MUSIC, LANGUAGE, AND HUMAN EVOLUTION

Music, Language, and Human Evolution

Edited by
NICHOLAS BANNAN

OXFORD
UNIVERSITY PRESS

OXFORD
UNIVERSITY PRESS

Great Clarendon Street, Oxford, OX2 6DP,
United Kingdom

Oxford University Press is a department of the University of Oxford.
It furthers the University's objective of excellence in research, scholarship,
and education by publishing worldwide. Oxford is a registered trade mark of
Oxford University Press in the UK and in certain other countries

First Edition published in 2012

Impression: 3

British Library Cataloguing in Publication Data

Data available

Library of Congress Cataloging in Publication Data

Data available

ISBN 978–0–19–922734–1

Printed and bound by CPI Group (UK) Ltd, Croydon, CR0 4YY

Contents

Acknowledgements and Notes to the Accompanying CD vii
Contributors ix
Abbreviations x
Glossary xi

PART I: INTRODUCTION

1 Introduction: Music, Language, and Human Evolution 3
Nicholas Bannan

PART II: PERSPECTIVES FROM ANTHROPOLOGY AND ARCHAEOLOGY

2 Music and Mosaics: The Evolution of Human Abilities 31
Robert A. Foley

3 The Evolution of the Human Vocal Tract: Specialized for Speech? 58
Margaret Clegg

4 When the Words Dry Up: Music and Material Metaphors Half a Million Years ago 81
Clive Gamble

PART III: PERSPECTIVES ON THE EVOLUTIONARY PREREQUISITES FOR MUSICAL BEHAVIOUR

5 Hominin Physiological Evolution and the Emergence of Musical Capacities 109
Iain Morley

6 Vocal Traditions of the World: Towards an Evolutionary Account of Voice Production in Music 142
Trân Quang Hai and Nicholas Bannan

7 Found Objects in the Musical Practices of Hunter-Gatherers: Implications for the Evolution of Instrumental Music 173
Pedro Espi-Sanchis and Nicholas Bannan

PART IV: PERSPECTIVES FROM SOCIAL AND COGNITIVE PSYCHOLOGY

8 On the Evolutionary Function of Song and Dance 201
Robin Dunbar

9 The Vocal Learning Constellation: Imitation, Ritual Culture, Encephalization 215
Björn Merker

PART V: PERSPECTIVES FROM MUSICOLOGY

10 Music as an Emergent Exaptation 263
Ian Cross

11 Musicians' Performance Prosody 277
Johan Sundberg

12 Harmony and its Role in Human Evolution 288
Nicholas Bannan

Index 341

Acknowledgements and Notes to the Accompanying CD

This book deals with a significant aspect of the human condition that is impossible to capture in words: the phenomenon of music. The theme around which its authors have presented their various arguments implies an evolutionary role for music that requires us to imagine its purpose and potential in the ascent of our species. Illustration of the kinds of music-making that may be associated with this developmental trajectory benefits from seeing and hearing, and the purpose of the accompanying CD is to provide some glimpses of the practice of music in cultures that may be unfamiliar to the reader, and to ways of listening to the human voice that reveal its intrinsic musicality.

Material on the CD resonates with proposals as to the purpose and practice of music throughout this book, but is especially relevant to the following chapters: Chapter 6 (Trân Quang Hai and Nicholas Bannan); Chapter 7 (Pedro Espi-Sanchis and Nicholas Bannan), and Chapter 12 (Nicholas Bannan). These chapters deal with the production and structure of musical sounds, and most benefit from the support of the audio and video examples provided.

The CD comprises three sections related to these three chapters. The first two draw on recorded material presented at the University of Reading at the research workshop that gave rise to this book. The CD was edited by Pedro Espi-Sanchis, who recorded further material in the field in South Africa. We are grateful to the following for their contributions: Mike McMillan and Brian Hayter, then of the Faculty of Education, University of Reading, for recording and editing video and audio material; Adrian Lowe, then Lay Clerk and cantor at Christ Church Cathedral, Oxford, for singing the Collect for Evensong; Jesse Stack, technician of the School of Music, University of Western Australia, for editing and recording video examples; members of The Winthrop Singers of UWA for performing vocalizations illustrative of points made in Chapter 12; Madosini, for performing material illustrative of South African practice. Excerpts of performances by the Dithlaka, the Dinaka, and the Tshikona Pipe

Ensembles at the 2004 Ethnomusicology Symposium are used with the kind permission of the International Library of African Music. We are grateful to Andrew Tracy for the opportunity to incorporate this material.

The phenomenon YouTube has revolutionized access to recorded material, and this has proved especially valuable in the field of ethnomusicology. New examples of the practices of music from across the globe are becoming available daily, and readers are encouraged to seek out further audio and video clips of musical examples mentioned in the text. In particular, Trân Quang Hai's website gives access to a wide selection of recordings he has made at various locations:

<http://tranquanghai.info/index.php?lang=en>.

I acknowledge the support of a great many colleagues around the world in achieving the completion of this book. The Faculty of Arts, Humanities and Social Sciences at the University of Western Australia provided financial support towards the creation and editing of the CD without which it could not have been included.

Above all, I cannot sufficiently express my gratitude to my wife, Valerie, who has lived with this project since its inception, and generously given time and encouragement to it at every stage.

Nicholas Bannan
School of Music
University of Western Australia

Contributors

Nicholas Bannan is a composer and choral conductor at the School of Music, University of Western Australia.

Margaret Clegg is Head of the Human Remains Unit, Natural History Museum, London.

Ian Cross is Reader in Music & Science, and Director of the Centre for Music and Science, University of Cambridge.

Robin Dunbar is Professor of Evolutionary Anthropology at the Institute of Cognitive & Evolutionary Anthropology, University of Oxford.

Pedro Espi-Sanchis is a freelance musician and storyteller based in Cape Town, South Africa.

Robert Foley is Professor of Human Evolution and Co-Director of the Leverhulme Centre for Human Evolutionary Studies, University of Cambridge.

Clive Gamble is Professor in the Department of Archaeology, Faculty of Humanities, University of Southampton.

Björn Merker is a neuroscientist and independent scholar working in Kristianstad, Sweden.

Iain Morley is Lecturer in Palaeoanthropology and Human Sciences, School of Anthropology, University of Oxford.

Johan Sundberg is retired Professor of Music Acoustics, Department of Speech Music Hearing, School of Computer Science and Communication, KTH (Royal Institute of Technology), and visiting Professor at the University of London and Universität Freiburg.

Trân Quang Hai is an honorary member of CREM (Centre de Recherche en Ethnomusicologie) of the National Centre for Scientific Research, Université Paris Ouest, France.

Abbreviations

AD	adult-directed
AMH	anatomically modern human
BP	before present
EEG	electroencephalography
fMRI	Functional Magnetic Resonance Imaging
ID	infant-directed
IDS	infant-directed speech
MYA	million years ago
PAG	periaqueductal grey matter
PET	Positron Emission Tomography
WAM	Western art music

Glossary

Altriciality. Literally 'requiring nourishment', and referring to the condition of the young of a species whose parents invest in their protection and upbringing, without which care they would not survive.

Anagenetic change. The process in evolution whereby an entire population acquires sufficient fixed mutations that it represents a separate species to the ancestor population.

Autapomorphy. A derived trait in a species that is neither shared with relatives within the species, clade, or family, nor with the ancestor species from which it evolved: a uniquely evolved characteristic.

β-endorphin levels. Beta-endorphin is a naturally occurring neurotransmitter that dulls pain and produces a feeling of well-being.

Exaptation. A trait that may have evolved to serve a particular function that, due to a change in the organism's interaction with its environment, can be co-opted to serve a different function. Sometimes referred to as a 'pre-adaptation' for the new characteristic.

Formant. Formants were defined in 1960 by voice scientist and acoustician Gunnar Fant as 'the spectral peaks of the sound spectrum of the voice'. The word is now generally used as a synonym for vocal tract resonance. Formants are determined by vocal tract shape and are decisive to vowel, consonant, and personal voice quality.

FOXP2. FOXP2 is an abbreviation of *Forkhead box protein P2*, and applies to the human version of a protein located on human chromosome 7. Versions of FOXP2 exist in a variety of distantly related species including mice and songbirds. In humans, mutations in FOXP2 cause severe disorders in speech and language functions, suggesting that adaptation for language descends from a modification in this gene.

Handicap principle. A term coined by Zahavi in 1975 to describe the runaway sexual selection whereby apparently disadvantageous traits, such as large antlers or the peacock's tail, operate instead as an 'honest signal' of genetic fitness. An organism able to flourish with such an attribute ensures that it will be inherited: if it were, within its environment, a true handicap, the organism would not survive to reproduce.

Homeostasis. The maintenance of a steady internal condition in response to external stimuli, whether necessary to the survival of the organism (food, air, water, warmth, love) or a threat to it (trauma, extremes of temperature, hostility, contaminants).

Hominid. A member of the *Hominidae*, the group comprising humans, chimpanzees and bonobos, gorillas, their shared ancestors and related species.

Hominin. A member of the *Hominini*, the group comprising modern humans and related species since the divergence of the human line from the African great apes. In very simple anatomical terms, hominins are all bipedal but not all hominids are.

K-selection promotes traits that allow organisms to compete effectively for limited resources in populations at or near environmental carrying capacity. This often involves heavy investment in few offspring with protracted development, reflected in large body size, longevity, encephalization, and extensive post-natal learning, rendering individuals into precious repositories of knowledge and skill. Examples: elephants, humans, and whales.

Mode 1, 2, and 3 technologies. Classifications of levels of technological advance in the working of stone implements in hominin evolution. Mode 1 represents the stone cores and flakes associated with the earliest technology first appearing in eastern Africa 2.4–2.3 Mya; Mode 2 with the next phase around 1.6–1.4 Mya in which cores begin to be bifacially worked to a standard shape; Mode 3 embraces the more refined, smaller artefacts that appeared between 250,000 BP and 50,000 years BP.

Megadonty. Possessing large teeth; adaptation to dietary needs dependent on the possession of large teeth either for predation or for the consumption of tough vegetation.

Neoteny. Preservation into adulthood of the characteristics of infancy: for example, the retention of infant-like proportions of the skull in humans compared to chimpanzees; the continuation of play behaviour into adulthood of domesticated silver foxes, and of humans.

Ontogeny. The development of an organism within its own life-span, from embryo to mature adult.

Phylogeny. The development of a species from the perspective of its evolutionary history.

Plesiomorphic. Literally 'old-feature', the retention of characteristics shared with related species and derived from a common ancestor.

Prognathic. Having a projecting lower jaw (see also Megadonty, above).

Recto-tono. Chanted on a single pitch or *reciting note.* Monotone.

Source-filter theory. The accepted model of vocalization that analyses speech and song as depending on a sound source (the vocal folds) modified by filtering through control of the vocal tract (pharynx, soft palate, tongue, and lips).

Part I

Introduction

1

Introduction: Music, Language, and Human Evolution

Nicholas Bannan

The genesis of this book was a meeting sponsored by the European Science Foundation and supported by the British Academy that brought together researchers from a variety of disciplines to explore the role of music in human evolution and its relationship to the development of language. The meeting took place at the University of Reading, UK, in the autumn of 2004, and was hosted by Professor Steven Mithen and Nicholas Bannan. The chapters of this book were all written subsequently by invited contributors to this event. Unusually for an academic conference, the meeting embraced two practical workshops, led by Trân Quang Hai and Pedro Espi-Sanchis, that permitted live and intimate experience of musical practices from various parts of the world intended to illuminate the issues under scrutiny in more conventional spoken presentations. Steven Mithen had proposed a structure for the meeting that alternated what we termed, by way of shorthand, 'archaeological' and 'musicological' contributions, intended to support a close interpenetration of content and interpretation arising from presentations that would not normally be heard in such proximity. The selection of chapters that make up this book has been arranged differently, in order to convey its themes more sequentially. It has been the intention of contributors, though, to try and capture the sense of mutual support and interest that made the Reading meeting memorable (see Balter 2004).

The relationship between music and language has concerned academics ever since Jean Jacques Rousseau and his eighteenth-century contemporaries considered how these may have originated (Levman 2000; Thomas 1995). The treatment of this topic has, however, tended to be discipline-specific, predominately reflecting the concerns of particular individuals, whether they be Charles Darwin approaching the field from his theory of evolution in 1871, his contemporary Herbert Spencer in *Mind* (1860), the linguist Otto Jespersen (1922), John Blacking from the perspective of ethnomusicology in his 1973 book *How Musical is Man?*, or John Sloboda from that of cognitive psychology in his 1985 book *The Musical Mind*. The necessary dialogue between academics with different specialisms but similar concerns has been lacking.

The need for such dialogue has intensified in recent years owing to two factors. First are the remarkable developments in numerous fields, especially the cognitive neuroscience of music, child development, and our understanding of human evolution, all of which throw new light on how music and language are related, as well as archaeological discoveries that are posing new questions that require music to be taken into account. Second, is that many of those working on the evolution of language have neglected the significance of music, failing to appreciate the inherent relationship between the two media despite speculation on this from Rousseau via von Humboldt (1836), Helmholtz (1863), and Darwin (1871) onwards. A principal reason for this is that, as all disciplines have developed in their theories, methods, and terminologies, it has become increasingly difficult for academics to remain sufficiently knowledgeable about other fields to achieve the degree of interdisciplinarity that is essential for understanding the relationship between music and language, and how these evolved. For the reader new to these issues, it is worth reviewing the accelerated trajectory that research and theory have traced in the relatively recent past.

The extraordinary advances in the various branches of enquiry on which speculations on the origins of music and language are founded have taken place precisely because they have to an extent isolated themselves as distinct specialisms that created a demand for expert training. Systematic musicology as defined by Adler (1885), semantics (Saussure, 1916), psychoacoustics (Helmholtz, 1863), and other disciplines essential to this topic all have their origins in the second half of the nineteenth century, and have developed and to some extent fragmented as they became enmeshed in the expansion of the university sector internationally during the twentieth century. But the consequences of increasing specialization led to trends in the application of knowledge that favoured progress in some fields over others.

Similarly, research in one field misapplied in another led to failed opportunities to devise convincing theories:

> For most of the century after Darwin wrote about expression, his views were rejected or simply ignored. The intellectual and scientific world was dominated by those who saw culture as determining every important aspect of our behaviour. As influential an anthropologist as Margaret Mead [1975] claimed that facial expressions differ from culture to culture as much as language, customs, attitudes and values . . . A few scientists went on to claim that the very idea of emotions was an invention of Western culture. (Ekman, 1998, p. xxiii)

Interdisciplinary exchange in the field of language origins had a head start on a similar approach to music, perhaps reflecting the convergence traceable from Saussure and Peirce to Chomsky that has sought to identify and explain linguistic universals. By comparison, ethnomusicology has tended to present cultures as so individually different that detecting universals has been taken to betray a prejudiced viewpoint, especially given the religious and ritual purposes associated with music in many cultures, and matters of taboo and researcher sensitivity that these determined:

> That music exists as a universal phenomenon is probably true. But this depends . . . on how music is defined. . . . To my mind the most universal characteristic of music is its non-universality as a means of communication. Whatever it communicates is communicated to the members of the in-group only, whoever they may be (List 1971, p. 399).

List's remarks are taken from his contribution to a plenary discussion on universals in music at the Seattle conference of the Society for Ethnomusicology in November 1970. He was replying to the then controversial proposal by McAllester that the topic of universality in music required attention, for all the problems involved:

> Any student of man must know that somewhere, someone is doing something that he calls music but nobody else would give it that name (McAllester, 1971, p. 379)

A landmark in the study of language origin was the publication of the proceedings of the New York Academy of Sciences symposium devoted to the field (Harnad, Steklis, and Lancaster 1976), 110 years after discussion of the topic was banned by the Société Linguistique de Paris. Building on foundations laid in the work of Chomsky (1957), Hockett (1960), and Lenneberg (1967), contributors to the

New York meeting began to represent the interdisciplinary methodology required to make progress, and the universalist agenda to which this could be applied. Concurrently, Lieberman and Crelin (Lieberman and Crelin 1971; Lieberman 1975) attempted to see a pattern in the comparative anatomy that they hoped would yield convincing data. In this they were followed by Johanson and Edgar (1996), as well as by a variety of researchers who focused on diverse specific features of anatomy in living species and those represented in the fossil record (see Morley, Bannan, Clegg, this volume; Fitch, 2000).

The field of *biolinguistics* was mapped out to Chomskyan specifications by Jenkins (2000). Other monographs on the acquisition of the capacity for language were published by Bickerton (1991), Pinker (1994), and Aitchison (1996). While such research in linguistics embraced a universalist stance, this did not necessarily determine the acceptance of evolutionary interpretation on Darwinian lines. A parallel contribution to the investigation of human communication arose in the work of researchers who represented an overtly Darwinist position within the complementary disciplines of cognitive and experimental psychology, socio-biology, and philosophy. Plotkin (1994) illustrated how the acquisition of knowledge could be set out hierarchically according to the capacity for learning of organisms at different stages of development; Donald (1991) devised a corresponding, stage-based sequence for the evolution of culture; Damasio (1994) argued the necessity of accounting for emotion as the complement of cognition in the processes involved in learning and memory; Deacon (1997) examined the neurological conditions for symbolization and their consequences for socialization and communication; Dennett reviewed the literature of post-Darwinian approaches to consciousness (1991), communication, social organization, and morality (1995); E. O. Wilson (1975) traced patterns of social behaviour across diverse species; and Dawkins (1976; 1986) published overarching explanations for the evolution of anatomy and behaviour, pioneering the new sub-discipline of memetics, proposing the *meme* as an extension into the analysis of cultural change of the role of the gene in evolutionary biology (see also Blackmore 1999).

Concurrent with the emergence of this agenda within the evolutionarily central disciplines of zoology and anatomy, work emerged in the humanities that took a chronologically long view of human development. One research method applied in both language and music study has been the analysis and translation of ancient notations, and the attempt to find patterns that might explain language evolution through plotting rates of change and distribution in languages themselves. From such processes, Ostler (2005) derived principles for language development and distribution—what one might term 'fitness'—that traced emergence, hybridization, decay, and extinction analogous to the life cycles of species. Bonvillian, Garber and Dell (1997) reported early historical attempts in Classical and Central Asian literature to account for language acquisition. Speculation on the distribution of languages prior to written evidence gained new support in examining the correlation between the global distribution of DNA and spoken languages (Cavalli-Sforza 1991). Bowra (1962) sought parallels between song forms in early Classical literature and the themes and structure of song in the repertoires of surviving Southern African and Australian Aboriginal hunter-gatherers, though his focus was more on text than music. However, analysis of the earliest evidence for notated intentions in music has been published by Hickmann (2001), Franklin (2002), and Davison and Apel (1946).

Providing a significant parallel to work on the evolution of human language and culture has been analysis of the communication of non-human species. This is central to Darwin's original speculations both in his account of human language as arising due to sexual selection for vocal ability in *The Descent of Man* (1871/2004, pp. 635–639), and throughout the argument of *The Expression of Emotion in Man and Animals* (1872). An overview of research in non-human communication and its biological development has been provided by Hauser (1997). Foremost in the quantity and detailed analysis published on other species has been research into birdsong, in which the leading figures have been Nottebohm (1989), Marler (1977; 2000), and Slater (2000). Description and analysis of vocal communication in aquatic mammals have confirmed that humpbacked whales engage in vocal learning (Payne 2000), as do pinnipeds, of which some harbour seals have been known to imitate human speech (Ralls et al. 1985). Fitch (2006) has proposed criteria to define what should and should not properly be referred to as song in non-human species. Nevertheless, the sub-discipline of *zoomusicology* has emerged to embrace the study of animal communication exhibiting behaviour parallel in structure or intention to human music-making (Martinelli 2008; Taylor 2008; see also 'Animal Music', Slater 2009; and Mâche 2000).

Unsurprisingly, attention has focused on the behaviour, anatomy, and capacity for communication of species evolutionarily closest to humans, and to comparisons that illuminate human evolution. Fossey (1983) initiated research into wild gorillas in their natural habitat, as did Goodall for chimpanzees (1986). De Waal's behavioural studies (2000) of chimpanzees and bonobos in captivity provide a useful complement to the fieldwork of these pioneers. Schwartz (1987; see also Lawton 2009) has asserted a keen advocacy for the anatomical and behavioural characteristics of orang-utans, despite generally accepted evidence for their more distant relationship to humans. The research of Seyfarth and Cheney into vervet monkey alarm calls (1980) presented data widely cited as an evolutionary threshold for the first word-like utterances.

As comparative research progressed, a key paradox became increasingly apparent, with significant implications for the evolution of human language and music: our closest genetic cousins, in terms of their respiratory anatomy and oral configuration, are poorly equipped for speech, while distant relatives such as Hoover the Harbour Seal (Deacon 1997, pp. 225–30) and Alex the Grey Parrot (Pepperberg 2002) possess the aural perception, neural processing and vocal mechanism to mimic human utterances convincingly. As this limitation on the vocal potential of non-human primates proved insurmountable, the ape communication agenda shifted towards alternative means by which the intelligence of these species could manifest itself in theories of mind and the capacity for problem-solving and the use of symbols, including sign language (Savage-Rumbaugh et al. 1978; Premack 1986). Related to this approach, and taking into account the revealed linguistic characteristics of deaf signing, the proposal has been considered that speech has its origins in complex manual gestures (Pollick and de Waal 2007; Gentilluci and Corballis 2006; Stokoe 2001; Corballis 1992) as much as in the development of the vocal system.

Nevertheless, the nature of human song and speech as orally generated and aurally perceived (Hockett 1960) remains the prime arena for evolutionary investigation, and the vocalizations of primate species less closely related to humans have continued to fascinate for what they may reveal (Marler 2000). Geissmann (2000) and Merker (2000a) researched the role and development of gibbon 'song', originally cited in Darwin (1871, pp. 590, 633, 639) as a useful source for comparison; and Richman (1980) explored the role of social vocalization in geladas. Both Richman (2000) and Merker (2000b) proposed coordinated, song-like communication in intermediate primate species as a bridge between animal vocalization and human language.

Other aspects of animal behaviour have received attention as contributing to the conditions in which the emergence of symbolic communication and the systems it depends on could be explained. Significant within this field has been Dunbar's (1997) theory of 'gossip' as a descendant of grooming behaviour, where expansion in group size determined the effectiveness of flexible vocalization that could be carried out 'at a distance', as well as the number of participants that could be vocally 'groomed' simultaneously. Dunbar (this volume) recognizes that a grooming-to-song model is no less valid. A parallel to this, focusing on the mother-infant relationship, has been proposed by Falk (2004), and accords to maternal singing a role in nursing while freeing the hands to carry out other tasks.

A crucial conceptual Rubicon separates scholars whose work recognizes the potential role played by musical behaviour in human evolution, and those who accept an adaptationist explanation for the origins of language, but are doubtful that music can either have played a part in this, or have possessed any adaptive advantages of its own. Pinker (1997) is most closely associated with a sceptical view of the role of music, seeing it as a side-benefit conferred by the discovery that skills acquired for other purposes could be employed to give pleasure. The proof Pinker's position demands is evidence that music was ever necessary. Bickerton (1991) outlined a different challenge in the 'continuity paradox' whereby it is impossible to capture in language the nature of pre-linguistic communication. Mithen (2005) responded to this in his speculation on the musicality of Neanderthal communication, and some form of description of what proto-language (or proto-music) may have been like, consistent with the anatomical, neurological, and archaeological evidence, is essential to providing an answer to Pinker. Forensic analysis of biological and geological data enables the visual aspects of life in Palaeolithic Europe or Africa to be depicted in artists' impressions: we can no longer pretend, when viewing these, that our ancestors lived in silence.

But speculation on what the intermediate species between our last common ancestor with the chimpanzee and ourselves may have been able to sing or say is fraught with difficulty. The example of Eggar

(1946) is a salutary reminder of the potential traps, her address to the Royal Musical Association on 'The Musical Condition of Man' referring as it did to the musical potential of the Piltdown relics prior to their exposure as a forgery. Livingstone (1973), whose brave proposal that the Australopithicenes may have sung represented a landmark in the reinvigoration of debate on the musical quality of human origins, faced considerable criticism in the issue of *Current Anthropology* that dissected his case.

A problem with thinking about music in relation to language that pervades the literature is that terminology can be misleading, and the conceptual framework within which comparisons are attempted can be uneven. We tend, for instance, to know what 'language' is, and are well supported by definitions in the literature: but defining 'music' is much trickier. Livingstone's paper leant on an understanding of theories of language prevailing over the previous decade that had descended from Hockett's (1960) proposal for the design features of communication (see also Fitch 2006 and Bannan, this volume). But Hockett's paper contains a partial account of the nature of music that is historically constrained to modern European practice. Similarly, the theory of musical generativity Lerdahl and Jackendoff (1983) derive from a Chomskyan combinatorial approach is confined in its application to only a relatively small (if, for many, highly significant) sub-section of mankind's musical output. The question of universals again arises: to speculate on what ancestral music-making may have been like, it does not help to try to export into the past customized impressions of our own favourite musical works.

Universals have emerged as accepted and useful concepts in anthropology as a Darwinist approach to explaining behaviour in terms of biological evolution has prevailed. Brown (1991; 2004) has developed a framework for thinking about universals, both of behaviour and in culture, that embraces music and dance, while Ekman (1973) collected and analysed data that laid old myths about the cultural derivation of facial expression. Ethnomusicologists responded to these and earlier influences, and two special issues of a leading journal, *The World of Music*, were devoted to the topic of musical universals (1977; 1984), initiating a more open and well-informed debate than that represented at the 1971 conference cited above. The 1977 publication featured articles by Blacking, Harrison, Hood, Lomax, Nattiez, Nettl, and Trân Van Khê; while the 1984 issue included Boiles, Brook and Bain, Gourlay, Koskoff, List, Nketia, and

Schneider. Significantly, both special issues also embraced proposals for universals in dance—Kurath 1977; Hannah 1984. Notably, these publications included papers by the Vietnamese musicologist Trân Van Khê (1977) and the Ghanaian scholar J. H. Kwabena Nketia (1984) that contributed to leavening the often Eurocentric slant of previous discussion in this field (Trân Van Khê is father of Trân Quang Hai whose work is featured in this volume). The challenge to archaeologists, psychologists, and linguists, as well as many musicians involved in speculation on the origins and role of music, is to ensure that their experience is widened to embrace the musical styles beyond the Euro-American mainstream, including those of surviving hunter-gatherers whose traditions may represent the best possible links we have to the remote past. The Australian composer Percy Grainger attempted such a 'commonsense view of all music' in a series of radio broadcasts in 1934 (Blacking 1987), but music education provision around the world has yet to deal satisfactorily with the cultural, epistemological, and pedagogical ramifications of a universalist approach to music.

In parallel with the awakening of universalist trends in ethnomusicology, the new discipline of music archaeology emerged as a branch of the International Council for Traditional Music, holding initial conferences in Cambridge (1982) and Stockholm (1984). In a wide-ranging review of the principal influences on the field, Schneider (1986) set out to place music archaeology in a historical context:

> Nearly one hundred years ago . . . Rudolf Virchov . . . complained about a lack of interest in the prehistory of music while he stated that prehistoric linguistics and even mythology had gained reasonable insights in the more remote ages of the past (Virchov 1886). Strange enough, it was a philosopher and psychologist, Carl Stumpf, who devoted a major book to the study of the *Anfänge der Musik* (Beginnings of Music, see Stumpf 1911), and some of the most interesting contributions with respect to the origins of music have been made by philosophers, linguists, psychologists, for example, Spencer (1890), Torrefranca (1907), Révész (1941). (Schneider, 1986, p. 196)

He went on to review the syntheses arising from work built upon these, as he saw them, essentially *comparative* foundations, citing in particular studies by the Berlin School of ethnomusicologists: Sachs (1962) and Hornbostel (pub. posth., 1973), and its East German post-war continuation in the work of Knepler (1977). Schneider then prescribed roles for the fields of bioacoustics and biocybernetics in filling in the missing links in the historical evidence, illustrating how the hypothesis forwarded by Tembrock (1978) that singing precedes speech in human evolution cannot be formulated by reference to archaeological evidence alone.

Schneider's identification of the key role played by psychologists in sustaining interest in this topic during the mid twentieth century heralds the significant contribution to research and debate made over the last fifty years within the relatively new specialist field of music psychology. The sub-discipline owes much to the work of Leonard Meyer, whose key publication on music, emotion, and meaning (1956) appeared just a year prior to Chomsky's seminal text on syntactic structures in language (1957). The two could be taken to define the complementary themes in the study of human communication that have influenced most subsequent work.

Music has featured increasingly in the development of both cognitive and social psychology. The multi-authored volume edited by Deutsch (1999) reviewed trends across most aspects of the field, and remains a valuable introductory text. Storr's monograph of 1992 defined an evolutionary context for psychological investigation; Levitin's of 2006 focuses especially on recent research in music cognition. In the important field of music and emotion, Scherer (1992) analysed similarities between the structure of animal calls across a variety of species that suggest ancient neural responses govern their production and perception. Research with human subjects by Juslin and Laukka (2004) and Gabrielsson (2001) has illuminated aspects of memory, affect, and social behaviour elicited through musical experience.

A significant contribution to theory and practice that exemplifies the instinctive, biological nature of human response to music has emerged in the field of music therapy. Many of the cases studied in Sacks (2007) illustrate the capacity of music to affect both emotion and cognition in ways that suggest an adaptive function underlying such responses. Malloch and Trevarthen's edited volume exploring *Communicative Musicality* (2009) embraces the full range of cognitive, social, developmental, and therapeutic features of musical experience.

From research related to the acquisition of language and music that developed independently in a wide range of disciplines and locations, a tradition of enquiry thus emerged strongly over the last decades of the twentieth century which drew on interdisciplinary studies in a variety of international centres to provide new data and syntheses able to supply missing pieces to the jigsaw. A pioneer of this movement was Wallin (1991), who established a Centre for Biomusicology[1] in Sweden that mounted the conference in Fiesole, Italy, in 1997 which led to the book *The Origins of Music* (Wallin, Merker and Brown, 2000). Daniel Schneck chaired a conference at Virginia Polytechnic Institute, Music in Human Adaptation (1997), whose proceedings include his own paper on the neurology and perception of music and one by Bannan (1997) on the adaptive features of human singing. In an early example of the increasingly useful medium of e-journals, Vaneechoutte and Skoyles (1998) published their synthesis of existing research supportive of the prior manifestation of human singing as a prerequisite for language. Benzon (2001) attempted an overview of the evolutionary nature of music from a perspective largely rooted in his experience of jazz improvisation.

Further cross-disciplinary and synthetic research initiatives also broadened the field during the 1990s. A conference sponsored by the Max Planck Institute on non-verbal vocal communication gave rise to an influential volume of proceedings (Papousek, Jürgens and Papousek, 1992). Besson and Friederici co-edited an issue of *Music Perception* (1998) devoted to comparison between research in music and language; Cross published his provocative paper 'Is Music the Most Important Thing We Ever Did?' (1999); and two further conferences presented work relevant to the field. The International Conference on Music Perception and Cognition (Keele, 2000) included papers by Bannan, Hodges, Tolbert, and Cross (Woods, Luck, Brochard, Seddon and Sloboda 2000) submitted for the sub-theme 'Music and Evolution', as well as others on related topics by Cox (2001) and Trevarthen (2000). In 2001, the New York Academy of Sciences sponsored a conference entitled 'The Biological Foundations of Music', leading to a volume of proceedings edited by Zatorre and Peretz (2001).

Research in music archaeology has given rise to a new interest in the acoustic properties of ritual spaces and prehistoric sites (Scarre

[1] Despite its usefulness to the current study and the strong case made by Wallin, the term *biomusicology* has yet to be accepted by, for instance, the New Grove. On the other hand, an entry on *Archaeomusicology* by Hickmann appeared in the New Grove Online (2001).

and Lawson 2006; Cross 2000a; Watson and Keating 1999; Devereux and Jahn 1996; see also Muecke and Zach 2007, and Gamble, this volume); the role of dance in the organization of society (Garfinkel 2003) that may be associated with large-scale developments such as irrigation and agriculture; and evidence for symbolic thinking associated with artefacts claimed to have been designed as musical instruments (Conard, Malina and Münzel 2009; d'Errico et al. 2003; Kunej and Turk 2000).

The discovery of musical instruments from 35,000 years ago (Conard, Malina and Münzel 2009) provides evidence for the role that music may have played in human evolution that cannot be ignored. The difficult technology of working ivory and bone so as to allow the performance of pitched sounds independent of those produced by the human voice conveys that these objects must have been accorded considerable symbolic and ritual significance. Nevertheless, what such technology achieves remains a far more recent development than the range of vocalization in our species that it extends and adapts. In comparison with these material objects, the precedence of speech and song acquired over a still more extensive biological past leaves tantalizingly little evidence.

Nevertheless, the scale of research and interest in the field of vocal evolution, embracing comparisons with other species as well as the cybernetic modelling of communication, has developed momentum in the first decade of the new millennium. A key paper by Hauser, Chomsky and Fitch (2002) set out to redefine the distinctly human features of language, identifying a crucial role for recursion. Justus and Hustler (2005) provided a complementary attempt to define the cognitive boundaries of human musicality and their origins, to which McDermott and Hauser added a concurrent response (2005). Fitch followed (2006) with a comprehensive review of these arguments set within a broader cross-species perspective.

The most wide-ranging and daring attempt to model a coherent theory out of the myriad interdisciplinary data that have informed this topic is the book *The Singing Neanderthals* by Steven Mithen (2005). Well received by many musicians and general readers, Mithen's synthesis of research in support of a mimetic, musical stage in vocal evolution prior to the development of the full faculty of language in *Homo sapiens sapiens* has also attracted some impassioned rejoinders and reviews, both pro and contra including Benzon (2008), Fitch (2005), Bispham (2006), Gamble (2006), Morley (2006),

Tallerman (2006), Wray (2006), Avorgbedor (2008), and Smith (2008). The book has been subjected to considerable scrutiny, and this has itself contributed to advancing the relationship between the ideas Mithen expresses and their capacity to influence future research. Even those unsympathetic to Mithen's viewpoint must concede that it is no longer wise to wish music away when speculating upon the acquisition and development of vocalization in humans.

Signs that the topic of music's nature and origins have entered the academic mainstream have included the 2006 issue of *Cognition* devoted to the topic, edited by Isabelle Peretz; and the inclusion of several papers on music in the Cradle of Language conference hosted at Stellenbosch University the same year. The latter, organised by Rudi Botha, informed his own commentary (2009) on Mithen's theory. A conference mounted at the University of Durham in 2007 by Bennett Zon focused on 'Music and Evolutionary Thought'.

Arising from this mass of new information and accelerated contact between researchers in different disciplines, a convergence has come about which is defining with increasing clarity the questions to be posed and the means that must be used to address them. It is only in the last fifteen years or so that research has focused on fully embracing the role of music into the evolutionary agenda, and encouraged speculation on the prehistoric soundscape and its function in shaping modern human abilities. This period has seen the discovery of crucial new evidence in the fossil record, as well as the development of new tools in neuroscience and forensic analysis, which together have accelerated interest in the phenomenon of music's origins and their relationship to language. Research informed by evolutionary speculation has in turn influenced developments in music therapy, pedagogy, child development, and ethnomusicology that is itself giving rise to additional data that is illuminating this debate (Malloch and Trevarthen 2009).

A common complaint of those addressing questions relating to human origins is the limited scope of the fossil evidence. Yet in the last 150 years, an enormous amount of material has come to light, and new discoveries can be anticipated that may provide key evidence. When one recalls that Darwin speculated on the vocalizations of the precursors of man prior to the discovery or examination of any early human fossil remains, the material to which we now have access represents a rich and growing resource. Within the first decade of the twenty-first century, startling new finds have stimulated lively debate regarding their significance and interpretation. The claimed *Homo floresiensis* discovered on an Indonesian island in 2004 divided scholars as to whether it represented a distinct species (Brown et al. 2004; Dixon 2004) or a genetically deformed *Homo sapiens* (Obendorf et al. 2008). The Dikika Child (Alemseged et al. 2006) painstakingly separated over several years from the dense matrix in which it had been preserved, represents a rare specimen of an Australopithecene child of about three years, including the hyoid bone *in situ*.

Far more remote in time and genetic lineage, the identification and analysis of the 'Ida' fossil, a form of lemur-like species from 47 Mya claimed to represent an ancestor of the primate line (Tudge 2009; but see Gibbons 2009), illustrates that ancient bones may survive with an astonishing level of intactness and detail (Beard 2009).

New evidence of early human technology able to construct musical instruments has come to light at Hohle Fels and Vogelherd (Conard, Malina, and Münzel 2009), adding to and predating that already known at Geißenklösterle (Conard et al. 2004) and Isturitz (d'Errico et al. 2003), and, in comparison with the Divje Babe I 'flute' (Kunej and Turk 2000), providing convincing evidence of the purpose of the artefacts. In their illustration of early modern human capacity for symbolization and ritual culture, finds such as these in Europe complement the ochre and shell objects unearthed at Blombos Cave in South Africa (Henshilwood et al. 2001) that suggest the full achievement of bodily and extrasomatic decoration by 70,000 years ago. While we can only speculate on what role music or language may have played in the life of the settlers of Blombos Cave, the existence of these artefacts provides a marker for the extended timescale in which the development of human communicative vocalization has occurred.

New research that complements this archaeological data and its anthropological interpretation is emerging as faster and more accurate equipment has become available in neurology. This is providing clearer understanding of the processing of pitch in vocalization and its perception (Braun 1999), and similarities and differences between the neural systems on which speech and song depend (Patel and Peretz 1997; Parsons 2001) as well as the nature of bodily movements coordinated to rhythm (Brown and Parsons 2008). A feature of brain organization first studied in macaques and dubbed the 'mirror neuron' system (Rizzolati and Arbib 1998) has excited considerable interest as an adaptation of potential significance to the capacity to imitate and empathize through the senses of sight and sound, including the role this could clearly play in music (Molnar-Szakacs and Overy 2006). Some commentators remain sceptical regarding exactly what mirror neurons explain, notably Hurford (2004), who analyses the nature of the contribution mirror neurons may make in suggesting limitations to the model; and Borg (2007), who queries 'If Mirror Neurons Are the Answer, What Was the Question?'

A further approach to research in this field made possible by advances in technology is being developed through computer modelling, both in language (Kirby 1999; de Boer 2000) and music (Baroni et al. 1992; Miranda 1999; Cook 1996). Processes involved in the evolution of, for instance, the production and learning of vowel sounds, the development of the optimally resonant singing voice, the nature of a musical grammar that can generate melody, or the transition from randomly generated to meaningful utterances within a virtual population, have been modelled with illuminating consequences for our understanding of communication in living organisms.

The authors who have contributed to this book each inhabit a different portion of the territory mapped in this introduction. This book presents their work as a valuable snapshot of the progress being made in the disciplines they represent, and the questions their approaches pose for the future. While no consensus exists even within these covers regarding the eventual resolution of the debate here engaged in regarding the origins of music and language, on one matter we would all be in broad agreement: whatever the difficulties in devising research to test the proposition, music would seem to have played a significant part in the evolution of our species that deserves serious attention.

REFERENCES

Adler, G. (1885). 'Umfang, Methode und Ziel der Musikwissenschaft', *Vierteljahrsschrift für Musikwissenschaft* pp. 5–20.

Aitchison, J. (1996). *The Seeds of Speech: Language Origin and Evolution.* Cambridge: Cambridge University Press.

Alemseged, Z., F. Spoor, W. H. Kimbe, et al. (2006). 'A Juvenile Early Hominin Skeleton from Dikika, Ethiopia', *Nature*, 443: 296–301.

Avorgbedor, D. (2008). Review of S. Mithen, 'The Singing Neanderthals', *Empirical Musicology Review* 3, 1: 22–35.

Balter, M. (2004). 'Seeking the Key to Music', *Science* 306, 5699: 1120–2.

Bannan, N. (2000). 'Prosody, Meaning and Musical Behaviour'. In: C. Woods, G. Luck, R. Brochard et al. (eds.), *Proceedings of the Sixth International Conference on Musical Perception and Cognition.* [CD-ROM]. Keele: Department of Psychology, University of Keele, UK.

——(1997). 'The Consequences for Singing Teaching of an Adaptationist Approach to Vocal Development'. In: D. Schneck and J. Schneck (eds.), *Music in Human Adaptation* pp. 39–46. Blacksburg, VA: Virginia Polytechnic Institute and State University Press.

Baroni, M., R. Dalmonte, and C. Jacobini (1992). 'Theory and Analysis of European Melody'. In: A. Marsden and A. Pople (eds.) *Computer Representations and Models in Music,* pp. 187–205. New York: Academic Press.

Beard, C. (2009). 'IDA's a Fine Fossil, But She's Not the Missing Link', *New Scientist* 201, 2710: 18–19.

Benzon, W. L. (2008). 'Synch, Song and Society: A Review of "The Singing Neanderthals" by Steven Mithen', *Human Nature Review*, 5: 66–86.

——(2001). *Beethoven's Anvil: Music in Mind and Culture.* New York: Basic Books.

Besson, M. and A. D. Friederici (1998). 'Language and Music Processing', *Music Perception*, Special Issue. Berkeley, CA: University of California Press.

Bickerton, D. (1991). *Language and Species.* Chicago: University of Chicago Press.

Bispham, J. (2006). '"Music" means nothing if we don't know what it means'. Lead review of *The Singing Neanderthals* by S Mithen. *Journal of Human Evolution*, 50: 587–93.

Blacking, J. (1987).*'A Commonsense View of All Music': Reflections on Percy Grainger's Contribution to Ethnomusicology and Music Education.* Cambridge: Cambridge University Press.

——(1977). 'Can Musical Universals be Heard?' *The World of Music*, 19, 1/2: 14–29.

——(1973). *How Musical is Man?* Seattle, WA: University of Washington Press.

Blackmore, S. (1999). *The Meme Machine.* Oxford: Oxford University Press.

Boiles, C. (1984). 'Universals of Musical Behaviour: A Taxonomic Approach', *The World of Music* 26, 2: 50–65.

Bonvillian, J. D., A. M. Garber, and S. B. Dell (1997). 'Language Origin Accounts: Was the Gesture in the Beginning?' *First Language* 17, 51: 219–39.

Borg, E. (2007). 'If Mirror Neurons Are the Answer, What Was the Question?' *Journal of Consciousness Studies* 14, 8: 5–17.

Botha, R. (2009). 'On Musilanguage/"Hmmmmm" as an Evolutionary Precursor to Language', *Language & Communication*, 29, 1: 61–76.

Bowra, C. M. (1962). *Primitive Song.* London: Weidenfeld and Nicolson.

Braun, M. (1999). 'Auditory Midbrain Laminar Structure Appears Adapted to f0 Extraction: Further Evidence and Implications of the Double Critical Bandwidth', *Hearing Research* 129: 71–82.

Brook, B. S. and D. Bain (1984). 'Music in the Life of Man: Theoretical and Practical Foundations for a World History', *The World of Music* 26, 2: 120–36.

Brown, D. E. (2004). 'Human Universals, Human Nature and Human Culture', *Daedalus* 133, 4: 47–54.

——(1991). *Human Universals.* New York: McGraw-Hill.

Brown, P., T. Sutikna, M. J. Morwood et al. (2004). 'A New Small-Bodied Hominin from the Late Pleistocene of Flores, Indonesia', *Nature* 431: 1055–61.

Brown, S. and L. M. Parsons (2008). 'The Neuroscience of Dance', *Scientific American* 299, 1: 78–83.

Cavalli-Sforza, L. L. (1991). 'Genes, People, and Languages', *Scientific American* 265: 104–10.

Chomsky, N. (1957). *Syntactic Structures.* The Hague: De Gruyter Mouton.

Cook, P. R. (1996). 'Singing Voice Synthesis: History, Current Work, and Future Directions', *Computer Music Journal*, 20, 3: 38–46.

Conard, N., M. Malina, S. Münzel et al. (2004). 'Eine Mammutelfenbeinflöte aus dem Aurignacien des Geissenklösterle: neue Belege für eine musikalische Tradition im Frühen Jungpaläolithikumauf der Schwäbischen Alb', *Archäologisches Korrespondenzblatt* 34, 4: 447–62.

—— —— —— (2009). 'New Flutes Document the Earliest Musical Tradition in Southwestern Germany', *Nature* 460: 737–40.

Corballis, M. C. (1992). 'On the Evolution of Language and Generativity', *Cognition* 44, 3: 197–26.

Cox, A. (2001). 'The Mimetic Hypothesis and Embodied Musical Meaning', in *Musica Scientiae* 5, 2: 195–212.

Cross, I. (2000a). 'Lithoacoustics—Music in Stone', preliminary report. <http://www.mus.cam.ac.uk/~ic108/lithoacoustics/>.

Cross I (2000b). 'Music in Human Evolution', unpublished paper. Abstract in C. Woods, G. Luck, R. Brochard et al. (eds.) *Proceedings of the Sixth International Conference on Musical Perception and Cognition.* [CD-ROM]. Keele: Department of Psychology, University of Keele, UK.

——(1999). 'Is Music the Most Important Thing We Ever Did? Music, Development and Evolution'. In: Y. Suk Won, (ed.), *Music, Mind and Science* pp. 10–39. Seoul, S. Korea: Seoul National University Press.

Damasio, A. (1994). *Descartes' Error: Emotion, Reason and the Human Brain.* New York: Avon Books.

Davison, A. and W. Apel (1946). *Historical Anthology of Music*, i. Cambridge MA: Harvard University Press.

Dawkins, R. (1986). *The Blind Watchmaker.* London: Longman.

——(1976). *The Selfish Gene.* Oxford: Oxford University Press.

Darwin, C. R. (1872/1999). *The Expression of the Emotions in Man and Animals*, ed. P. Ekman. London: HarperCollins.

——(1871/2004). *The Descent of Man.* ed. with an introduction by J. Moore and A. Desmond. London: Penguin.

Deacon, T. (1997). *The Symbolic Species.* London: Allen Lane.

de Boer, B. (2000). 'Self-Organisation in Vowel Systems', *Journal of Phonetics*, 28: 441–65.

Dennett, D. (1995). *Darwin's Dangerous Idea.* London: Allen Lane.

——(1991). *Consciousness Explained.* London: Allen Lane.

d'Errico, F., C. Henshilwood, G. Lawson et al. (2003). 'Archaeological Evidence for the Emergence of Language, Symbolism, and Music—An Alternative Multidisciplinary Perspective', *Journal of World Prehistory* 17, 1: 1–70.

Deutsch, D. (1999). *The Psychology of Music.* London: Academic Press.

Devereux, P. and R. G. Jahn (1996). 'Preliminary Investigations and Cognitive Considerations of the Acoustic Resonances of Selected Archaeological Sites', *Antiquity* 70: 665–6.

De Waal, F. (2000). *Chimpanzee Politics: Power and Sex Among Apes.* Baltimore, MD: Johns Hopkins University Press.

Dixon, P. W. (2004). 'The Hawaiian Menehune in Myth and Paleontology', *Human Evolution* 19, 4: 287–91.

Donald, M. (1991). *Origins of the Modern Mind: Three Stages in the Evolution of Culture and Cognition.* Cambridge, MA: Harvard University Press.

Dunbar, R. (1997). *Grooming, Gossip and the Evolution of Language.* Cambridge, MA: Harvard University Press.

Eggar, K. (1946). 'The Musical Condition of Man', *Proceedings of the Royal Musical Association,* 73rd Session, pp. 13–23.

Ekman, P. (1998). Introduction, afterword and commentaries to Charles Darwin, *The Expression of the Emotions in Man and Animals.* London: HarperCollins.

——(1973). *Darwin and Facial Expression.* London: Academic Press.

Falk, D. (2004). 'Prelinguisic Evolution in Early Humans: Whence Motherese?' *Behavioural and Brain Sciences* 27: 491–503.

Fitch, W. T. (2006). 'The Biology and Evolution of Music: A Comparative Perspective', *Cognition* 100: 173–215.

——(2005). 'Dancing to Darwin's Tune', *Nature*, 438: 288.

——(2000). 'The Evolution of Speech: A Comparative Review', *Trends in Cognitive Sciences*, 4, 7: 258–67.

Fossey, D. (1983). *Gorillas in the Mist.* Boston: Houghton Mifflin.

Franklin, J. (2002). 'Musical Syncretism in the Greek Orientalizing Period'. In: E. Hickmann and R. Eichmann (eds.) *Archäologie früher Klangerzeugung und Tonorinungen. Studien zur Musikarchäologie III, Orient-Archäologie 10,* pp. 441–51.

Gabrielsson, A. (2001). 'Emotion Perceived and Emotion Felt: Same or different?' *Musicae Scientiae* Special Issue, 123–47.

Gamble, C. (2006). 'Musical chairs'. Commentary on *The Singing Neanderthals* by S Mithen. *Cambridge Archaeological Journal*, 16: 108–9.

Garfinkel, Y. (2003). *Dancing at the Dawn of Agriculture.* Austin, TX: University of Texas Press.

Geissmann, T. (2000). 'Gibbon Songs and Human Music'. In: N. Wallin, B. Merker, and S. Brown (eds.), *The Origins of Music*, pp. 103–23. Cambridge, MA: MIT Press.

Gentilluci, M. and M. Corballis (2006). 'From Manual Gesture to Speech: A Gradual Transition', *Neuroscience & Biobehavioral Reviews*, 30, 7: 949–60.

Gibbons, A. (2009). 'Celebrity Fossil Primate: Missing Link or Weak Link?' *Science*, 324, 5931: 1124–5.

Goodall, J. (1986). *The Chimpanzees of the Gombe: Patterns of Behaviour.* Cambridge, MA: Belknap/Harvard University Press.

Gourlay, R. A. (1984). 'The Non-Universality of Music and the Universality of Non-Music', *The World of Music*, 26, 2: 25–39.

Grainger, P. (1934/1987). 'A Commonsense View of all Music'. Transcripts of lectures broadcast by the ABC, Australia in 1934. Appendix A of J. Blacking, *A Commonsense View of all Music,* pp. 151–80. Cambridge: Cambridge University Press.

Hanna, J. L. (1984). 'Towards Discovering the Universals of Dance', *The World of Music*, 26, 2: 88–103.

Harnad, S., H. Steklis, and J. Lancaster (1976). *Origins and Evolution of Language and Speech.* Annals of the New York Academy of Sciences, 280.

Harrison, F. (1977). 'Universals in Music: Towards a Methodology of Comparative Research', *The World of Music*, 19, 1/2: 30–42.

Hauser, M. (1997). *The Evolution of Communication.* Cambridge, MA: MIT Press.

——N. Chomsky, and T. Fitch (2002). 'The Faculty of Language: What Is It, Who Has It, and How Did It Evolve?' *Science* 298: 1569–79.

Helmholtz, H. von (1863). *Die Lehre den Tonempfindungen als physiologische Grundlage für die Theorie der Musik.* Braunschweig: F. Vieweg und Sohn. Trans. A. J. Ellis, *On the Sensations of Tone* (1954). New York: Dover.

Henshilwood, C., J. Sealy, R. Yates et al. (2001). 'Blombos Cave, Southern Cape, South Africa: Preliminary Report on the 1992–1999 Excavations of the Middle Stone Age', *Journal of Archaeological Science*, 28: 421–48.

Hickmann, E. (2001). 'Archaeomusicology', ed. L. Macy. New Grove Online. <http//www.grovemusic.com>.

Hockett, C. F. (1960). 'The Origin of Speech', *Scientific American*, 203: 89–96.

Hodges, D. A. (2000). 'Why Are We Musical? Support for an Evolutionary Theory of Human Musicality'. In: C. Woods, G. Luck, R. Brochard et al. (eds.) *Proceedings of the Sixth International Conference on Musical Perception and Cognition*. [CD-ROM]. Keele: Department of Psychology, University of Keele, UK.

Hood, M. (1977). 'Universal Attributes of Music', *The World of Music*, 19, 1/2: 63–75.

Hornbostel, E. M. v (1973). 'Geburt und erste Kindheit der Musik', posthumously published in *Jahrbuch für musikalische Volks- und Völkerkunde*, 7: 9–17.

Humboldt, W. v (1836/1972). *Linguistic Variability and Intellectual Development*, tr. G. Buck and F. Raven. Philadelphia, PA: University of Pennsylvania Press.

Hurford, J. R. (2004). 'Language Beyond Our Grasp: What Mirror Neurons Can, and Cannot, Do for Language Evolution'. In: K. Oller, U. Griebel, and K. Plunkett, (eds.), *Evolution of Communication Systems: A Comparative Approach*, pp. 297–313. Boston, MA: MIT Press.

Jenkins, L. (2000). *Biolinguistics*. Cambridge: Cambridge University Press.

Jespersen, O. (1922). *Language: Its Nature, Development and Origin*. London: Allen & Unwin.

Johanson, D. C. and B. E. Edgar (1996). *From Lucy to Language*. New York: Simon & Schuster.

Juslin, P. and P. Laukka (2004). 'Expression, Perception, and Induction of Musical Emotions: A Review and a Questionnaire Study of Everyday Listening', *Journal of New Music Research*, 33, 3: 217–38.

Justus, T. and J. Hustler (2005). 'Fundamental Issues in the Evolutionary Psychology of Music'. *Music Perception*, 23, 1: 1–27.

Kirby, S. (1999). 'Learning, Bottlenecks and Infinity: A Working Model of the Evolution of Syntactic Communication'. In: K. Dautenhahn and C. Nehaniv (eds.), *Proceedings of the A/SB '99 Symposium on Imitation in Animals and Artifacts*, pp 121–9. Society for the Study of Artificial Intelligence and the Simulation of Behaviour, Brighton.

Knepler, G. (1977). *Geschichte als Weg zum Musik verständnis*. Leipzig: Philipp Reclam Verlag.

Koskoff, E. (1984). 'Thoughts on Universals in Music', *The World of Music*, 26, 2: 66–87.

Kunej, D. and I. Turk (2000). 'New Perspectives on the Beginnings of Music: Archaeological and Musical Analysis of a Middle Paleolithic Bone "Flute"'. In: N. Wallin, B. Merker, and S. Brown (eds.), *The Origins of Music*, pp. 235–68. Cambridge, MA: MIT Press.

Kurath, G. P. (1977). 'Universals in Dance'. *The World of Music*, 19, 1/2: 43–62.

Lawton, G. (2009). 'Could the Orang-Utan Be Our Closest Relative?' *New Scientist*, 202 (2713): 6–7.

Lenneberg, E. H. (1967). *The Biological Foundations of Language*. New York: Wiley.

Lerdahl, F. and R. Jackendoff (1983). *A Generative Theory of Tonal Music*. Cambridge, MA: MIT Press.

Levitin, D. (2006). *This Is Your Brain on Music: The Science of a Human Obsession*. New York: Dutton Books.

Levman, B. (2000). 'Western Theories of Music Origin, Historical and Modern', *Musicae Scientiae*, 4, 2: 185–212.

Lieberman, P. (1975). *On the Origins of Language*. New York: Macmillan.

——and E. S. Crelin (1971). 'On the Speech of Neanderthal Man', *Linguistic Inquiry*, 2: 203–22.

List, G. (1984). 'Concerning the Concept of the Universal and Music', *The World of Music*, 26, 2: 40–9.

——(1971) 'On the Non-Universality of Musical Perspectives', *Ethnomusicology*, 15, 3: 399–402.

Livingstone, F. B. (1973). 'Did the Australopithecenes Sing?' *Current Anthropology*, 14, 1/2: 25–9.

Lomax, A. (1977). 'Universals in Song', *The World of Music*, 19, 1/2: 117–41.

Mâche, F.-B. (2000). 'The Necessity Of and Problems With a Universal Musicology'. In: N. Wallin, B. Merker, and S. Brown (eds.), *The Origins of Music* pp. 473–80. Cambridge, MA: MIT Press.

Malloch, S. and C. Trevarthen (2009). *Communicative Musicality: Exploring the Basis of Human Companionship*. Oxford: Oxford University Press.

Marler, P. (2000). 'Origins of Music and Speech: Insight from Animals'. In: N. Wallin, B. Merker, and S. Brown (eds.), *The Origins of Music*, pp. 31–48. Cambridge, MA: MIT Press.

——(1977). 'The Structure of Animal Communication Sounds'. In: T. H. Bulloch (ed.), *Recognition of Complex Acoustic Signals*, pp. 17–35. Berlin: Springer-Verlag.

Martinelli, D. (2008). 'Zoomusicology and Musical Universals: The Question of Processes', *Revista Transcultural de Música 12*. <http://www.sibetrans.com/trans/56/how-to-cite-trans>.

McAllester, D. (1971). 'Some Thoughts on 'Universals' in World Music', *Ethnomusicology*, 15, 3: 379–80.

McDermott, J. and M. Hauser (2005). 'The Origins of Music: Innateness, Uniqueness, and Evolution', *Music Perception*, 23: 29–59.

Mead, M. (1975). Review of P. Ekman 'Darwin and Facial Expression', *Journal of Communication*, 25: 205–13.

Merker, B. (2000a). 'Synchronous Chorusing and Human Origins'. In: N. Wallin, B. Merker, and S. Brown (eds.), *The Origins of Music*, pp. 315–27. Cambridge, MA: MIT Press.

——(2000b). 'The Birth of Music in Synchronous Chorusing at the Hominid Chimpanzee Split'. In: C. Woods, G. Luck, R. Brochard et al. (eds.), *Proceedings of the Sixth International Conference on Musical Perception and Cognition.* [CD-ROM]. Keele: Department of Psychology, University of Keele, UK.

Meyer, L. B. (1956). *Emotion and Meaning in Music.* Chicago, IL: Chicago University Press.

Miranda, E. (1999). 'The Artificial Life Route to the Origins of Music', *Scientia*, 10, 1: 5–33.

Mithen, S. (2005). *The Singing Neanderthals.* London: Weidenfeld & Nicholson.

Molnar-Szakacs, I. and K. Overy (2006). 'Music and Mirror Neurons: From Motion to 'E'motion', *Scan*, 1: 235–41.

Morley, I. (2006). 'Name That Tune'. Commentary on *The Singing Neanderthals* by S Mithen, *Cambridge Archaeological Journal*, 16: 101–3.

Muecke, M. W. and M. S. Zach (2007). *Essays on the Intersection of Music and Architecture.* Raleigh, NC: Lulu.com.

Nattiez, J.-J. (1977). 'Under What Conditions Can One Speak of the Universals of Music?' *The World of Music*, 19, 1/2: 92–116.

Nettl, B. (1977). 'On the Question of Universals'. *The World of Music*, 19, 1/2: 2–13.

Nketia, J. H. K. (1984). 'Univeral Perspectives in Ethnomusicology'. *The World of Music*, 26, 2: 3–24.

Nottebohm, F. (1989). 'From Bird Song to Neurogenesis', *Scientific American*, 260: 56–61.

Obendorf, P. J., C. E. Oxnard, and B. J. Kefford (2008). 'Are the Small Human-Like Fossils Found on Flores Human Endemic Cretins?' *Proceedings Of The Royal Society of London B*, 275: 1287–96.

Ostler, N. (2005). *Empires of the Word: A Language History of the World.* London: HarperCollins.

Papousek, H., U. Jürgens, and M. Papousek (1992). *Nonverbal Vocal Communication: Comparative and Developmental Approaches.* Paris and Cambridge: Editions de la Maison des Sciences de l'Homme and Cambridge University Press.

Parsons, L. (2001). 'Exploring the Functional Neuroanatomy of Music Performance, Perception, and Comprehension'. *Annals of the New York Academy of Sciences*, 930: 211–31.

Patel, A. and I. Peretz (1997). 'Is Music Autonomous from Language? A Neuropsychological Appraisal'. In: I. Deliège and J. A. Sloboda (eds.), *Perception and Cognition of Music*, pp. 191–215. Hove: Psychology Press.

Payne, K. (2000). 'The Progressively Changing Songs of Humpback Whales: A Window on the Creative Process in a Wild Animal'. In: N. Wallin, B. Merker, and S. Brown (eds.), *The Origins of Music*, pp. 135–50. Cambridge, MA: MIT Press.

Pepperberg, I. (2002). *The Alex Studies: Cognitive and Communicative Abilities of Grey Parrots.* Cambridge, MA: Harvard University Press.

Pinker, S. (1997). *How the Mind Works.* New York: W. W. Norton.

——(1994). *The Language Instinct.* London: Allen Lane.

Plotkin, H. (1994). *Darwin Machines and the Nature of Knowledge.* London: Allen Lane.

Pollick, A. and F. de Waal (2007). 'Ape Gestures and Language Evolution'. *Proceedings of the New York Academy of Sciences*, 104, 19: 8184–9.

Premack, D. (1986). *'Gavagai!' or the Future History of the Animal Language Controversy*. Cambridge, MA: MIT Press.

Ralls, K., P. Fiorelli, and S. Gish (1985). 'Vocalizations and Vocal Mimicry in Captive Harbor Seals, Phoca Vitulina', *Canadian Journal of Zoology*, 63: 1050–6.

Révész, G. (1941). 'Der Ursprung der Musik', *International Archiv für Ethnographie*, 40: 70–75.

Richman, B. (2000). 'How Music Fixed 'Nonsense' into Significant Formulas: On Rhythm, Repetition, and Meaning'. In: N. Wallin, B. Merker, and S. Brown (eds.), *The Origins of Music*, pp. 301–14. Cambridge, MA: MIT Press.

——(1980). 'Did Human Speech Originate in Coordinated Vocal Music?' *Semiotica*, 32, 3/4: 233–44.

Rizzolati, G. and M. Arbib (1998). 'Language within Our Grasp', *Trends in Neuroscience*, 21: 188–94.

Sachs, C. (1962). *The Wellsprings of Music.* The Hague: Martinus Nijhoff.

Sacks, O. (2007). *Musicophilia: Tales of Music and the Brain.* London: Picador.

Saussure, F. de (1916/1959). *Course in General Linguistics*, tr. W. Baskin. New York: McGraw-Hill.

Savage-Rumbaugh, E. S., D. M. Rumbaugh, and S. Boysen (1978). 'Linguistically Mediated Tool Use and Exchange by Chimpanzees (Pan Troglodytes)'. *Behavioral and Brain Sciences*, 4: 539–54.

Scarre, C. and G. Lawson (2006). *Hearing Secret Harmonies: Archaeoacoustics.* Cambridge: McDonald Institute for Archaeological Research.

Scherer, K. (1992). 'Vocal Affect Expression as Symptom, Symbol and Appeal'. In: H. Papousek, U. Jürgens, and M. Papousek (eds.), *Nonverbal Vocal Communication: Comparative and Developmental Approaches,* pp. 43–60. Paris and Cambridge: Editions de la Maison des Sciences de l'Homme and Cambridge University Press.

Schwartz, J. (1987). *The Red Ape: Orang-Utans and Human Origins.* New York: Houghton Mifflin.

Seyfarth, R. M. and D. Cheney (1980). 'The Ontogeny of Vervet-Monkey Alarm-Calling Behavior: A Preliminary Report', *Zeitschrift für Tierpsychologie*, 54: 37–56.

Slater, P. (2009). 'Animal Music'. <http://www.oxfordmusiconline.com/subscriber/article/grove/music/47872>.

——(2000). 'Birdsong Repertoires: Their Origins and Use'. In: N. Wallin, B. Merker, and S. Brown (eds.), *The Origins of Music*, pp. 49–63. Cambridge, MA: MIT Press.

Sloboda, J. (1985). *The Musical Mind: The Cognitive Psychology of Music.* Oxford: Oxford University Press.

Schneck, D. J. (1997). 'A Paradigm for the Physiology of Human Adaptation'. In: D. Schneck and J. Schneck (eds.), *Music and Human Adaptation*, pp. 1–22. Blacksburg, VA: Virginia Polytechnic Institute and State University Press.

Schneider, A. (1986). '*ΑΡΧΑΙΟΛΟΓΕΩ*. Some comments on methods and sources in Music Archaeology'. In: C. S. Lund (ed.), *Second Conference of the ICTM Study Group on Music Archaeology*, pp. 195–224. Stockholm: Royal Swedish Academy of Music.

Schneider, M. (1984). 'The Birth of the Symbol in Music', *The World of Music*, 26, 3: 3–21.

Smith, N. (2008). 'The Evolution of Music and Language', *Current Anthropology*, 49, 3: 529–30.

Spencer, H. (1890). 'The Origin of Music', *Mind*, 60: 449–68.

Stokoe, W. C. (2001). *Language in Hand: Why Sign Came Before Speech.* Washington, DC: Gallaudet University Press.

Storr, A. (1992). *Music and the Mind.* London: HarperCollins.

Stumpf, C. (1911). *Anfänge der Musik.* Leipzig: J.A. Barth.

Tallerman, M. (2006). 'Abracadabra! Early Hominin for "I Think My Humming's Out of Tune With the Rest of the World!"' Commentary on *The Singing Neanderthals* by S. Mithen. *Cambridge Archaeological Journal*, 16: 106–7.

Taylor, H. (2008). 'Decoding the Song of the Pied Butcherbird: An Initial Survey', *Revista Transcultural de Música*, 12. <http://www.sibetrans.com/trans/a98/decoding-the-song-of-the-pied-butcherbird-an-initial-survey>.

Tembrock G (1978). 'Bioakustik, Musik und Sprache', *Sitzungsberichte der Akademie der Wissenschaften der DDR :N, Mathematik, Naturwissenschaften, Technik,* Nr 1/N, Berlin.

Thomas, D. A. (1995). *Music and the Origins of Language: Theories from the French Enlightenment.* Cambridge: Cambridge University Press.

Tolbert, E. (2000). 'Music and Meaning: An Evolutionary Story', In: C. Woods, G. Luck, R. Brochard et al. (eds.), *Proceedings of the Sixth International Conference on Musical Perception and Cognition.* [CD-ROM]. Keele: Department of Psychology, University of Keele, UK.

Torrefranca, F. (1907). 'Le origini della musica'. *Rivista musicale italiana* XIV: 555–94.

Trân Van Khê (1977). Is the Pentatonic Universal? A Few Reflections on Pentatonism', *The World of Music,* 19, 1/2: 76–91.

Trevarthen, C. (2000). 'Timing and Expression of Emotion in Natural Musicality'. In: C. Woods, G. Luck, R. Brochard et al. (eds.), *Proceedings of the Sixth International Conference on Musical Perception and Cognition.* [CD-ROM]. Keele: Department of Psychology, University of Keele, UK.

Tudge, C. (2009). *The Link: Uncovering Our Earliest Ancestor.* London: Brown, Little.

Vaneechoutte, M. and J. R. Skoyles (1998). 'The Memetic Origin of Language: Modern Humans as Musical Primates', *Journal of Memetics - Evolutionary Models of Information Transmission, 2.* <http://www.cfpm.org/jom-emit/1998/vol2/vaneechoutte_m&skoyles_jr. html>.

Virchov, R. (1886). Presidential address, *Zeitschrift für Ethnologie,* 18 (*Verhandlungen*).

Wallin, N. (1991). *Biomusicology: Neurophysiological, Neuropsychological and Evolutionary Perspectives on the Origin and Purposes of Music.* New York: Pendragon Press.

——, B. Merker, and S. Brown (2000). *The Origins of Music.* Cambridge, MA: MIT Press.

Watson, A. and D. Keating (1999). 'Architecture and Sound: An Acoustic Analysis of Megalithic Monuments in Prehistoric Britain', *Antiquity,* 73: 325–36.

Wilson, E. O. (1975). *Sociobiology: The New Synthesis.* Cambridge, MA: Harvard University Press.

Wray, A. (2006). 'Joining the Dots: The Evolutionary Picture of Language and Music'. Commentary on 'The Singing Neanderthals' by S. Mithen. *Cambridge Archaeological Journal,* 16: 103–5.

Zatorre, R. J. and I. Peretz (2001). *The Biological Foundations of Music (Annals of the New York Academy of Sciences, V.930).* New York: New York Academy of Sciences.

Part II

Perspectives from anthropology and archaeology

2

Music and Mosaics: The Evolution of Human Abilities

Robert A. Foley

INTRODUCTION

While there has been an explosion in the range of disciplines interested in human evolution in recent years, it remains the case that the historical record preserved in fossils and archaeology provides the core historical framework within which Darwinian models must be formulated and tested. This chapter provides a brief overview of the major radiations that occurred over the course of hominin evolution, and suggests their adaptive bases. The evolution of human abilities, including that of producing and appreciating music, is considered in this context, and in terms of the adaptive basis for these abilities. It is suggested that the period prior to the evolution of modern humans (500,000 to 200,000 years ago) may have been a time of major evolutionary change. Developments occurring less than 150,000–100,000 years ago are unlikely to be part of any universal human genetic heritage.

If there is anything that characterizes studies of human evolution in the past two decades or so, it is the expansion of approaches. For most of the last century, the source of all information about human evolution was the fossil record, perhaps supplemented here and there by the stone tools and other debris of the archaeological record. Given this fossil basis, human evolution was, of course, largely seen as an anatomical journey—upright posture, bipedal striding, manual dexterity, facial reduction, and so on—topped off, of course, with encephalization. This last was the major trend, and the one from which the non-anatomical parts could be perhaps gleaned—usually speculations about the evolution of culture or language, and a vague model of how increased intelligence was resulting in greater flexibility, adaptability, or complexity, in the human lineage.

This has now all changed. Occasionally to the chagrin of the anatomically-based anthropologist, human evolution is now accessible and interesting to a wide range of disciplines. Molecular geneticists were perhaps the first to make a significant impact (Cann et al. 1987; Sarich 1971; Underhill and Kivisild 2007). While this was originally in terms of reconstructing phylogenetic history from neutral markers (Bowcock et al. 1994; Gonder et al. 2007), such work has now expanded markedly to include models of ancient demography (Nei and Takahata 1993; Quintana-Murci et al. 2008), discussion of reproductive and kinship parameters (Quintana-Murci et al. 2008), and increasingly, the identification of the functional genes underlying the human phenotype, as well as estimates of its genetic history (Enard et al. 2002; Tishkoff 2007). Next in line have been archaeologists, who have, in recent years, considered human Pleistocene prehistory within a much more explicit evolutionary framework than before (Foley 1987b; 2002; Klein 1992; 2009; Mellars 1992). Nowhere has this been more marked than in the debates over the role of behaviour in the origin of modern humans, and in particular the development of ideas about a cultural or symbolic revolution (d'Errico 2003; Klein 2000; Mellars 1991). And the most recent major input has come from the evolutionary psychologists (Barkow et al. 1992; Barrett et al. 2002). Based on the idea that the human mind is an adaptive organ, and therefore shaped by selection, evolutionary psychologists have developed experiments that can identify the adaptive characteristics of the mind, and then developed hypotheses that attempt to explain how or why these may have been favoured in the past, during the course of human evolution. This work has not only made psychology more explicitly evolutionary, it has also resulted in a much more broad-based reconstruction of human evolution than is found in the bones alone; the mind is now accessible as well.

While the specialists who have entered the field of human evolutionary studies have been mainly from these disciplines, there are many others—linguists, cognitive scientists, primatologists and ethologists, physiologists and clinicians (Darwinian medicine (Nesse et al.

2010)) who have also both taken an interest in the field and contributed to it. Musicians have now joined the fray.

Not that this is necessarily new. Darwin considered the evolution of musical abilities in *The Descent of Man*.

> When we treat of sexual selection we shall see that primeval man, or rather some early progenitor of man, probably first used his voice in producing true musical cadences, that is in singing, as do some of the gibbon-apes at the present day; and we may conclude from a widely-spread analogy, that this power would have been especially exerted during the courtship of the sexes,—would have expressed various emotions, such as love, jealousy, triumph,—and would have served as a challenge to rivals. It is, therefore, probable that the imitation of musical cries by articulate sounds may have given rise to words expressive of various complex emotions. (Darwin 1871)

Darwin, as he often did with human traits, found it hard to identify a strong, environmentally adaptive basis for music, and proposed that it was an outcome of sexual selection. In doing so he identified a number of themes that have persisted in considerations of the evolution of music—its links to the emotions, the importance of tonal variation and prosody, and thus its continuity with speech, and the difficulty of establishing whether speech may have formed the basis for song, or whether song preceded speech. These remain major research issues, and represent themes that are further examined in the subsequent chapters of this book.

Having said that there are now many diners at the table of human evolution, the role of this chapter is to return to the core of evolutionary studies. Ultimately geneticists, cognitive scientists, linguists and psychologists want to establish an explanation—either in terms of function or mechanism, and additionally a historical pathway—by which musical abilities were incorporated into the human capacity. However, the core of a Darwinian approach is that adaptations—be they the product of natural or sexual selection—can only be understood in the context in which they evolved. This means that if some hominin had a mutant set of vocal cords that enhanced its reproductive success, this occurred in a particular context, at a particular time and place. This means that we need to establish when and where that might be, as a basis to considering what its function—in Darwinian terms—might be.

Thus, in this chapter, I shall set out to do two things. The first is to provide a brief overview of our current understanding of human evolution, based on genetic, fossil, and archaeological evidence, so that we can establish the appropriate framework for considering the evolution of language and musical potential. The second will be to consider the issue of what I shall refer to as 'human abilities' from an evolutionary perspective. The key question I shall be considering is that when we look at music and/or language, we are selecting out one or two of the abilities that humans possess and use. How do these relate to others? The central issue here is whether we have reasons to expect a concerted, synchronised pattern of evolution—a great leap forward model—or whether different abilities may have evolved in a more mosaic fashion.

The pattern of human evolution

A paradigm shift in human evolutionary studies

Twenty years ago or so it was relatively easy to give an overview of hominin evolution, as it was considered to be a straightforward process of largely anagenetic change—that is, gradual evolution within a few lineages. As this is no longer the case, it is worth perhaps starting by emphasizing three key findings. First, the hominin subfamily is remarkably diverse—there are currently some 26 candidate hominin species, belonging to six genera (see Foley 2005a for an up-to-date list). This implies a pattern of cladogenesis—evolution by branching—and that the process of human evolution is similar to that of other lineages (Figure 2.1). Hominin evolution is therefore a series of adaptive radiations. It is worth emphasizing at the outset one important implication of this diversity: each of these species is different in some way, and lived in a particular context. For anyone attempting to reconstruct a particular adaptive trajectory—be it music or any other feature—it is necessary to recognize the specificity of each taxon and lineage. Very often such reconstructions are based on un-named hominins, or else on a crude contrast between modern humans and the rest.

Second, through both genetics and the fossil record, we have pinpointed very closely the origin of the hominin clade. Results based on molecular phylogeny suggest that hominins are the sister

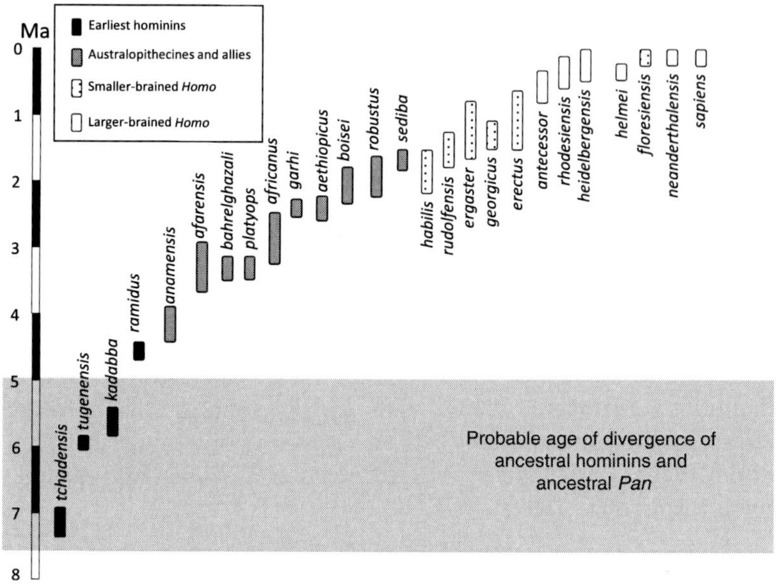

Figure 2.1. The extinct hominins: the chronological distribution of the taxa recognized in the hominin fossil record. The grey shaded area shows the probable age of the last common ancestor of hominin and chimpanzees/bonobos, based on genetic evidence.

clade of *Pan* (the chimpanzee) and that divergence between the hominin and this lineage occurred in the Late Miocene, probably sometime between six and eight million years ago (Williams and Goodman 1989). This allows us to refer to an important creature fairly precisely—the last common ancestor or LCA (Figure 2.1). Based on what we know of the first hominins, and of the living chimpanzees, we can make estimates of the nature of the LCA (although recent interpretations of *Ardipithecus ramidus* may weaken the strengths of such inferences (Lovejoy 2009)), which provide us with a basis for determining the 'starting point' for human and hominin uniqueness. Put another way, we can be sure that there was no singing six million years ago—after that, we cannot be so sure.

The third important discovery is that although there is this vast array of early hominins stretching back over the last five or more million years, the history of our own species—*Homo sapiens*—is only a small and recent twig on that branch (Stringer 2002). Evolutionary genetics has shown that all living humans are descended from a small ancestral population that lived in Africa between 150,000 and 200,000 years ago—and by small, I mean very small, probably less than 50,000 (Atkinson et al. 2008). This is important as it means that our own species has a very localized and specific origin, and that all living humans share a remarkable number of their characteristics—we are a very homogenous species (Gagneux and Varki 2001). The significance of this is worth returning to later with regard to abilities.

Having outlined these major points, which provide an important set of constraints on how we reconstruct language or musical history, let me now briefly outline the major pattern of hominin evolution as it is currently indicated by the fossil and archaeological record. It should be remembered that as these are all extinct species, genetics can say nothing about them (with the exception of Neanderthals, for whom there is now ancient DNA (Green et al. 2006; Krings et al. 2000; Krings et al. 1999)).

HOMININ ADAPTIVE RADIATIONS

As described above, the pattern of hominin evolution is best viewed as a series of adaptive radiations. An adaptive radiation is the diversification of a series of descendent forms from a common ancestor. It is commonly associated with a relatively high rate of diversification and a shared adaptive theme. It should be remembered that to some extent adaptive radiations are arbitrary events imposed analytically, but they serve as a useful analytical tool. The following represent the major adaptive radiations of the hominins (Figure 2.2, based on Foley 2002).

The radiation of the African apes in the later Miocene: this may have been a response to the invasion of Africa from Asia of an ancestral lineage, with the African apes and the hominins being the outcome of these dispersals (Stewart and Disotell 1998). The taxa of *Sahelanthropus tchadensis*, *Orrorin tugenensis*, *Ardipithecus kadabba*, and *Ar. ramidus* are placed in this radiation very tentatively. Given the absence of a fossil record for chimpanzees and gorillas, and the contentious positions of all of these relative to each other (White et al. 2009), little can be said about any pattern of dispersal and diversification except on the basis of inference from the living taxa. Recent evidence from *Ar. ramidus* would caution against too strong a

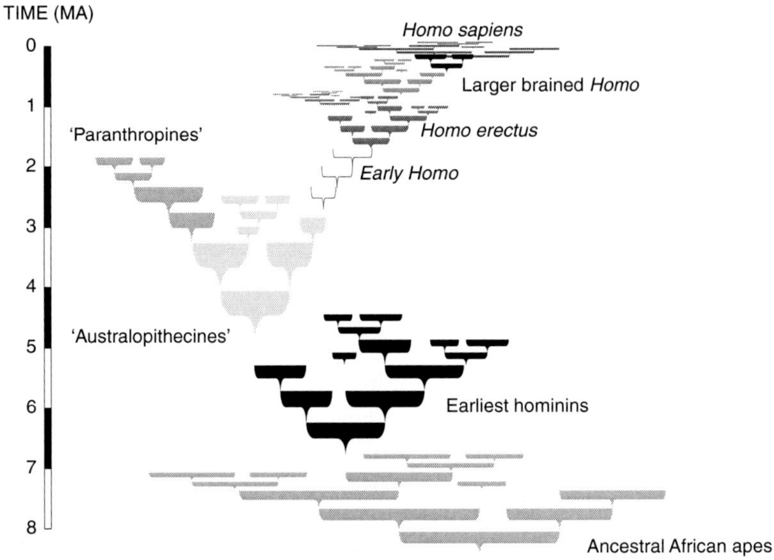

Figure 2.2. Adaptive radiations described in the text.

dependence upon chimpanzees as a model for the last common ancestor, and overall there is evidence that the early hominins and their closest allies were considerably diverse.

The radiation of the bipedal apes (early australopithecines): all this group, occurring in the Pliocene, have bipedal adaptations, and this seems to be the basis for a widespread, pan-African (Chad—Eastern—Southern Africa) series of dispersals. All retain a number of generalized hominin cranial characters, although they also display a trend towards increasing megadonty. The taxa included in this group are *Au. anamensis, Au. afarensis, Au. africanus, Au. garhi,* and *Au. bahrelghazali. Kenyanthropus platyops* is probably part of this group, although its affinities remain uncertain (see below).

The radiation of megadontic specialists is the dominant trend of the latest Pliocene and early Pleistocene. This seems to be a radiation of more savanna-dwelling hominins, with adaptations towards the processing of coarse fibrous plant material. It is unclear whether this was a monophyletic radiation, or part of a more general series of trends among the australopithecines, all of which show elements of megadonty. Indeed, this may be the trend which characterizes all the early (australopithecine) hominins, and may involve considerable homoplasy. The taxa included in this radiation are sometimes placed in their own genus—*Paranthropus*—and sometimes lumped with the Australopithecines *Au. (P.) aethiopicus, Au. (P.) robustus*, and *Au. (P.) boisei.*

The radiation of earliest Homo: this is perhaps the most problematic of all the radiations as its phylogenetic position is extremely uncertain (Wood and Collard 1999b), and there is considerable variation within the group (Wood and Lonergan 2008; Wood 1991). *Homo rudolfensis* retains megadontic adaptations, while sharing with *Homo habilis* an element of brain enlargement. Whether there is more that characterizes this radiation than just a variant on the australopithecine themes of the preceding few million years remains to be seen. The basis for placing it as a radiation lies solely on the fact that it may be a group with a presence in both eastern and southern Africa (SK53, Malawi) (Bromage and Schrenk 1999; Grine 2005). It could arguably be placed within later *Homo* or among the australopithecines (Wood and Collard 1999a). Furthermore, it is unclear what features link the specimens placed in this group adaptively. Some would also argue that the recently described taxon *Kenyanthropus platyops* would fall within this radiation, although its earlier date—3.5 Mya—would greatly extend in time the range of this group.

The radiation of Homo: there is considerable debate on the taxonomic level involved here, with some distinguishing a number of species, including the early African form *Homo ergaster*, and others incorporating all the diversity within forms of *Homo erectus*. There is a strong case, regardless of nomenclature, that this radiation represents a major grade shift among hominins, shown in cranial and dental morphology, the post-cranium, behaviour and technology. This radiation may actually be considered to be two sub-events; the first of these is the diversification and dispersal into the warmer parts of Eurasia in excess of 1.0 Mya, and the second is the radiation of later forms (especially *Homo heidelbergensis*) in the period around 0.5 Mya (Rightmire 1998). The first is largely associated with Mode 1 technology, the second with Mode 2 (Acheulean). This second part is more geographically extensive, and includes the more cold temperate parts of Eurasia.

The radiation of larger-brained Homo: the last half million years or so have seen the development of larger-brained hominins, often with very robust morphologies, and associated with the expansion of Acheulean (Mode 2) and prepared core (Mode 3) technologies, as well as evidence for more human-like behaviour. These radiations occur in the context of major glacial–interglacial climatic cycles, and so the underlying biogeographical patterns may have led to repeated dispersals overlain on each other, giving rise to the complexity of the later hominin fossil record—descendents of *Homo helmei* derived from Africa leading to the evolution in Europe of Neanderthals, and in Africa of *Homo sapiens.*

The radiation of Homo sapiens: this is a very recent event (last 100 Kyr) and is different from preceding dispersals as it involved far less morphological diversification, no speciation and resulted in the first complete global colonization.

THE ADAPTIVE BASIS

These radiations describe the timing, range, and diversity of the hominin lineage, but obviously they are just that—descriptions. For those whose minds turn more to questions of behaviour, these radiations require a further level of interpretation. As stated earlier, adaptive radiations only make sense in terms of the adaptive basis for them—in other words, what it is that causes the lineage to diversify and disperse at these particular times and places. Here we move into the realm of interpretation and inference, but a number of fairly well substantiated claims can be made.

Very little can be said about the radiation of the very earliest hominins, such as *Sahelanthropus*, if indeed they are hominins. Although there are some claims that can be made that they (*Sahelanthropus*, *Ororrin*, and *Ardipithecus*) had some bipedal adaptations, it is more likely that they retained a relatively generalized semi-terrestrial environment and way of life, and that what we are seeing here is the beginnings of the modern African ape adaptations. Rather more can be said about the next radiation, which is that of the australopithecines. Two major points should be made here—firstly, they are all clearly bipedal in the sense that they are more fully adapted to a terrestrial, bipedal way of life than any known ape, even if that bipedalism did not conform to the full striding gait associated with humans and their more recent ancestors. What may be important here is to recognize that if there are prerequisites for speech and song that are dependent upon an upright stance (Aiello

1996), then these were established well over four million years ago. However, the second point is in contradistinction of this—namely that although the australopithecines were bipedal, they show no other indications of behavioural conditions similar to those of modern humans. In particular, they do not exhibit encephalization, they do not have any major change in life history strategy (delayed maturation, etc.), and there is no, or at least very little, associated technology. I think from this we can infer that in most respects the australopithecines can be treated as bipedal apes.

The radiations of the genus *Homo* present a different story. Although one can argue about where the boundary between *Homo* and other hominins may lie (and see Berger et al. 2010, for a recent find (*Au. sediba*) which blurs the distinction yet further), there is a general consensus that from a morphological and adaptive perspective the key unifying and unique trait of *Homo* is the enlargement of the brain. It goes without saying that this also implies major cognitive and behavioural changes. A difficulty arises in the fact that it is brain size that is important here and that this changes in a continuous manner. Although one can look at the rate and see that it is not constant, nonetheless we are in a situation where we have no idea, on morphology alone, whether a brain size of 600 is significantly worse than one of 700, and so on. Furthermore, such calculations are strongly influenced by changes in body size.

One can conclude that although the changes in the brain are the most consistent element of evolution in *Homo*, it is through the cognitive and behavioural manifestations that it can best be perceived.

This is not the place to describe in detail the evidence—and indeed the qualifications—for the adaptive bases of the various radiations of *Homo* that were identified above. Each one, though, does seem to be associated with some major changes which can be seen in the palaeobiological record.

Early Homo *and* Homo ergaster: the early part of this radiation is obscure and indeed questionable; *Homo habilis* and *Homo rudolfensis* show few major differences from the australopithecines, and indeed some would remove them from the genus (Wood and Collard 1999b). If, however, there is anything which might be broadly associated with the time period in which these early creatures evolved, it is the appearance of stone tools. Although some tool-use and even toolmaking would almost certainly have been present among australopithecines and was probably plesiomorphic with chimpanzees, stone tools broadly coincide with the first *Homo*. The issue of *Au. garhi*, which is also associated with stone tools, however, casts some doubt on this simple association (Asfaw et al. 1999).

Perhaps more significant are the changes that occur with *Homo ergaster* sometime around 2.0 Mya. The most notable of these is the expansion of geographical range, which results in the first hominins beyond Africa, and the occurrence of a pan-tropical Old World lineage for the first time. The most likely adaptive basis for this event—not one event, probably, but a repeated succession of such expansions and dispersals, of which we can see only the bare bones—is meat-eating. The reason for saying this is partly empirical—there is, for the first time, major evidence for processed animal bones—and partly theoretical: geographical expansions with relatively low levels of speciation are a characteristic of carnivores (Foley 1987a, 2001). As I shall discuss later, meat-eating also has implications in terms of other aspects of behaviour and cognition.

Homo heidelbergensis: hominin populations probably ebbed and flowed under the influence of climatic and other changes for much of the period from 2.0 to 0.7 Mya, creating a lattice of species and subspecies across the warmer parts of the Old World. However, as far as we can tell, more northerly latitudes were never colonized for other than short warmer periods. From around 0.7 Mya, though, there is evidence for more northerly dispersals, associated most strongly with the Mode 2 or Acheulean industries. Although it is normally said that this technology appeared at around 1.5 Mya, there is clear evidence that the classic Acheulean is generally much later, and coincides with the radiation of *Homo heidelbergensis*. The regularization, standardization, and sophistication (and indeed the over-production) of the hand-axe imply a cognitive shift. Perhaps more significantly, it also is perhaps associated with the first persistent use of fire and the nutritional benefits of cooking (Wrangham et al. 1999).

Larger-brained Homo: many authorities would not see this as a clearly identifiable clade and radiation, preferring to see the Neanderthals and modern humans as homoplasic continuities of the archaic hominins of Europe and Africa (Hublin 1996; Stringer and Hublin 1999). However, there are some grounds, both archaeological and morphological, to suggest that the Mode 2 technologies were replaced by Mode 3 ones in Africa, associated with a larger-brained form of hominin—*Homo helmei*—which is ancestral to both Neanderthals and modern humans (Foley and Lahr 1997; Lahr and Foley

1998, 2001). If this is the case, then there is a dispersal event leading to adaptive diversity (cold adapted in Europe, tropical in Africa), in the last part of the Middle Pleistocene. It can be argued, on the basis of the shared traits found in Neanderthals and modern humans (see below), that there are major cognitive and behavioural changes, perhaps associated with the point-based and prepared core technology, perhaps more subtly cognitive and social. However, the chronology of the fossil evidence of the early stages of both human and Neanderthal lineages makes resolution of these issues difficult, although genetic evidence suggests it was not more than 500,000 years ago (Green et al. 2006).

Homo sapiens: the radiation of modern humans or *Homo sapiens* has to be characterized in a slightly different way, in that it is a massive geographical dispersal, but it does not involve speciation; all modern humans are remarkably similar, although it may have been the case that within the clade there were more different populations which are now extinct. As such, the evolutionary scale of modern humans is less than that of earlier hominins, and indeed this somewhat different evolutionary signal may itself be evidence for a change in hominin behaviour. There have been many claims for what forms the basis for modern human dispersals—language, symbolic thought, upper Palaeolithic technology, decreased cognitive modularity, and so on (Davidson and Noble 1991; Klein 1995; Mithen 1996). None is entirely satisfactory. Most explanations for the geographical expansions of the hominins tend to fall foul of the difficulties of chronology. Some of the traits appear too late to be part of a major biological change; this is especially the case for the Upper Palaeolithic. Others, which could be rooted at an earlier time at the base of the modern human clade, suffer from the problem that we do not find evidence for these at a very early date. There is often—and this can be very significant—a major chronological gap between the appearance of modern humans as a clade (at between 150 and 200,000 years ago (McDougall et al. 2005)) and the evidence for truly modern behaviour. The upper Palaeolithic changes of Eurasia are later than 50,000 years ago, but there is evidence for the presence of similar behaviours in southern Africa from 140,000 years ago (Marean et al. 2007). However, this evidence is intermittent rather than continuous, and it is only much later that such behaviours persist. As this is much younger than the earliest anatomically modern humans, a major issue in recent human evolution is what exactly was going on in those 'empty' millennia.

Be that as it may, one possibility concerning the ecological basis for the spread of modern humans lies in the first genuine evidence for significant use of marine resources. Exploitation of aquatic resources may have been significant for a number of reasons—access to a particularly rich source of food, with long-chained fatty acids that may have been critical to brain development; higher density populations along coasts, leading to greater social complexity; and the opportunities for dispersal that arise from a marine technology.

MULTIPLE LINEAGES, MULTIPLE EVENTS, MULTIPLE ADAPTATIONS

A recitation of the intricacies of the archaeological and fossil records is probably something that most can live without. However, details do matter, for they are the framework on which any reconstruction of a particular evolutionary trait—such as music and language—must be built. It is pointless on the one hand arguing that musical abilities are fixed in the australopithecines five million years ago, or that a gene for music spread with the Upper Palaeolithic. Neither model has any empirical basis. The chronology, geography, and demography of extinct hominin populations, as well as what we can infer about their adaptations, must act as a straitjacket on reconstructive models. In the light of the evidence I have outlined briefly above, I would emphasize five important elements of that straitjacket.

Homo *specificity*: leaving aside the difficulties of deciding whether the so-called earliest *Homo* really do belong to a monophyletic clade, it is clear that the cognitive and behavioural shifts we associate with human evolution belong specifically to this clade, and thus major changes lie somewhere between approximately 2.0 Mya and the appearance of anatomically and behaviourally modern humans (say before the last 100,000 years). This is not necessarily to imply that all australopithecines and the other genera were 'just like chimpanzees', for they may well have developed their own evolutionary trajectories, some of which may have been behavioural and cognitive.

Graded variation: despite the emphasis placed here upon multiple species, it is clear from both the nature of evolutionary mechanisms and the empirical fossil evidence that distinctions between clades are not always clear-cut. Having divided up the fossil world into species, we should not fall into the trap of assuming they are all clearly demarcated in all traits. There were surely differences between taxa, but they all remain large, complex, intelligent primates.

Lineage, not time, specific: at virtually all points in the last five million years there have been more than one clade (species) of hominin in existence, sometimes several. This was the case even 30,000 years ago, long after modern humans evolved. If this is the case, then linkages between abilities and hominins must be based on their phylogenetic lineage, not a particular time. During the Middle Pleistocene, for example, there may have been several hominin species, some of which may have been much more musical than others.

Complexity: recent discoveries have shown that human evolution is far more complex than was previously thought, and the rate at which the nature of this complexity has emerged in the last decade is staggering. We should therefore remember that simple reconstructions of behaviour and cognition, of the emergence of human abilities, must reflect this complexity and the areas which remain relatively unknown.

Not all periods are equally interesting, but . . .: in recent years it has been the case that much attention has focused, in discussions of modern humans, on the emergence of the modern species and the appearance of so-called fully modern behaviour. In its more extreme versions, the whole of human evolution has been encapsulated in this event. Leaving aside the fact that just these two may be separated by 100,000 years, we should not neglect the fact that, while significant changes may have occurred at the origins of *Homo sapiens*, these were built on capacities that had accumulated over the previous two million years. Without arguing for a return to simple gradualism, we should recognize that there is more than one important period of change in human evolution, and that the evolution of complex traits such as musical abilities was likely to have been more than one event.

A WORD ABOUT NEANDERTHALS

It is probably not too far from the truth that the main cornerstone of most non-specialists in talking about human evolution is the Neanderthals. There are good reasons for this, as they are both the best known of all the extinct hominins and also the one most closely related to modern humans. As a result they have often been used as the reflective mirror from which human uniqueness could be gauged. It is worth briefly considering exactly where Neanderthals fit into human evolution, a prerequisite for determining exactly what they might tell us.

Neanderthals are a particular lineage of extinct hominin that lived in Europe, the Middle East and possibly Central Asia. They probably evolved sometime in the last 350,000 years (Lahr and Foley 2001), although there is considerable controversy, with some arguing for a longer evolutionary history (Hublin 1996). However, the classic and full Neanderthal phenotype is much later in time. They last shared a common ancestor with modern humans between 700,000 and 350,000 years ago (based on genetic and fossil evidence (Green et al. 2006)), and almost certainly became extinct by 25,000 years ago, leaving little or no genetic trace in the human population (Hublin et al. 1996). Morphologically they are very large-brained hominins, with faces and body proportions indicating specialist adaptation to cold. In other ways they are remarkably like modern humans, albeit a robust and muscular form. Behaviourally there is considerable debate about the extent to which they are or are not like modern humans (Fig. 2.3). Some have claimed that a chasm exists between Neanderthals and modern humans, with the former lacking speech, symbolic thought, cross-modular thought, organized foraging, art, bone tools, and, of course, music (Mellars 1999). Such a stark contrast is unwarranted by the evidence. There are several key points to hold in mind about the Neanderthals in considering what they may or may not show us about modern humans. First, they are not antecedent to *Homo sapiens*, but are contemporary. Second, they are, in any analysis, more closely related to modern humans than to any other hominin, and therefore are not part of a general archaic clade relative to *sapiens*. Third, Neanderthals share many characteristics with modern humans—the same Mode 3 technology, burials, hunting behaviour, the use of ochre, to name only those that are least contested. That is not a matter of serious dispute, and suggests that if there are differences in cognitive or behavioural ability, they are subtle ones. The real issue with regard to them is whether these shared traits are plesiomorphic (Lahr and Foley 2001), homoplasic (d'Errico 2003; d'Errico, et al. 1998), or the result of cultural contact (Mellars 1999)—in other words, to understand what Neanderthals can tell us about modern humans we need to know how they came to be so similar in their behaviour and cognition.

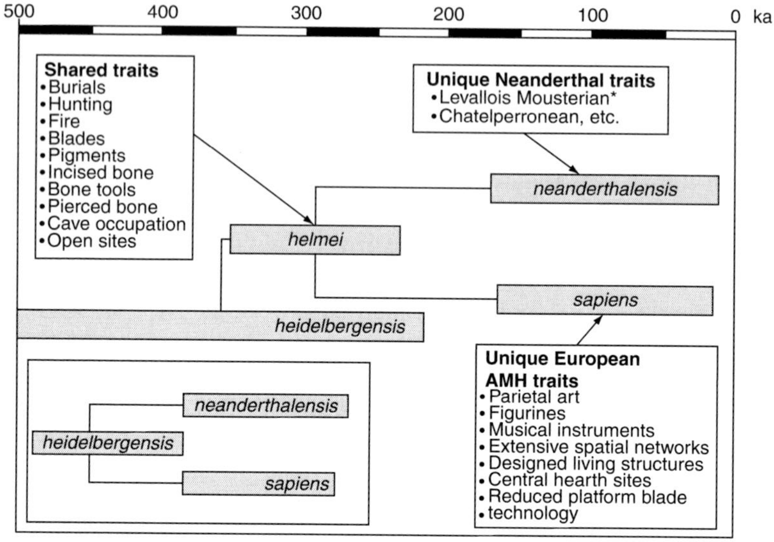

Figure 2.3. Neanderthals and modern humans: the distribution of behavioural characteristics assuming a recent common ancestor; the figure indicates what traits may have been present in the last common ancestor for humans and Neanderthals under this model. The box shows a more conventional model in which *Homo heidelbergensis* is the last common ancestor, in which case many of the traits shown as shared ancestral traits would be interpreted as homoplasies or occurrences of convergent evolution.

Human abilities: mosaics, modules and hopeful musicians

Against this background we can now consider the evidence for how the human abilities that presumably underlie these adaptive radiations may have themselves evolved. The key question central to all the chapters of this book is when and why musical abilities may have become part of the human biological capacity. I would stress that these are general ideas from the perspective of palaeoanthropology, rather than specific empirical conclusions. As indicated above, they arise as suggestions from the straitjacket of the pattern of hominin history and its Darwinian context.

From an evolutionary perspective, we can perhaps consider two key issues here. The first is whether there is a synchronized, integrated evolution of human abilities, such that, at some point in our evolutionary past, all the features that make us human come into operation, effectively as a single package. This can be thought of as the punctuated model, which in practical terms in recent years has been associated with the idea that modern humans have a late and relatively sudden origin. This model has come under a number of names, usefully summed up as the symbolic or human revolution (Mellars and Stringer 1989). The alternative to this idea is that different human abilities have evolved at different periods of time, and that human evolution is the cumulative outcome of these more independent events. I would term this a 'mosaic model'. The second issue is the classic Darwinian algorithm: what are the costs and benefits of a trait—such as music—that might explain why it evolved *in the particular context in which it did.*

THE PATTERN OF EVOLUTION OF HUMAN ABILITIES

Human abilities can be thought of very informally as the things which we can do which are unique to us, and which we can argue provide an important adaptive basis to humanity. A long list can be constructed—language, tool-making, music, ethnic marking, labour organization, planning, and so on. For each of them it is usually possible to find similar abilities in other animals that may be considered either as plesiomorphies (e.g. tool-use in chimpanzees), or as homoplasies (song in some birds), and furthermore such traits, when dissected out, can often be seen to be continuations of those found in other primates.

Those issues have been discussed endlessly, often to no great conclusion, but I will focus here on a slightly different aspect. If we look specifically at the traits which can be argued a) to relate to cognitive and behavioural capacity and b) to be comparatively inferred to some extent in the palaeobiological record, then it should be possible to construct a pattern of cognitive evolution, either through time, or across extinct lineages. Figure 2.4 shows the resulting pattern. The variables on which these graphs are based are technology, life history pattern, and brain size (Foley 1996 and forthcoming). As can be seen, these clearly show that there is not a 'single package' pattern of evolution.

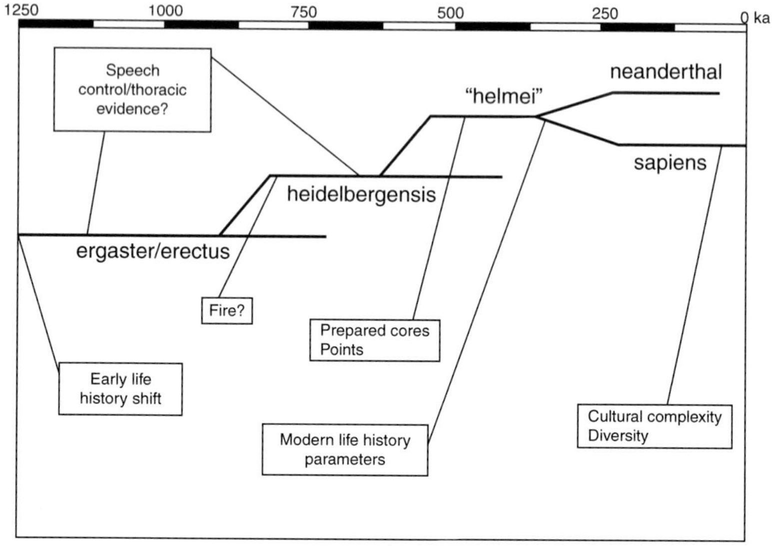

Figure 2.4. A chronology and phylogeny of the evolution of the major behavioural and biological traits of humans.

The features examined are, of course, not directly related to music, but to brain enlargement and to changes in development, in addition to technology. We can add to them the suggestion that the earliest evidence for the controlled breath associated with speech (or song) lies between 1.6 and about 0.3 Mya, and that various other lines of evidence might place the origins of significant linguistic abilities in the range of time between 0.5 and 0.3 Mya in either *Homo heidelbergensis* or *Homo helmei*. Broadly speaking, we could also incorporate the earliest 'symbolic' behaviour in the African Middle Stone Age, most certainly by 70,000 (Henshilwood et al. 2004) and possibly as early as 250,000 years ago (Barham 1998). However, against this, the serious fluorescence of these attributes in the archaeological record does not occur until less than 30,000 years ago, and in many cases it can be argued, not until less than 20,000 years ago. We can summarize these general patterns in schematic form (Figure 2.4). Three general observations can be drawn from this:

1. There is little doubt that there is a mosaic pattern to the evolution of human abilities, in the sense that they do not all appear suddenly, as a package, at a particular time, and certainly not at the time of the proposed demographic bottleneck associated with the origins of

modern humans; indeed, at that time, nothing much may be said to have happened.

2. If, as one tends to assume, there is a strong genetic basis to musicality, then for it to be universally present in the human population it must have been in place more than 150,000 years ago, and more probably more than 200,000 years ago; this means that later attributes cannot represent universal biological evolution, but cultural processes relating to its expression in particular contexts, or population-specific, frequency-based, traits.

3. The key period appears to be between 0.5 and 0.2 Mya. Regrettably, this is a period with relatively few fossil remains. It is also a period where there were probably several clades of hominin present, and it remains to be seen whether the traits we associate with developed human abilities occurred in all of these lineages (and therefore may have been plesiomorphic), or whether they were unique to the African lineage leading to modern humans.

DARWINIAN MUSIC

If the preceding sections have focused broadly on the general pattern of hominin evolution—the context, as it were, that constrains interpretation and hypothesis—it remains to consider briefly and inexpertly the question that lies at the centre of this book: why should an ability to produce, listen to, and appreciate music, in particular forms, have evolved? In general, two types of evolutionary answer to this question can be provided. One is that this particular trait—musicality, we can call it—has no adaptive function, but is the by-product of the evolution of something else (the cheesecake model (Pinker 1997)). For musicality, the usual hypothesis is that music may simply be the outcome of a large and complex brain that is good at generating novel phenomena. This would be the non-adaptive, non-Darwinian, explanation, and I will, as a matter of both heuristic device and principle, say no more about it.

A modified version of this would, though, be that musicality arose non-adaptively but, once acquired, had reproductive consequences which thus made it a focus of selection. This is the concept of exaptation (Gould and Vrba 1983). There may be a case for looking at music in this way, as we are very limited in how we can view the details of its origin, selective functions remain important and we can consider the issue adaptively.

The second type of explanation is that musicality provides individuals with a reproductive advantage, such that the properties of the individual that make him or her able to produce or respond to music are actively selected for. (As an aside, it may be worth remembering that these two aspects—production and appreciation—may be separate to some extent, even if they have a co-evolutionary element.)

Three major models for the emergence of musicality have been proposed. One, originating with Darwin himself, is that the primary selective basis for music is sexual selection; in particular, through either male-male competition or female choice, musical ability plays a part in mate choice and subsequent reproductive success. Miller (2000) has developed this most forcefully. The primary reasons for believing in this hypothesis are: that song in birds is also largely the product of sexual selection (a functional convergence); that males are more likely to be highly proficient in the production of music, while the ability to appreciate music can be found in both sexes; and that much music production relates to mating behaviour.

The second model is one of group cohesion. Much musical activity—most?—occurs in group contexts, and although there may be strong sexual elements, these are not necessarily the strongest ones. Much music serves to provide a shared emotional experience among a group or sub-group. Dunbar (2004), among others, has argued that this aspect of music may be the primary one in shaping musical ability. Support for it may come from the extensive group nature of music, from the existence of martial music in many societies, and from the ethnic specificity of much music, both at genre level and individual songs.

The third model can be thought of as a form of general language model—namely that music is a means of information transfer or signalling in the same way in which language is; although there is overlap with the previous two models, in the sense that are also signalling models, the distinction lies in the fact that the content has greater significance (in the sexual selection model, it may not matter what you sing as long as you are in tune; in the cohesion model, it may not matter what you sing as long as you are in time!).

I do not intend to provide a formal test of the adequacy of these models. Instead I want to highlight briefly some key elements that are important for defining the selective context of the evolution of musicality.

1. *Language, song, and music*: the evidence from brain size and structure, and from the thoracic vertebral foramina, indicates a middle Pleistocene point for the evolution of some speech-based linguistic component. If the key element of this is vocal communication, selected in relation to social complexity and scale as Dunbar (Aiello and Dunbar 1993) has argued, then it could be that in practice the distinction between singing and language may be a later derived one, resulting from selection for two different levels of prosody. This may well avoid the difficulty of the discussions of whether language preceded song, or vice versa. Furthermore, as I have argued elsewhere (Foley 2005b), it would allow for an important distinction to be made between the evolution of thought processes and the evolution of those traits which allow thoughts (and emotions) to be communicated. The selective forces for these are likely to be very different.

2. *Rhythm and emotion*: there is considerable evidence that music is strongly linked to those parts of the brain associated with the emotions, and endless anecdotal evidence concerning the emotional force of music. While this may be associated with mating behaviour, there are at least as strong grounds for linking it to community affiliation and coordination of group behaviour. Some suggested contexts for these may well be greater emphasis on hunting, which requires coordination, and warfare. Here the role of rhythm may be important, and may perhaps be independent of any selection for melodic capacity. In empirical terms, the evidence both for violent traumas in later hominin skeletons and the evidence for greater hunting dependence in the later Pleistocene may be significant.

3. *Life history*: a key element in the evolutionary patterns described above is the evolution of human life history patterns. These are clearly distinct from those of apes and australopithecines, and although they are more complex than this simple summary, can be broadly considered to be focused on increased longevity, extended development, and a later age of first reproduction. Much of what humans are capable of cognitively may well be linked into this critical change in development and the context in which adults spend their lives, including their mortality schedule. Although music is clearly associated in our own culture with the 'higher' forms of art, it is important

to consider how it occurs very much in the context of a particular schedule of development—for example, the role of prosody and melody in infant development, and the frequency of song and music associated with death and major life history transitions.

CONCLUSIONS

In this chapter, I have outlined the broad pattern of hominin evolution. I have argued that since the divergence with the African apes some five or six million years ago, there has been a series of successive radiations, each of which can be characterized by particular adaptive trends. Only within Homo do these trends show the changes in life history, behaviour or brain size which we might associate with higher cognitive abilities and the evolution of abilities such as music. Furthermore, the evidence of the fossil and archaeological records suggests that modern human abilities may have evolved in a mosaic fashion over a period of half a million years or so. It is likely that these were all in place by the time of the appearance of modern humans, although there is a very substantial gap between that point and the frequent expression of these abilities—a gap of more than 150,000 years or 6–7,000 generations (30 times the length of 'civilization').

This pattern provides the context in which the evolution of human abilities should be considered. Genetics, psychology, and many other disciplines have much to contribute to the mechanisms and functions of those abilities, but as the actual evolutionary process is a historical one, in the end it is into that historical context that these models must be placed.

Finally, I have looked very generally at human abilities, rather than specifically at music. There are two reasons for doing this. The first is that it is important to remember that musicality must be integrated with other abilities, and that its adaptive value is in the context of those other traits. For example, it could be that the adaptive value of music is greatly increased in the context of a language ability in which one can say how much one likes or dislikes a particular song. In evolution, the context is everything.

The second reason is that there have been many debates on the evolution of particular human traits—language, hunting, toolmaking, deception, mind-reading, fire-control, cooking, and so on. Looked at in isolation, it is very tempting for particular proponents to place that one trait or another centre stage, and ignore the rest. Often the same arguments are used for each trait. In the end, what is perhaps most striking about the human species is the breadth of its abilities—its ability to generate diversity across many theatres, of which music is only one.

REFERENCES

Aiello L. (1996). 'Terrestriality, Bipedalism and the Origin of Language'. *Proceedings of the British Academy*, 88: 269–89.

Aiello L. and R. Dunbar (1993). 'Neocortex Size, Group Size and the Origin of Language in the Hominids', *Current Anthropology*, 34: 184–93.

Asfaw, B., T. White, O. Lovejoy et al. (1999). '*Australopithecus garhi*: A New Species of Early Hominid from Ethiopia', *Science*, 284: 1623–3.

Atkinson, Q. D., R. D. Gray, and A. J. Drummond (2008). 'mtDNA Variation Predicts Population Size in Humans and Reveals a Major Southern Asian Chapter in Human Prehistory', *Molecular Biology and Evolution*, 25: 468–74.

Barham, L. S. (1998). 'Possible Early Pigment Use in South-Central Africa', *Current Anthropology*, 39: 703–10.

Barkow, L., L. Cosmides, and J. Tooby (1992). *The Adapted Mind*. New York: Oxford University Press.

Barrett, L., R. Dunbar, and J. Lycett (2002). *Human Evolutionary Psychology*. Princeton, NJ: Princeton University Press.

Berger, L. R., D. J. de Ruiter, S. E. Churchill et al. (2010). '*Australopithecus sediba*: A New Species of *Homo*-Like Australopith from South Africa', *Science*, 328: 195–204.

Bowcock, A. M., A. Ruiz-Linares, A. Tomfohrde et al. (1994). 'High Resolution of Human Evolutionary Trees with Polymorphic Microsatellites', *Nature*, 368: 455–7.

Bromage, T. and F. Schrenk (1999). *African Biogeography, Climatic Change, and Hominid Evolution*. Oxford: Oxford University Press.

Cann, R. L., M. Stoneking, and A. C. Wilson (1987). 'Mitochondrial DNA and Human Evolution', *Nature*, 325: 31–5.

Darwin, C. (1871). The *Descent of Man, and Selection in Relation to Sex*. London: John Murray.

Davidson, I. and W. Noble (1991). 'The Evolutionary Emergence of Modern Human Behaviour: Language and Its Archaeology, *Man*, 26: 223–54.

d'Errico, F. (2003). 'The Invisible Frontier. A Multiple Species Model for the Origin of Behavioral Modernity', *Evolutionary Anthropology*, 12: 188–202.

——J. Zilhao, M. Julie, D. Baffier et al. (1998). 'Neanderthal Acculturation in Western Europe? A Critical Review of the Evidence and Its Interpretation', *Current Anthropology*, 39: S1–S44.

Dunbar, R. (2004). *The Human Story*. London: Faber & Faber.

Enard, W., M. Przeworski, S. Fisher et al. (2002). 'Molecular Evolution of FOXP2, a Gene Involved in Speech and Language', *Nature*, 418: 869–72.

Foley, R. A. (forthcoming). 'The Emergence of Culture in the Context of Hominin Evolutionary Patterns'. In: S. Levinson and P. Jaisson (eds.), *Evolution and Culture*. Boston, MA: MIT Press.

——(2005a). 'Species Diversity in Human Evolution: Challenges and Opportunities', *Transactions of the Royal Society of South Africa*, 60: 67–72.

——(2005b). 'The Emergence of Culture in the Context of Hominin Evolutionary Patterns'. In: S. Levinson and P. Jaisson (eds.), *Evolution and Culture*, pp.53–78. Boston, MA: MIT Press.

——(2002). 'Adaptive Radiations and Dispersals in Hominin Evolutionary Ecology', *Evolutionary Anthropology*, 11 (Suppl. 1): 32–7.

——(2001). 'The Evolutionary Consequences of Increased Carnivory in Hominids'. In: C. B. Stanford (ed.), *The Early Human Diet: the Role of Meat*, pp. 305–31. Oxford: Oxford University Press.

——(1996). 'Measuring the Cognition of Extinct Hominids'. In: P. Mellars and K. Gibson (eds.), *Modelling the Early Human Mind*, pp. 57–66. Cambridge: MacDonald Institute for Archaeological Research Monograph.

——(1987a). *Another Unique Species: Patterns of Human Evolutionary Ecology*. Harlow: Longman.

——(1987b). 'Hominid Species and Stone Tools Assemblages: How Are They Related?' *Antiquity*, 61: 380–92.

——and M. M. Lahr (1997). 'Mode 3 Technologies and the Evolution of Modern Humans', *Cambridge Archaeological Journal*, 7: 3–36.

Gagneux, P. and A. Varki (2001). 'Genetic Differences between Humans and Great Apes', *Molecular Phylogenetics and Evolution*, 18: 2–13.

Gonder, M. K., H. M. Mortensen, F. A. Reed et al. (2007). 'Whole-mtDNA Genome Sequence Analysis of Ancient African Lineages', *Molecular Biology and Evolution*, 24: 757–68.

Gould, S. J. and E. S. Vrba (1983). 'Exaptation—a Missing Term in the Science of Form', *Paleobiology*, 8: 4–15.

Green, R. E., J. Krause et al. (2006). 'Analysis of One Million Base Pairs of Neanderthal DNA', *Nature*, 444: 330–6.

Grine, F. E. (2005). 'Early *Homo* at Swartkrans, South Africa: a Review of the Evidence and an Evaluation of Recently Proposed Morphs', *South African Journal of Science*, 101: 43–52.

Henshilwood, C. S., F. d'Errico, M. Vanhaeren et al. (2004). 'Middle Stone Age Shell Beads from South Africa', *Science*, 384: 404.

Hublin, J. J. (1996). 'The First Europeans', *Archaeology*, 49: 36–44.
——F. Spoor, M. Braun et al. (1996). 'A Late Neanderthal Associated with Upper Palaeolithic Artefacts', *Nature*, 381: 224–6.
Klein, R. G. (2009). *The Human Career*. Chicago, IL: Chicago University Press.
——(2000). 'Archeology and the Evolution of Human Behavior', *Evolutionary Anthropology*, 9: 17–36.
——(1995). 'Anatomy, Behaviour and Modern Human Origins', *Journal of World Prehistory*, 9: 167–98.
——(1992). 'The Archaeology of Modern Human Origins', *Evolutionary Anthropology*, 1: 5–14.
Krings, M., C. Capelli et al. (2000). 'A View of Neandertal Genetic Diversity', *Nature Genetics*, 26: 144–6.
——H. Geisert, R. W. Schmitz et al. (1999). 'DNA Sequence of the Mitochondrial Hypervariable Region II from the Neandertal Type Specimen', *Proceedings of the National Academy of Sciences of the United States of America*, 96: 5581–5.
Lahr, M. M. and R. A. Foley (2001). 'Mode 3, *Homo helmei*, and the Pattern of Human Evolution in the Middle Pleistocene'. In: L. Barham and K. Robson Brown (eds.), *Human Roots: Africa and Asia in the Middle Pleistocene*, pp. 23–40. Bristol: Westbury Press.
————(1998). 'Towards a Theory of Modern Human Origins: Geography, Demography, and Diversity in Recent Human Evolution', *Yearbook of Physical Anthropology*, 41: 137–76.
Lewin, R. and R. A. Foley (2003). *Principles of Human Evolution*. Oxford: Blackwell.
Lovejoy, C. O. (2009). 'Reexamining Human Origins in Light of *Ardipithecus ramidus*', *Science* 326, 5949: 74, 74el–74e8.
Marean, C. W., M. Bar-Matthews et al. (2007). 'Early Human Use of Marine Resources and Pigment in South Africa during the Middle Pleistocene', *Nature*, 449: 905–8.
McDougall, I., F. Brown, J. Fleagle (2005). 'Stratigraphic Placement and Age of Modern Humans from Kibish, Ethiopia', *Nature*, 433: 733–6.
Mellars, P. A. (1999). 'CA* Forum on Theory in Anthropology—The Neanderthal Problem Continued', *Current Anthropology*, 40: 341–64.
——(1992). 'Archaeology and the Population-Dispersal Hypothesis of Modern Human Origins in Europe', *Philosophical Transactions of the Royal Society of London B*, 337: 225–34.
——(1991). 'Cognitive Changes and the Emergence of Modern Humans in Europe', *Cambridge Archaeological Journal*, 1: 63–76.
——and C. B. Stringer (1989). *The Human Revolution: Behavioural and Biological Perspectives on the Origins of Modern Humans*. Edinburgh: Edinburgh University Press.

Miller, G. F. (2000). 'Evolution of Human Music through Sexual Selection'. In: N. Wallin, B. Merker, and S. Brown (eds.), *The Origins of Music*, pp. 329–60. Boston, MA: MIT Press.

Mithen, S. (1996). *The Prehistory of the Mind.* London: Thames & Hudson.

Nei, M. and N. Takahata (1993). 'Effective Population Size, Genetic Diversity and Coalescence Time in Sub-Divided Populations', *Journal of Molecular Evolution*, 37: 240–4.

Nesse, R. M., C. T. Bergstrom et al. (2010). 'Making Evolutionary Biology a Basic Science for Medicine', *Proceedings of the National Academy of Sciences, USA*, 107, 1: 1800–7.

Pinker, S. (1997). *How the Mind Works.* New York: Norton.

Quintana-Murci, L., H. Quach et al. (2008). 'Maternal Traces of Deep Common Ancestry and Asymmetric Gene Flow between Pygmy Hunter-Gatherers and Bantu-Speaking Farmers', *Proceedings of the National Academy of Sciences, USA*, 105: 1596–601.

Rightmire, G. P. (1998). 'Human Evolution in the Middle Pleistocene: The Role of *Homo heidelbergensis*', *Evolutionary Anthropology*, 6: 218–27.

Sarich, V. M. (1971). 'A Molecular Approach to the Question of Human Origins'. In: P. Dolinhow and V. M. Sarich (eds.), *Background for Man*, pp. 60–81. Boston, MA: Little, Brown.

Stewart, C. B. and T. R. Disotell (1998). 'Primate Evolution—in and out of Africa', *Current Biology*, 8, R: 582–8.

Stringer, C. B. (2002). 'Modern Human Origins: Progress and Prospects', *Philosophical Transactions of the Royal Society of London, B*, 357: 563–79.

——and J. J. Hublin (1999). 'New Age Estimates for the Swanscombe Hominid, and their Significance for Human Evolution', *Journal of Human Evolution*, 37: 873–7.

Tishkoff, S. A. (2007). 'Convergent Adaptation of Human Lactase Persistence in Africa and Europe', *Nature Genetics*, 39: 31–40.

Underhill, P. and T. Kivisild (2007). 'Use of Y Chromosome and Mitochondrial DNA Population Structure in Tracing Human Migrations', *Annual Review of Genetics*, 41: 539–64.

White, T. D., B. Asfaw, Y. Beyene et al. (2009). '*Ardipithecus ramidus* and the Paleobiology of Early Hominids', *Science*, 326: 75–86.

Williams, S. A. and M. Goodman (1989). 'A Statistical Test that Supports a Human/Chimpanzee Clade Based on Non-Coding DNA Sequence Data', *Molecular Biology and Evolution*, 2: 338–46.

Wood, B. and M. Collard (1999a). 'Anthropology—The Human Genus', *Science*, 284: 65–71.

————(1999b). 'The Changing Face of Genus *Homo*', *Evolutionary Anthropology*, 8: 195–207.

——and N. Lonergan (2008). 'The Hominin Fossil Record: Taxa, Grades and Clades', *Journal of Anatomy*, 212: 354–76.

——(1991). *Koobi Fora Research Project*, iv: *The Hominid Cranial Remains*. Oxford: Clarendon Press.

Wrangham, R. W., J. H. Jones, G. Laden et al. (1999). 'Cooking and the Ecology of Human Origins', *Current Anthropology*, 40: 567–94.

3

The Evolution of the Human Vocal Tract: Specialized for Speech?

Margaret Clegg

INTRODUCTION

The human upper respiratory tract, or vocal tract, appears to have undergone a major reorganization during our evolution. This reorganization has resulted in a wide separation of the soft palate and epiglottis and a consequently low laryngeal position compared to other mammals (Fig. 3.1). This difference is usually presented as both the result of and the reason for our ability to produce the sounds associated with human speech. Furthermore, this low laryngeal position in humans is regarded as conferring a functional disadvantage on us in the form of choking to death on the food we eat.

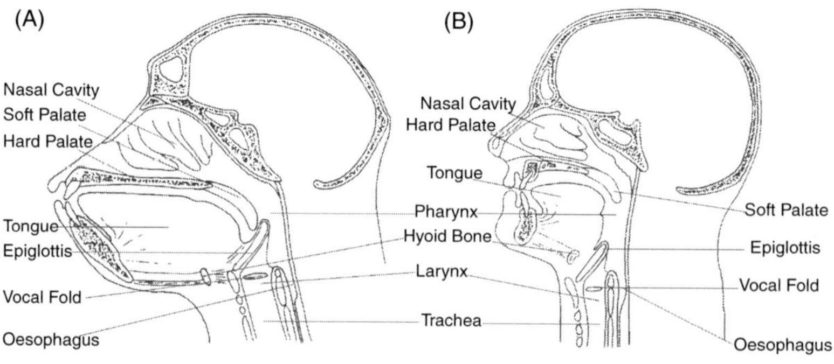

Figure 3.1. Sagittal section of the chimpanzee (A) and human (B) vocal tract. The diagram shows the position and main features of the vocal tract.

The descent of the larynx during human evolution is usually considered to be a gradual process, with the modern configuration being present in archaic *Homo sapiens*. The Neanderthals, however, are regarded as having had a high laryngeal position and less developed speech capabilities. In effect, according to some authors, the Neanderthals reversed the trend to a low laryngeal position.

Evidence will be presented to question these and other commonly held assumptions regarding the morphology of the vocal tract and possible selection pressures resulting in the descent of the human larynx.

Since earliest times, we have been fascinated by why humans alone of all animals have the ability to speak. With the discovery of fossil hominins, people became interested in how and why this capability evolved. The Neanderthals have always been the hominin group that has generated the most controversy and debate concerning the evolution of speech. This chapter presents a range of evidence to support the hypothesis that Neanderthals had vocal tracts similar to those of modern humans.

The earliest dissections of non-human primates showed major similarities between the vocal structures of these primates and modern humans (Cunningham 1997; Hill 1969). So similar are, for example, human and chimpanzee vocal apparatus that many researchers in the eighteenth and nineteenth centuries could not provide an anatomical reason why chimpanzees could not speak (Youati 1835; Hill 1969). They concluded, as did Darwin (1871), that the lack of speech in chimpanzees related to either the size or organization of the brain. However, researchers continued to look for signs of speech on human bones, the 'stigmata' of speech (Spuhler 1977; Keith 1915; see also Dubrul and Reed 1960). They felt that such an important faculty must give rise to skeletal markers that would allow the evolution of this unique character to be distinguished on fossil remains. Unfortunately, many of the skeletal indicators chosen varied between modern individuals and were therefore unreliable (Hooten 1946; Edgar 1952; Vallois 1961).

One major distinction that does exist between humans' and other primates' vocal tracts is the difference in the position of the larynx (see Fig 3.1). Primates, in common with many other mammals, have the larynx positioned high in the pharynx with the soft palate and epiglottis touching. In contrast, modern humans have a different configuration; the human larynx is in a low position with a wide separation between the soft palate and epiglottis (Negus, 1923, 1949; Lieberman et al. 1972, 1982; Liebermann 1985; Laitman et al. 1977, 1978, 1982; Crelin 1987; Harrison 1995; Falk 1975; Larson and Herring 1996; and for modern humans, see Kirchner, 1993; Wind 1976, 1983). This difference in position is easily observed in X-rays, CT scans, and dissections.

The vocal tract, or more correctly the upper respiratory tract, evolved for respiration and ingestion. Vocal abilities in mammalian species use mechanisms that evolved for eating and breathing (Negus 1949; Harrison 1995; Kirchner 1993; Fitch 1997). The muscles that move the larynx to enable breathing and swallowing to occur are therefore the same muscles used in vocalization in mammalian species, including humans.

These laryngeal muscles suspend the larynx from the cranial base via the hyoid bone (Fig. 3.2). The origins and insertion of these muscles are very similar in all primates, including humans. However, there are differences in the relationships between the muscles that take origin on the cranial base (Dean 1984). These relationships are modified in response to changes in the length and breadth of the cranial base (Dubrul and Laskin 1961; Reidenberg and Laitman 1991; Dean 1984). These changes would have had consequences for the action of the muscles if the laryngeal position were to remain unchanged. Some of the laryngeal muscles take origin or insert on the mandible and sternum, so changes in position or orientation of either of these regions would also have had consequences for muscle action.

In the 1970s Lieberman made an innovative proposal that focused on the difference in the position of the larynx in humans and non-human primates and the association between laryngeal position and cranial base length and flexion (Lieberman and Crelin 1971). He also realized that those ontogenetic changes which occur in humans in the position of the larynx and flexion of the cranial base might relate to the development of speech in humans.

Comparison of the vocal tracts of an adult human male, an adult chimpanzee, and a human neonate showed similarities between the chimpanzee and the human neonate, both of whom lacked speech (Lieberman and Crelin 1971). These two had flat cranial bases, a particular angulation of the styloid process and a high laryngeal position. Latex models of both vocal tracts were used to reproduce the range of sounds that chimpanzees and human babies can make. These ranges were then compared to that produced by an adult

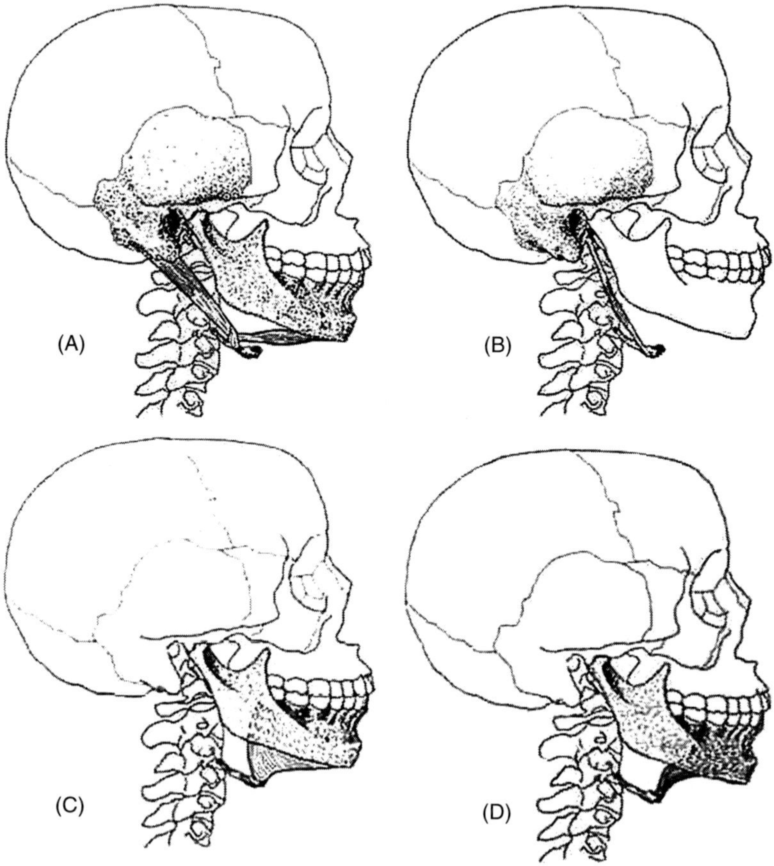

Figure 3.2. The muscle attachment sites on the hyoid bone. A = digastric muscle. B = stylohyoid muscle. C = mylohyoid muscle. D = geniohyoid muscle.

human male. Only the adult human, with his low laryngeal position, had the superlaryngeal space necessary, according to Lieberman, to produce all the sounds of speech (Lieberman and Crelin 1971; Lieberman et al. 1972; Lieberman 1992).

He also proposed that this expanded superlaryngeal space came at a price, that of choking to death on food. This view is widely accepted and is used to illustrate how important speech is to us; if it is correct, then the acquisition of speech compromised our ability to swallow safely and thus jeopardized our own survival, potentially before we could reproduce.

Lieberman then applied this modelling technique to fossil remains. He chose to model the Neanderthal vocal tract first, using the La Chapelle-aux-Saints cranium. He proposed that the similarity of this Neanderthal adult to the adult chimpanzee and human neonate in its cranial base morphology meant that the larynx must have been in a high position. The Neanderthals, he concluded, could not produce the full range of sound associated with human speech (Lieberman and Crelin 1971; Lieberman et al. 1972).

This work was extended by Laitman and colleagues, producing evidence, wholly based on the length and flexion of the cranial base, of the laryngeal position of a wide variety of earlier hominin species. They proposed a gradual descent of the larynx during human evolution, with a child-like rather than neonatal position of the larynx for the later hominins (Laitman et al. 1978, 1979; Laitman 1985; Reidenberg and Laitman 1991; Laitman et al. 1992). However, both Lieberman (1992) and Laitman et al. (1979, 1982) argued that the laryngeal position in the Neanderthals was higher than in earlier hominins or modern humans. In effect they argue that Neanderthals may have reversed the trend for a lower laryngeal position and consequently lost the ability to produce all the sounds of speech.

The discovery in 1983 of the Kebara hyoid bone, with its human-like morphology, suggested that Neanderthals were more like modern humans than previously thought (Arensburg et al. 1989). However, both Lieberman (1990) and Laitman et al. (1990) were dismissive of the importance of this bone, calling it a singular discovery. They also argued that the evidence for its human-like morphology was based on an analysis which could not distinguish species-specific hyoid bone morphology. However, more recent discoveries of hyoid bones from several other species (*Homo neanderthalensis*, *Homo heidelbergensis*, and *Australopithecus afarensis*) have helped to undermine their dismissal.

The models proposed by both Lieberman and Laitman have implicit assumptions underpinning them. These assumptions are:

- The needs of speech production, especially clarity, caused the descent of the larynx.
- Only humans have laryngeal descent and have the necessary space in the pharynx to produce the sounds of speech.
- The low position of the human larynx is costly.

- Neanderthals had a reversal in laryngeal position.
- Neanderthal vocal tracts are shorter than modern human and more child-like in their morphology.
- A single hyoid bone is unlikely to give much information about laryngeal position.

This chapter sets out to look at each of these assumptions and to present evidence, either from the literature or from my own research, that suggests that the model of vocal tract evolution is not as clear-cut as previously claimed. The Lieberman and Laitman models have become the accepted view within anthropology. Although many researchers have questioned many aspects of this work (e.g. Siegal and Carlisle 1974; Burr 1976; Falk 1975; Dubrul 1977; Houghton 1993; Duchin 1990; Arensburg et al. 1989, 1990), these views are normally given less space in textbooks and in the general biological anthropology literature. It is often assumed that the Lieberman/Laitman models are the definitive ones.

What evidence can I present to question these widely held beliefs? It comes in several forms—medical and animal studies, veterinary research and morphological analysis—and often has implications for more than one assumption.

THE DESCENT OF THE LARYNX

Why the larynx descends

It is important to remember that the position of the larynx is determined by several factors. Most of these factors have little to do with speech. The larynx is suspended by a sling of ligaments and muscles from the cranial base and mandible. The positioning of these muscles and their relationship to each other are important in maintaining the ability to swallow safely (Falk, 1975). During the course of human evolution, changes have occurred in both the cranial base, due to bipedalism and brain expansion, and to the face, in the reduction of prognathism, possibly due to dietary changes (Strait 1999; Camperio-Ciani and Chiarelli 1991; Bilsborough and Wood 1988; Gould 1977). These alterations in the relationship between the muscles that control laryngeal movement would have consequences for the action of the muscles if the larynx were to remain in the same high position found in other mammals. Many different sets of circumstances can cause the relationship between these muscles to change.

Permanent changes in head posture in both humans and other mammals, including primates (Adams and Moore 1975) and rats (Reidenberg and Laitman 1991), are associated with a permanent lowering of the larynx compared to their normal counterparts. In humans, such changes can be caused by breathing problems, while in other mammals they may be found where animals have been surgically altered or continually forced to walk bipedally. Differences in facial proportion can also produce a lower laryngeal position. In humans, a variety of medical conditions that lengthen the face also produce a lower laryngeal position compared to those with normal proportions (Joseph et al. 1998). Even a larger than usual mandible will cause a lower laryngeal position than found in individuals with small mandibles (Peng and Cooke 1999). Further evidence for this association is found in experimental rats in whom part of the cranial base is oblated during growth (Dubrul and Laskin 1961; Reidenberg and Laitman 1991). The rats all had lower larynges than normal. Indeed, the removal of a portion of the cranial base also produced many of the cranial changes seen in humans during their evolution, including a globular cranium and flatter face. Those animals in which these changes were incomplete died primarily from problems in swallowing (Reidenberg and Laitman 1991).

These studies all show that alterations in the regions to which the pharyngeal and laryngeal muscles attach influence the position of the larynx regardless of species. The need to swallow safely is the driving factor in the change in laryngeal position. This, rather than the needs of speech, appears to be the major cause of laryngeal descent.

The ontogenetic descent of the larynx

It is often assumed that humans are the only species in which laryngeal descent occurs. In humans there are two periods of laryngeal descent. The first occurs at around two years of age and gives the first permanent loss of contact between the soft palate and epiglottis (Negus 1949; Westhorpe 1987). The second descent in humans happens at puberty (Negus 1949; Westhorpe 1987; Fitch and Giedd 1999; Eveleth and Tanner 1990). However, there is marked sexual dimorphism in the degree of this secondary descent (Negus 1949;

Westhorpe 1987; Fitch and Giedd 1999; Eveleth and Tanner 1990). Boys have a much bigger change in laryngeal position than do girls. The descent in boys is needed to accommodate the much larger larynx that develops at the time the voice 'breaks'. This change is so great that adult human males no longer have a direct relationship between vocal tract length and body size (Fitch and Giedd, 1999). We can all think of examples of men whose voice indicates a large individual but when seen are much shorter than expected.

The assumption that other species do not have laryngeal descent has not been supported by studies on several primate and other mammalian species (Flugel and Rohen 1991; Fitch 2000; Nishimura 2003). Changes in mandibular position and head orientation during growth also alter the relationship between the muscles of the larynx in other animals. Many species seem to show a difference in laryngeal position between juveniles and adults (Avril 1963; Jordan 1971; Flugel and Rohen 1991; Fitch 2000; Nishimura 2003).

During growth the relationship between the hyoid and the mandible needs to be maintained (Flugel and Rohen 1991; Falk 1975) to ensure safe swallowing. Otherwise the changes due to growth would interfere with the action of the muscles controlling swallowing. At a minimum, the hyoid must be below the mylohyoid line (Falk 1975; Aiello and Dean 1990). This is the position at which the muscles between the mandible and hyoid bone attach to the mandible.

We have already seen that human males have a massive secondary growth of the pharynx, which places the larynx in a very low position. This change does not appear to confer any benefit with respect to clearer articulation. No one usually regards men as better at talking than women. Why then should Lieberman's model use the male vocal tract as the human standard? The reason for this is historic. In anatomical studies, men have been regarded as the perfect examples of human anatomy. For this reason, when Lieberman used source-filter theory (Müller 1848; Fant 1960) to explain how the sounds of speech would be produced by the various vocal tract models, he followed accepted practice and used a male vocal tract.

It is the comparisons with this model that produce Lieberman's interpretation. However, as the male vocal tract has such a massive increase in size without adding to articulation, it is perhaps not the best model to use (Pickett 1999). The lower laryngeal position in men seems to be a sexually selected characteristic, either through female choice (women like men with deep voices or with large body size), or male/male competition (trying to convince your opponent you are big). As all human males have this secondary growth of the larynx and pharynx, it seems that the male vocal tract may be for deception rather than any articulatory advantage. There is a further reason for comparisons with male vocal tracts: most researchers in the field are male.

The human female vocal tract may be a better model for earlier hominins, as we have no idea when the extra growth spurt in the pharynx and larynx occurred. Furthermore, the human female vocal tract is not a just scaled-down version of the male vocal tract. It seems to follow a longer and slower growth pattern more in common with other primates (Laver 1991).

Comparisons of the vocal tract lengths estimated for the Neanderthals by Lieberman, Crelin, and Laitman all show that these hominins had vocal tracts shorter than those of modern human males. But more recent work by Clegg (2004) has shown that, although estimates of Neanderthal vocal tract length based on Lieberman's and Laitman's work do not fall within the range for modern human males measured in her study, they are within the range for modern human females. (As both Lieberman and Laitman suggested, the Neanderthals are unlike modern human males. They do however, fall within the range of human variation.) Clegg's study does not model the sounds the Neanderthal vocal tract could produce, as there is no suggestion that human females are limited in the range of sounds they are capable of producing. Therefore, the hypothesis that the Neanderthals had high laryngeal positions and could not produce the sounds of speech cannot be supported.

Functional descent of the larynx

Furthermore, evidence exists that other animals do have the necessary space within the pharynx to produce a wide range of sounds. Evidence from red deer, roaring cats, and primates shows that, to produce certain sounds, the larynx is lowered within the pharynx to expand the pharyngeal space, allowing louder or different ranges of sounds to be produced (Owen 1834; Hast 1989; Fitch 2000). This ability also occurs in children before the larynx reaches its adult position. Children lower the larynx to produce many of the sounds associated with human speech (Crelin 1987). Indeed, human adults do so under many circumstances to change register or increase the volume of the voice (Riordan 1977). This ability is widely used in acting and operatic performances (P. Rodenburg pers. comm.), and also during musical performance on wind instruments (Thomas et al. 1988). One could suggest that it is the ability to manipulate the vocal tract that is more important in the range of sounds produced than simply the resting position of the larynx.

The cost of speech

The low position of the human larynx is often regarded as having a price: that is, speech is costly. There are two aspects to consider when assessing how costly speech really is. The first is that speech is a cooperative activity: it has no point without a listener. Speech is generally a two-way flow of information—conversation is passed back and forth with each party taking turns. This is even seen in mother/child interactions before speech has developed in the child (Jaffe et al. 2001, 1990; Trevarthen and Aitken 1994; Trevarthen 1979; Bruner 1975; Sinclair 1970). We seem adapted to this mechanism. If speech is cooperative, it does not need to be costly.

The second aspect is: what price do we pay for our low larynx? This is generally presented as the risk of choking on the food we eat (Darwin 1871; Lieberman 2000; Wind 1976; Stringer and Gamble 1993). But we seem to have mechanisms in place to ensure we swallow safely. In most normal individuals, swallowing is a smooth, fluid motion, and the relationships between the parts of the vocal tract—particularly the descent of the epiglottis and the rapid rise in the larynx—ensure safe swallowing (Gay et al. 1994; Hiiemae and Crompton 1985). Why then has this problem been proposed? Most studies that discuss speech mention this cost (e.g. Stringer and Gamble 1993; Lieberman 2000; Schepartz 1993; Fitch 1994). It was originally suggested that human swallowing was unique—that we were the only species to swallow across the larynx and all other species had the food stream round an upright epiglottis. (Negus 1923, 1949; Darwin 1871; Lieberman et al. 1972, 1992). However, this was extrapolated from dissections, not from life. It has long been known in the veterinary literature that all adult mammals swallow in the same way, across the larynx (e.g. Getty 1975). Videofluoscopies have now shown this to be so in a wide variety of mammals (Kobora-Mates et al. 1995; German et al. 1991; Hiiemae and Crompton 1985).

The low larynx has also been proposed as a 'design' flaw (Lieberman 2000; Titze and Talkin 1979). This maladaptation is seen as putting us at risk of food dropping into and blocking the larynx. It is always assumed that this causes many deaths every year. One of the commonest statistics quoted in this context is that 3,000 people choke to death on food in the USA (Kitay and Shafer 1989). Much effort, especially in the USA, is directed at preventing these deaths, through teaching the Heimlich manoeuvre (Epperly 1990; Choa et al. 2006).

Remarkably, then, no one until recently has looked at the mortality from choking on food in the context of laryngeal descent. When we examine the figures, we see some surprising results. The overall death rate averages 0.63 per 100,000 (Clegg and Aiello 2000). This is very similar to the results of a medical study by Mittleman and Wetli (1982), who found a consistent death rate from choking on food of 0.66 per 100,000 over a twenty-year period in the USA.

Choking on food is a rare event, and the majority of deaths occur in two groups. In the very young—under one year old, before the larynx descends—they are due to such factors as congenital abnormalities and early infant weaning. The other group are the elderly, for whom death by choking on food is a more recent phenomenon, found in individuals who have survived strokes (Clegg and Aiello 2000). Older individuals would not contribute to gene selection for two reasons: first, they are past the age when death from choking would be important in altering reproductive success; second, they would not have survived the illness that caused the swallowing problems. In those healthy adults who do die, two factors are particularly important: the consumption of alcohol and the use of psychiatric drugs, both of which impair swallowing competence (Bonte and Jacob 1990; Strange 1992). The greatest risk factor for death from choking on food in healthy adults is therefore to be male and drunk, primarily because men drink far more heavily than do women (Johnson and Hosking 1991). Most deaths are therefore due to cultural practices, including infant feeding, consumption of alcohol, and medical intervention, rather than any functional disadvantage in modern humans (Harris et al. 1984; Mittleman and Welti 1982; Fioritti et al. 1997; Hogstel and Robinson 1989). These factors are unlikely to have exerted much in the way of selection pressure on humans. One could argue that alcohol-induced deaths might exert selection pressure; certainly there are anecdotal examples of other animals choking after eating fermented fruit. However, the low level of deaths in modern humans would seem to militate against this argument.

One might also ask what sort of risk these figures imply. One method of assessing mortality risk is through comparison with the Calman chart. This chart (Calman 1996) grades risk into seven bands, 'High' (>1:100), 'Moderate' (1:100–1:1000), 'Low' (1:1000–1:10,000), 'Very Low' (1:10,000–1:100,000), 'Minimal' (1:100,000–1:1,000,000), and 'Negligible' (<1:1,000,000). A death in a railway accident would be classed as minimal risk, while being killed by lightning is a negligible risk. Most of the risk of asphyxiation on food is in the minimal to negligible range. As a cause of death, it is therefore unlikely to have a major impact on overall death rates. It would seem that we do not pay the price suggested for our ability to speak.

Although the evidence is counter-intuitive, we should not really be so surprised. If humans have extended the trend seen in other primates for laryngeal descent and speech as a cooperative activity, then the cost in terms of deaths from choking would be unlikely to be high.

What the hyoid bone can tell us

Further evidence for laryngeal position might be determined from the relationship of the hyoid bone to the cranium and mandible. The hyoid bone is the only bone in the vocal tract and, as soft tissue does not usually fossilize, it is likely to present the only direct evidence of earlier hominin vocal tracts. The hyoid bone has many of the pharyngeal muscles attached to it, and consequently has relationships with various parts of the cranium (Dean 1984, 1985; Bibby and Preston 1981). This allows inferences about hyoid position to be drawn.

The hyoid bone also has species-specific differences in morphology (Saber and Hofman 1985). These distinctions are illustrated in Figure 3.3, where the differences between the human, chimpanzee, and neanderthal hyoids can be clearly seen. The relationship of the various muscles that attach to the hyoid from the cranial base and mandible may be implicated in these morphological differences.

If we look at the relationships within the hyoid bone of two species, humans and chimpanzees, we can see that measurements from the bone produce a different pattern of relationships in the two species (Clegg and Aiello 2002). For example, humans have a greater total maximum breadth (TMB) than total maximum length (TML), but the opposite is true for chimpanzees (Fig 3.4).

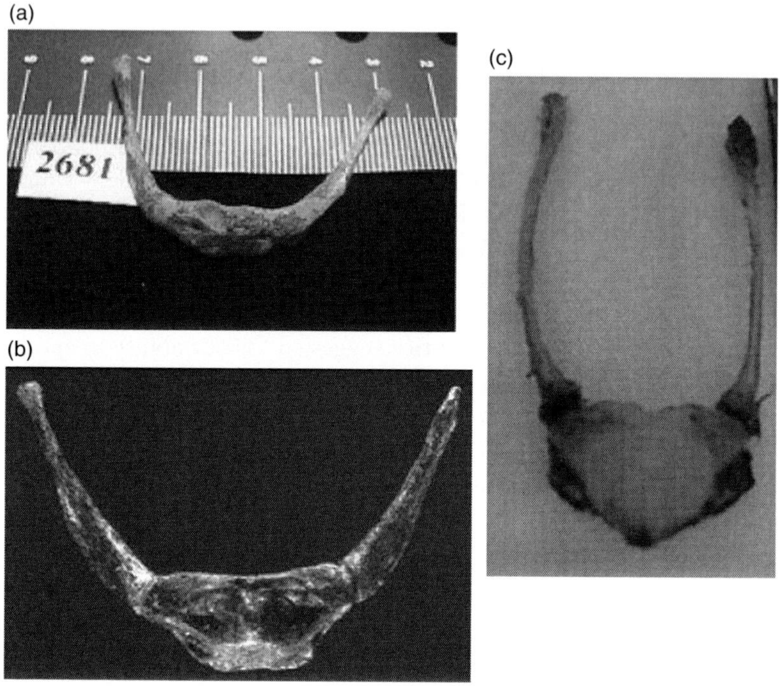

Figure 3.3. The hyoid bone in (a) humans, (b) Neanderthals, and (c) chimpanzees.

If these differences held true across hominin species, then we might expect two lines of evidence: first, that the relationships within the hyoid bone might be more human-like or more ape-like; second, that the relationship between the cranium and the hyoid was either more like a human or more like that of an ape. This would be relatively straightforward to test if we had a good fossil record of hyoid bones. However, the fossil evidence is sparse. Examination of only five complete hyoid bones has so far been published: the Kebara 2 Neanderthal (Arensburg et al. 1989); a second Neanderthal hyoid body (Rodríguez et al. 2003); a juvenile *Australopithecus afarensis* (Alemseged et al. 2006); and two *Homo heidelbergensis* hyoid bones from Sima de los Huesos (Martínez et al. 2008).

The Kebara hyoid bone looks human-like in its morphology, and falls within the measured range of modern human variation (Arensburg et al. 1990). However, the original analysis was criticized as

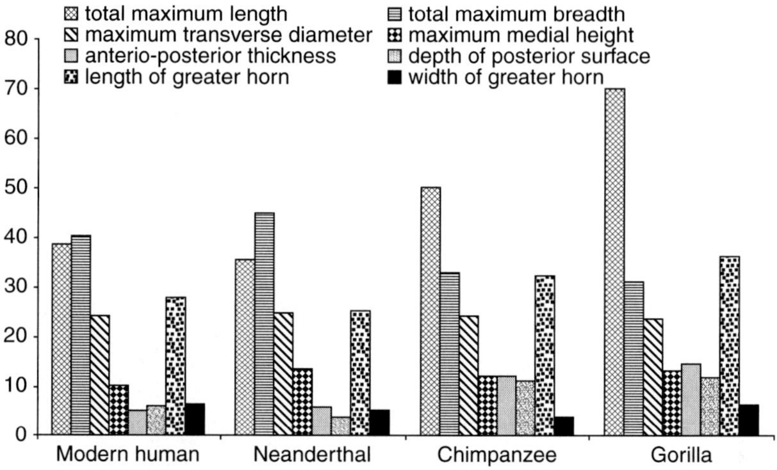

Figure 3.4. Mean hyoid bone measurements in humans, Neanderthals, chimpanzees, and gorillas. The Neanderthal measurements are those observed on the Kebara 2 hyoid bone. Humans and Neanderthals have a greater TMB than TML; the opposite is true for the African apes.

not distinguishing species-specific morphology (Laitman et al. 1990). So to use the hyoid bone to get information about earlier hominin vocal tracts, ideally we need more fossil evidence, preferably from more than a single hominin species. Failing this, we need to find ways to get more information from the evidence we already possess.

One possible method is to determine the relationship between the hyoid bones and crania of humans and chimpanzees, and use this analysis to predict the form of the hyoid bone of earlier hominins and its relationship to the cranium (Clegg and Aiello, 2002). This analysis produced different sets of regression equations for humans and chimpanzees. The Kebara hyoid followed the human pattern, and when a variety of Neanderthal crania were used in the regression equations, only the human equation predicted a hyoid bone similar to the Kebara hyoid bone. Other early hominins, including *P. boisei, A. africanus, Homo erectus,* and an archaic *Homo sapiens,* were also tested to see the type of crania that this Neanderthal hyoid bone might be associated with. Only those from the genus Homo predicted a Kebara-like hyoid bone. The conclusions of this study clearly show that the Kebara hyoid bone would not be associated with either a chimpanzee or australopithecine cranium.

We can see therefore that, early in human evolution, the relationship between the cranium and hyoid bone was similar to that in modern humans; it appears to be in place by the beginning of our own genus. The Neanderthals have the same human-like relationship between the hyoid bone and cranium found in both earlier and later species. The implication of this study is that a human-like vocal tract was in place early in human evolution.

The contribution of the brain to speech and language

Many researchers, including Darwin, have suggested that the most important area in the evolution of language and speech was the brain. Our species-specific ability to develop language was due to a unique reorganization of specific areas in the brain. This view is well founded in the case of children who, for a variety of reasons, lack the ability to produce sounds but can develop a language system such as sign language that they pick up with the ease that other children develop spoken language (e.g. Pinker 1994). It is therefore obvious that there is more to evolving language than having a particular respiratory tract anatomy. As with much of evolution, existing structures are often co-opted for an additional function, and it would seem that our existing respiratory tract was employed to allow our spoken language. It is therefore of interest to understand if other hominins had a similar configuration.

There is also a close relationship between areas of the brain associated with language and those associated with motor control, particularly fine motor control (Rizzolatti et al. 1996). Problems with areas of the brain such as Broca's have been shown to produce deficits in the ability to articulate sounds and control the fine movements of the respiratory tract necessary for speech (e.g. Dronkers 1996). Hand and mouth movements are also closely related and often occur simultaneously, and it seems probable that these language areas arose out of regions associated with fine motor control which is essential for the articulation of speech.

Music and the vocal tract

The simplest musical instrument to use is the human voice. It requires no special ability to make singing sounds. However, that we sound the way we do is due to our particular respiratory anatomy.

Singing was probably one of the earliest ways of making music, and seems to have arisen early in our evolution. Researchers such as Dunbar have suggested that this ability developed or was used to bring about group bonding as group size made the use of grooming impossible (Aiello and Dunbar 1993; Dunbar, this volume). Communal singing does have the ability to produce a feeling of unity in crowds such as those in football stadiums and at church services.

CONCLUSIONS

The evidence presented in this chapter does not support the implicit received assumptions referred to at the opening. The evolutionary descent of the larynx would appear to be due more to physiological factors relating to bipedalism, brain expansion and facial reduction than to the needs of speech. Ontogenetic descent of the larynx has been shown in all mammalian species studied so far. Furthermore, many species lower the larynx to extend the range of sounds produced. The ontogenetic and functional descent of the larynx calls into question the assertions that only humans produce the requisite type of vocal tract required to produce the sounds of speech, and that humans alone have an ontogenetic descent of the larynx. The anatomy of the vocal tract does not compromise our ability to swallow safely. This price of speech is a rare event, related more to cultural practices such as infant feeding, medical interventions, and the consumption of alcohol than to our anatomy or physiology.

Estimates of Neanderthal vocal tract lengths fall within the range of modern human variation. Although these estimates produce a vocal tract shorter than that found in human males, they fall within the range of modern human females. The Neanderthals would not, therefore, have a restricted range of speech sounds. It is unlikely that they reversed the trend for a low laryngeal position compared to earlier hominins. The hyoid bone has relationships with the cranium and mandible which allow estimates of both hyoid bone morphology and implications regarding laryngeal position to be drawn. The Neanderthal hyoid bone would not be associated with either chimpanzee or australopithecine crania. The morphology of the hyoid bone, and by implication the position of the larynx, has been human-like since the evolution of our genus. Therefore, the ability to produce the full range of human speech sounds, but not necessarily fully human speech or language, probably evolved early in human evolution.

If this line of reasoning is correct, then our ability to speak developed from our existing anatomy and capabilities. Humans sound the way they do by accident, not design. The anatomy of the vocal tract that allows us to produce the sounds of speech is fortuitous, and not a necessary prerequisite.

REFERENCES

Adams, L. M. and W. J. Moore (1975). 'A Biomechanical Appraisal of Some Skeletal Features Associated with Head Balance and Posture in Hominoidea', *Acta Anatomica*, 92: 580–94.

Aiello L. C. and C. Dean (1990). *Human Evolutionary Anatomy.* London: Academic Press.

—— and R. Dunbar (1993). 'Neocortek Size, Group Size and the Evolution of Language', *Current Anthropology*, 34: 184–93.

Alemseged, Z., F. Spoor, W. H. Kimbel et al. (2006). 'A Juvenile Early Hominin Skeleton from Dikika, Ethiopia', *Nature*, 443, 7109: 296–301.

Arensburg, B., L. A. Schepartz, A. M. Tillier et al. (1990). 'A Reappraisal of the Anatomical Basis for Speech in Middle Palaeolithic Hominids', *American Journal of Physical Anthropology*, 83: 137–56.

——A. M. Tillier, H. Duday et al. (1989). 'A Middle Palaeolithic Human Hyoid Bone', *Nature*, 338: 758–60.

Avril, C. (1963). 'Kehlkopf und Kehlsack des Schimpansen (*Pan troglodytes*)', *Gegenbaurs Morphologisches Jahrbuch*, 105: 74–129.

Bibby, R. E. and C. B. Preston (1981). 'The Hyoid Triangle', *American Journal of Orthodontics*, 80, 1: 92–7.

Bilsborough, A. and B. A. Wood (1988). 'Cranial Morphology of Early Hominids: Facial Region', *American Journal of Physical Anthropology*, 76: 61–86.

Bonte, W. and B. Jacob (1990). 'Death by Bolus Impaction in the Larynx', *Acta Medicinae Legalis et Socialis*, 40: 139–51.

Bruner, J. (1975). 'The Ontogenesis of Speech Acts', *Journal of Child Language*, 2, 1: 1–19.

Burr, B. D. (1976). 'Neanderthal Vocal Tract Reconstructions: A Critical Appraisal', *Journal of Human Evolution*, 5: 285–90.

Calman, K. C. (1978). 'Cancer, Science and Society and the Communication of Risk', *British Medical Journal* 313: 799–802.

Camperio-Ciani, A. and B. Chiarelli (1991). 'A Systematic Relationship between Brain Size Increase and Larynx Transformation during Hominization'. In:

J. Wind, B. Chiarelli, B. Bichakjian et al. (eds.), *Language Origins: A Multidisciplinary Approach*, pp. 51–66. Dordrecht: Kluwer Academic Publishers.

Clegg, M. (2004). 'A New Model for the Neanderthal Vocal Tract', *American Journal of Physical Anthropology*, S38: 76.

——and L. C. Aiello (2000). 'Paying the Price of Speech? Analysis of Mortality Statistics for Choking on Food', *American Journal of Physical Anthropology*, S30: 126.

————(2002). 'Estimating Hyoid Bone Morphology in Earlier Hominin Species', *American Journal of Physical Anthropology*, S34: 54.

Choa, M. H., I. C. Park, H. S. Chung et al. (2006). 'Internet-Based Animation for Instruction in Cardiopulmonary Resuscitation', *Journal of Telemedicine and Telecare*, 12SU, Suppl. 3: 31–3.

Crelin, E. S. (1987). *The Human Vocal Tract, Anatomy, Function, Development and Evolution*. New York: Vantage Press.

Cunningham, A. (1997). *The Anatomical Renaissance: The Resurrection of the Anatomical Projects of the Ancients*. Aldershot: Scolar.

Darwin, C. (1859). *On the Origin of Species by Means of Natural Selection*. London: John Murray.

——(1871). *The Descent of Man, and Selection in Relation to Sex*. London: John Murray.

Dean, M. C. (1984). 'The Comparative Myology of the Hominoid Cranial Base I: The Muscular Relations and Bony Attachments of the Digastric Muscle', *Folia Primatologica*, 43: 157–80.

——(1985). 'The Comparative Myology of the Hominoid Cranial Base II: The Muscles of the Prevertebral and Upper Pharyngeal Region', *Folia Primatologica*, 44: 40–51.

Dronkers, N. F. (1996). 'A New Brain Region for Co-ordinating Speech Articulation', *Nature* 384: 159–61.

Dubrul, E. L. and C. A. Reed (1960). 'Skeletal Evidence of Speech?' *American Journal of Physical Anthropology*, 18: 153–6.

——and D. M. Laskin (1961). 'Preadaptive Potentialities of the Mammalian Skull: An Experiment in Growth and Form', *American Journal of Anatomy*, 109: 117–32.

——(1977). 'Early Hominid Feeding Mechanisms', *American Journal of Physical Anthropology*, 47: 305–20.

Duchin, L. (1990). 'The Evolution of Articulated Speech: Comparative Anatomy of the Oral Cavity in Pan and Homo', *Journal of Human Evolution*, 19: 687–97.

Edgar, G. W. F. (1952). 'A Reconsideration of the Possible Relation between Articulate Speech and the Presence of Genioglossal Spines (Spinae m. genioglossi) in Man', *Proc. Konintel. Nederl. Akademic van Wetenschappen*, 55: 21.

Ekberg, O. and M. Feinberg (1992). 'Clinical and Demographic Data in 75 Patients with Near Fatal Choking Episodes', *Dysphagia*, 7, 4: 205–8.

Epperly, T. D. (1990). 'Teaching the Heimlich-Maneuver', *Postgraduate Medicine*, 88, 8: 31.

Eveleth, P. B. and J. M. Tanner (1990). *Worldwide Variations in Human Growth*, 2nd edn. Cambridge: Cambridge University Press.

Falk, D. (1975). 'Comparative Anatomy of the Larynx in Man and the Chimpanzee: Implications for Language in Neanderthal', *American Journal of Physical Anthropology*, 43: 123–32.

Fant, G. (1960). *Acoustic Theory of Speech Production.* The Hague: Mouton.

Fioritti, A., L. Gioccotto, and V. Melegal (1997). 'Choking Incidents among Psychiatric Patients: Retrospective Analysis of Thirty-One Cases', *Canadian Journal of Psychiatry*, 42, 5: 515–20.

Fitch, W. T. (2000). 'The Phonetic Potential of Nonhuman Vocal Tracts: Comparative Cineradiographic Observations of Vocalizing Animals', *Phonetica*, 57: 205–18.

——(1997). 'Vocal Tract Length and Format Frequency Dispersion Correlate with Bodysize in Rhesus Macaques', *Journal of the Acoustic Society of America*, 102: 1213–22.

——(1994). 'Vocal tract length and the evolution of language'. PhD thesis, Brown University. Ann Arbor, Michigan: UMI Dissertation Services.

——and J. Giedd (1999). 'Morphology and Development of the Human Vocal: A Study Using Magnetic Resonance Imaging', *Journal of the Acoustic Society of America*, 106, 3, 1: 1511–22.

Flugel, C. and J. W. Rohen (1991). 'The Craniofacial Proportions and Laryngeal Position in Monkeys and Man of Different Ages. (A Morphometric Study Based on CT-scans and Radiographs)', *Mechanisms of Ageing & Development*, 61: 65–83.

Gay, T., J. K. Rendell, J. Spiro et al (1994). 'Coordination of Oral Cavity and Laryngeal Movements during Swallowing', *Journal of Applied Physiology*, 77, 1: 357–65.

German, R. Z., A. W. Crompton, L. C. Levitch et al. (1991). 'The Mechanism of Suckling in Two Species of Infant Mammals: Miniature Pigs and Long Tailed Macaques', *Journal of Experimental Zoology*, 261: 322–30.

Getty, R. (1975). *Sisson and Grossman's The Anatomy of the Domestic Animals*, 5th edn. Philadelphia: Saunders.

Gould, S. J. (1977). *Ontogeny and Phylogeny*. Cambridge, MA: Harvard University Press.

Harris, C. S., S. P. Baker, G. A. Smith et al. (1984). 'Childhood Asphyxiation by Food: A National Analysis and Overview', *Journal of the American Medical Association*, 251, 1,7: 2231–5.

Harrison, D. F. N. (1995). *The Anatomy and Physiology of the Mammalian Larynx*. Cambridge: Cambridge University Press.

Hast, M. (1989). 'The Larynx of Roaring and Nonroaring Cats', *Journal of Anatomy*, 163: 117–21.

Hiiemae, K. M. and A. W. Crompton (1985). 'Mastication, Food Transport, and Swallowing'. In: M. Hildebrand, D. Bramble, K. Liem et al. (eds.), *Functional Vertebrate Morphology*, pp. 262–90. Cambridge, MA: Belknap Press.

Hill, W. C. O. (1969). 'The Discovery of the Chimpanzee'. In: G. Bourne (ed.), *The Chimpanzee*, i, pp. 1–21. Basel and New York: Karger.

Hogstel, M. O. and N. B. Robinson (1989). 'Feeding the Frail Elderly', *Journal of Gerontological Nursing*, 15, 3: 16–20.

Hooten, E. A. (1946). *Up from the Ape*. New York: Macmillan.

Houghton, P. (1993). 'Neanderthal Supralaryngeal Vocal Tract', *American Journal of Physical Anthropology*, 90: 139–46.

Jaffe, J., B. Beebe, S. Feldstein et al. (2001). 'Rhythms of Dialogue in Infancy: Coordinated Timing in Development—Introduction', *Monographs of the Society for Research in Child Development*, 66, 2: 1–132.

——S. W. Anderson et al. (1990). 'Conversational and Anticonversational Turntaking in Mother Infant Vocal Interactions', *Brain and Language*, 39, 4: 611–12.

Johnson, C. D. and S. Hosking (1991). 'National Statistics for Diet, Alcohol Consumption and Chronic Pancreatitis in England and Wales', *Gut*, 32, 11: 1401–5.

Jordan, J. (1971). 'Studies on the Organ of Voice and Vocalisation in Chimpanzees', *Folia Morphologica* (Warsz), 30: Pt 1, 97–126; Pt 2, 222–48; Pt 3, 323–40.

Joseph, A. A., J. Elbaum, G. J. Cisneros et al. (1998). 'A Celephometric Comparative Study of Soft Tissue Airway Dimensions in Persons with Hyperdivergent and Normodivergent Facial Patterns', *Journal of Oral and Maxillofacial Surgery*, 56, 2: 135–9.

Keith, A. (1915). *The Antiquity of Man*. London: Williams & Norgate.

Kirchner, J. A. (1993). 'The Vertebrate Larynx: Adaptations and Aberrations', *Laryngoscope*, 103, 10: 1197–201.

Kitay, G. and N. Shafer (1989). 'Café Coronary: Recognition, Treatment, and Prevention', *Nurse Practitioner*, 14, 6: 35–8, 43, 46.

Klatt, D. H. and L. C. Klatt (1990). 'Analysis, Synthesis, and Perception of Voice Quality Variations among Female and Male Talkers', *Journal of the Acoustic Society of America*, 87, 2: 820–57.

Kobora-Mates, M., J. A. Logemann, C. Larson et al. (1995). 'Physiology of Oropharyngeal Swallowing in the Cat: A Videofluroscopic Study', *American Journal of Physiology*, 268: 232–41.

Laitman, J. T. (1985). 'Evolution of the Hominid Upper Respiratory Track: The Fossil Evidence'. In: P. V. Tobias (ed.), *Hominid* Evolution: Past, Present and Future. New York: Alan R. Liss.

—— E. S. Crelin, and G. J. Conglogue (1977). 'The Function of the Epiglottis in Monkey and Man', *Yale Journal of Biology and Medicine*, 50: 43–8.

——P. J. Gannon, and J. S. Reidenburg (1989). 'Charting Changes in the Hominid Vocal Tract', *American Journal of Physical Anthropology*, 78, 2: 257–8.

—— and R. C. Heimbuch (1982). 'The Basicranium and Upper Respiratory System of the Australopithicines', *American Journal of Physical Anthropology*, 52, 2: 204.

————and E. S. Crelin (1978). 'Developmental Change in a Basicranial Line and its Relationship to the Upper Respiratory System in Living Primates', *American Journal of Anatomy*, 152: 467–82.

——————(1979). 'The Basicranium of Fossil Hominids as an Indicator of Their Upper Respiratory System', *American Journal of Physical Anthropology*, 51: 15–34.

——J. S. Reidenburg, P. J. Gannon et al. (1990). 'The Kebara Hyoid: What Can It Tell Us about the Evolution of the Hominid Vocal Tract', *American of Physical Anthropology*, 81, 2: 254.

————D. R. Friedland et al. (1992). 'The Demise of the Neanderthals: The Respiratory Specialisation Hypothesis', *American Journal of Physical Anthropology*, Supp 14: 104.

Larson, J. E. and S. W. Herring (1996). 'Movement of the Epiglottis in Mammals', *American Journal of Physical Anthropology*, 100: 71–82.

Laver, J. (1991). *The Gift of Speech: Papers in the Analysis of Speech and Voice*. Edinburgh: Edinburgh University Press.

Lieberman, P. (1979). 'Hominid Evolution, Supralaryngeal Vocal Tract Physiology and the Fossil Evidence for Reconstructions', *Brain and Language*, 7, 1: 101–26.

—— (1982). 'Can Chimpanzees Swallow or Talk? A Reply to Falk', American Anthropologist, 84: 148–52.

——(1983). 'On the Nature and Evolution of the Biological Basis of Language'. In: E. de Grolier, A. Lock, C. R. Peters et al. (eds.), *Glossogenetics: The Origin and Evolution of Language*, pp. 91–113. New York: Harwood Academic Press.

——(1985). 'On the Evolution of Human Syntactic Ability. Its Pre-Adaptive Base-Motor Control and Speech', *Journal of Human Evolution*, 14: 657–68.

——(1990). 'Hyoid Bone Position and Speech: Reply to Dr Arensberg', *American Journal of Physical Anthropology*, 94: 275–8.

——(1991). 'On the Evolutionary Biology of Speech and Syntax'. In J. Wind, B. Chiarelli, B. Bichakjian et al. (eds.), *Language Origins: A Multidisciplinary Approach*, pp. 409–29. Dordrecht: Kluwer Academic Publishers.

——(1992). 'On Neanderthal Speech and Neanderthal Extinction', *Current Anthropology*, 33, 4: 409–10.

——(2000). *Human Language and our Reptilian Brain: The Subcortical Bases of Speech, Syntax and Thought*. Cambridge, MA: Harvard University Press.

—— and E. S. Crelin (1971). 'On the Speech of Neanderthal Man', *Linguistic Inquiry*, 2: 203–22.

———— and D. H. Klatt (1972). 'Phonetic Ability and Related Anatomy of the Newborn and Adult Human, Neanderthal Man, and the Chimpanzee', *American Anthropologist*, 74, 3: 287–307.

——J. T. Laitman, J. S. Reidenberg et al. (1992). 'The Anatomy, Physiology, Acoustics and Perception of Speech: Essential Elements in Analysis of the Evolution of Human Language', *Journal of Human Evolution*, 23: 447–67.

Martínez, I., J. L. Arsuaga, R. Quam et al. (2008). 'Human Hyoid Bones from the Middle Pleistocene Site of the Sima de los Huesos (Sierra de Atapuerca, Spain)', *Journal of Human Evolution*, 54, 1: 118–24.

Mittleman, R. E. and C. V. Wetli (1982). 'The Fatal Cafe Coronary: Foreign Body Airway Obstruction', *Journal of the American Medical Association*, 247, 9: 1285–8.

Müller, J. (1848). *The Physiology of the Senses, Voice and Muscular Motion with Mental Facilities*, trans. W. Baly. London: Walton & Maberly.

Negus, V. E. (1923). *The Mechanism of the Larynx*. London: William Heinemann Medical Books.

——(1949). *The Comparative Anatomy of the Larynx*. London: William Heinemann Medical Books.

Nishimura, T. (2003). 'Comparative Morphology of the Hyo-Laryngeal Complex in Anthropoids: Two Steps in the Evolution of the Descent of the Larynx', *Primates*, 44: 41–9.

Owen, R. (1834). 'On the Anatomy of the Cheetah', *Transactions of the Zoological Society of London*, 1: 129–36.

Peng, L. and M. S. Cooke (1999). 'Fifteen-Year Reproducability of Natural Head Posture: A Longitudinal Study', *American Journal of Orthodontics and Dentofacial Orthopedics*, 116, 1: 82–5.

Pickett, J. M. (1999). *Acoustics of Speech Communication: Fundamentals, Speech Perception Theory, and Technology*. Boston, MA: Allyn and Bacon.

Pinker, S. (1994). *The Language Instinct*. New York: Harper Perennial.

Reidenberg, J. S. and J. T. Laitman (1991). 'Effect of Basicranial Flexion on Larynx and Hyoid Position in Rats: An Experimental Study of Skull and Soft Tissue Interactions', *The Anatomical Record*, 230: 557–69.

Riordan, C. J. (1977). 'Control of Vocal-Tract Length in Speech', *Journal of the Acoustical Society of America*, 62, 4: 998–1002.

Rizzolatti, G., L. Fadiga, V. Gallese et al. (1996). 'Premotor Cortex and the Recognition of Motor Actions', *Cognitive Brain Research*, 3, 2: 131–41.

Rodríguez, L., L. L. Cabo, and J. E. Egocheaga (2003). 'Breve nota sobre el hioides Neandertalense de Sidrón (Piloña, Asturias)'. In: R. Malgosa, Nogués. and M. Aluja (eds.), *Antropología y Diversidad*, i: 484–93. Barcelona: Edicions Bellaterra.

Saber, A. S. and R. R. Hofmann (1985). 'A Comparative Study of the Hyoid Bone (Os Hyoides) of Six Ruminant Species (Ruminantia Scopoli, 1777)', *Anatomischer Anzeiger*, 158, 1:43–50.

Schepartz, L. A. (1993). 'Language and Modern Human Origins', *Yearbook of Physical Anthropology*, 36: 1–126.

Siegal, M. I. and R. C. Carlisle (1974). 'Some Problems in the Interpretation of Neanderthal Speech Capabilities: A Reply to Lieberman', *American Anthropologist*, 76: 319–23.

Sinclair, H. (1970). 'The Transition from Sensory-Motor Behaviour to Symbolic Activity', *Interchange*, 1, 3: 119–26.

Spuhler, J. N. (1977). 'Biology, Speech and Language', *Annual Review of Anthropology*, 6: 509–61.

Strait, D. (1999). 'The Scaling of Basicranial Flexion and Length', *Journal of Human Evolution*, 37: 701–19.

Strange, P. G. (1992). *Brain Biochemistry and Brain Disorders*. Oxford: Oxford University Press.

Stringer, C. and C. Gamble (1993). *In Search of the Neanderthals: Solving the Puzzle of Human Origins*. London: Thomas & Hudson.

Thomas, L., G. J. Troup, and N. Ferris (1988). 'Vocal-Tract Shapes of a Flautist in Performance', *Journal of the Acoustical Society of America*, 84, 4: 1549–50.

Titze, I. R. and D. T. Talkin (1979). 'A Theoretical Study of the Effects of Various Laryngeal Configurations on the Acoustics of Phonation', *Journal of the Acoustical Society of America*, 66, 1: 60–74.

Trevarthen, C. (1979). 'Communication and Cooperation in Early Infancy: A Description of Primary Intersubjectivity'. In: M. Bullowa (ed.), *Before Speech: The Beginning of Interpersonal Communication*, pp. 321–47. Cambridge: Cambridge University Press.

——and K. J. Aitken (1994). 'Brain Development, Infant Communication and Empathy Disorders: Intrinsic Factors in Child Mental Health', *Development and Psychopathology*, 6: 599–635.

Vallois, H. V. (1961). 'The Social Life of Early Man: The Evidence of Skeletons'. In: S. Washburn (ed.), *Social Life of Early Man*, pp. 214–35. Chicago, IL: Aldine Publishing.

Westhorpe, R. N. (1987). 'The Position of the Larynx in Children and its Relationship to Ease of Intubation', *Anaesthesiology Intensive Care Medicine*, 15, 4: 384–8.

Wind, J. (1976). 'Phylogeny of the Human Vocal Tract', *Annals of the York Academy of Sciences*, 280: 612–30.

——(1983). 'Primate Evolution and the Emergence of Speech'. In: E. de Grolier, A Lock, C. R. Peters et al. (eds.), *Glossogenetics: The Origin and Evolution of Language*, pp. 15–35. New York: Harwood Academic Press.

Youati, W. (1835). 'Account of the Habits and Illness of the Late Chimpanzee', *The Lancet*, 26, 662: 202–6.

4

When the Words Dry Up: Music and Material Metaphors Half a Million Years ago

Clive Gamble

Musical performances force us to think about ways of being that prioritize not what people think but how they feel

Nichola Wood and Susan J. Smith (2004, p. 540)

We can best understand music by being in the middle of it

Nicholas Cook (1998, p. 84)

INTRODUCTION

Music provides a particular challenge to archaeologists interested in human evolution. In this contribution, the focus is moved away from identifying material proxies such as flutes to consider instead the significance of performance spaces. The importance of emotion in social life is stressed, and a model developed that relates music to the changing dominance of container metaphors based on the body. It is concluded that music has always been a part of hominin social life, but that during evolution it was co-opted to enhance positive emotions at hominin gatherings.

The business of becoming modern must, it seems, always be delayed. Universal declarations, such as that on the human genome and human rights (UNESCO 1997) speak of 'the heritage of humanity'. However, in its present cultural and genetic diversity, this modern project does not even extend back in time far enough to include recent hominin ancestors. Therefore, the process of modernity is regarded as delayed (Proctor 2003; Gamble 2007) and it is the business of the palaeoanthropologist working within this paradigm to uncover when and how.

Music is one facet of this universal heritage. It is often used, like language and other cultural capacities such as decoration, painting, sculpture, and notation, to raise fences between neighbouring ancestors. These elements, and the varied religious beliefs, moral, economic and political philosophies they support, comprise the human endowment that was hard won during evolution. It is as though a basket of eggs has been slowly filled over a long period until in modern times it is finally full.

In this contribution I take issue with the model of a universal human endowment that accrued in this manner because, in the process, humans are excluded from a proper evolutionary perspective. I will propose that the way to escape from the unresolvable, and unhelpful, debates about who exactly in the past was a 'modern human' (Ingold 2000; Gamble 2007) is to take a fresh perspective on what we regard as social life; to be less impressed by universal properties of the mind, and the endowment these created, and more with the experiences of the body and the actions they entailed. And it is here that music provides a key to unlock a door for archaeologists.

MUSIC AND MATERIAL PROXIES FROM THE PAST

For many palaeoanthropologists, the conceptual basket into which these evolutionary gifts are poured is the 'anatomically modern human' (AMH); a creature born from an understandable desire to counteract the political exploitation of human variation by locating our point of origin in some apparently neutral place in the remote past (Gamble 2007, Chapter 2). But AMH also bears witness to a deeper intellectual project where in all forms of investigation, including human ancestry, modernity requires evidence that it is always deferred (Thomas 2004).

Briefly, a combination of anatomical, archaeological, and genetic information now provides a minimum time depth of some 200,000 years ago for the origins of such AMH, *Homo sapiens sapiens*. After this time, the endowments mentioned above slowly accumulated so that eventually a not-fully-modern taxon was transformed into a fully-modern-person. Watching this process from the other side of the fence are similar, but different creatures—the Eurasian Neanderthals, Indonesian *Homo floresiensis* and the East Asian *Homo erectus*—who despite all being contemporaries and belonging to the genus *Homo* will never, it seems, attain a full basket of gifts.

Palaeoanthropologists, because they study populations, societies, and cultures, rather than individuals (Gamble and Porr 2005a), find it easy to subscribe to this set of human universals that accrued, and which include music and language among them. Therefore, their main task has been to discover when such aspects appeared, and the methods employed have been to use material forms as proxies.

Perhaps the two most familiar proxies are cave art and village architecture. The former has been used, and disputed, as a proxy for the appearance of language and symbolism (Noble and Davidson 1996). The latter is regarded as evidence for corporate social structures and the kind of built environment that a heightened, political symbolism feeds off (Watkins 2004). Music, as others have documented (D'Errico et al. 2003), is poorly served by comparable material proxies. The evidence for musical instruments is meagre and late in date. This circumstance contributes to the view that language was a prior necessity not only for music but also for ritual, ceremony, and dance. In fact everything that makes us human. Language came first, the rest followed; the cheesecake after the main course (Pinker 1997).

Not everyone agrees. Darwin inferred from the association between music and the strongest emotions—ardent love, rivalry and triumph—that

> the progenitors of man probably uttered musical tones before they had acquired the power of articulate speech; and that consequently, when the voice is used under any strong emotion, it tends to assume, through the principle of association, a musical character. (1872, p. 92)

Even so, it seems for many archaeologists that life without spoken language was always symbolically silent and conceptually limited, dependent instead on affordances from environmental cues rather than the cultural interpretation of signs. Such views explain the importance attached to material proof, and where in the absence of suitable proxies such as drums and flutes absence is taken as musical silence. But is language both a necessary and sufficient requirement for these other aspects of the human endowment? Has Darwin, yet again, got the order right?

To explore these questions I will examine other archaeological evidence and concentrate not on the instruments but rather on the performance spaces that arose from social gatherings in the past. These spaces related people and material culture and they survive particularly well, irrespective of age (Gamble and Porr 2005b). For the most part they are interpreted from a rational perspective as evidence for the manufacture of stone tools, the killing of large animals and their subsequent consumption. My aim is to show that we can do more with them than make these limited inferences. These performance spaces leave traces that can be interpreted as material metaphors, and in particular the experience of containment. Besides a rational assessment of the evidence we can now offer a relational interpretation, and where the basis of knowledge is embodied in emotions that would have been common to all hominins, much as Darwin indicated.

With the playing field now levelled between different hominin species we can then understand music as one means by which the emotional intensity of these gatherings was heightened. This opens up the potential to link changes in the means of performance to the selection pressure arising from complex social life on our embodied cognition (Dunbar 2003).

MUSIC AND PERFORMANCE

In *The Raw and the Cooked* Lévi-Strauss gets to the heart of the matter when it comes to music by opposing rational and relational approaches to understanding.

> But that music is a language by whose means messages are elaborated, that such messages can be understood by the many but sent only by the few, and that it alone among all the languages unites the contradictory character of being at once intelligible and untranslatable—these facts make the creator of music a being like the gods and makes music itself the supreme mystery of human knowledge. All other branches of knowledge stumble into it, it holds the key to their progress (in Leach 1970, p. 115).

He describes music and myth as 'machines for the suppression of time': by which he means that the last note of the symphony is implied in the first; 'myth and music thus appear as conductors of an orchestra of which the listeners are the silent performers (op.cit.)'. This is tantalizing stuff and as musicologist Nicholas Cook observes (personal communication), it's just at that moment when we think that we can say something about music that the words dry up.

That silence can be filled if we turn instead to a theory of practices, or non-representational theory (Nash 2000). The focus is the human body and its capacity to perform. However, non-representational theory does not focus solely on the body but rather on the body-practices of becoming, where 'the emphasis is on practices that cannot adequately be spoken of, that words cannot capture, that texts cannot convey—on forms of experience and movement that are not only or never cognitive' (Nash 2000, p. 655). What emerges is an emotional geography where different spatial contexts produce distinct affects as people perform within them. As Wood and Smith put it (2004, p. 540) the interest in music is that it prioritizes feelings above thoughts. We cannot be entirely rational about music if we are to follow up Lévi-Strauss's point about it being simultaneously 'intelligible and untranslatable'. To do that we must consider the relational impacts of music, and in an evolutionary context this emphasizes a neglected area, the importance of *performance*.

Thrift (2000) provides a useful definition of the concept:

> Performance can be thought of on two levels. On one, performance is a means of theorizing the day-to-day improvisations which are the means by which the now is produced. Arising especially out of the expressive powers of the body, these improvisations enable each moment to be open to possibility, even if . . . they often fall back into, indeed often form a part of the normative. At another level performance is the practical means by which these improvisatory skills are brought out and used to construct performances of various kinds. (Thrift 2000, p. 577)

The point about performance is that it unites the mundane with the extraordinary in a human life (Crang 1994). Music as a component of performance is similarly structured in scale; from the private performance on an air guitar to the closing axe-riff on the Pyramid Stage at Glastonbury. This elision is well illustrated in Goffman's dramaturgical metaphors for the study of interaction and where, irrespective of scale and seriousness of context, 'we look within an act for the involvement it seems to express; we look to the involvement for the regulations by which it is bound; and we look to these regulations as a sign of what is owed to the gathering and its social occasion as realities in their own right (Goffman 1963, p. 247).'

James (2003, p. 299) makes a similar case by concentrating on the ubiquitous presence in human cultures of ceremonially significant forms of practice based on the model of the game, the dance, or the drama. As a result, nothing starts from scratch. Everything, she points out, involves strategically significant motion that uses historical connections as the basis for understanding, rather than the prior delineation of abstract cultural phenomena such as styles of music, forms of religion, and artistic fashions.

As a consequence, rather than an aesthetics of style there exists an aesthetics of mobility. And with such performances comes the experience of power through embodiment. An example is provided by Cresswell's (2006, p. 143) account of the international tussle over the correct form of ballroom dancing in the post-war period. The bitter dispute resulted in victory for Victor Sylvester and the Imperial Society of Teachers of Dancing (ISTD). Their dispute was with free form dances such as the Shimmy that encouraged bodily improvisation. These steps had to be outlawed because to the ISTD they were primitive and barbaric, much like the reaction to jazz, pop, punk, and rap by each successive generation of parents. However, as Cresswell points out, this denial of movement in order to produce a norm of movement is not without its pleasures. Power is the production of pleasure itself (2006, p. 145) and this can come from strict regimentation of the body as well as allowing it to 'all hang out'.

EMOTION AND PERFORMANCE IN THE PAST

It is this aspect of performance, when the emotional charge is highest and the endorphins which create pleasure flow freely (Dunbar, this volume ch. 8), that has been ignored by palaeoanthropologists. In his account of the origins of music, Mithen (2005) makes only a brief mention of the power of performance and the contribution of music to that power. Archaeologists have dwelt instead on the rational, searching for examples of music-making artefacts to identify the origins of music. They have ignored the performance places where, according to non-representational theory, emotional relationships rather than rational connections should form the interpretive basis of the material traces.

But emotion has always been ignored in the study of past societies (Tarlow 2000; Hamilakis et al. 2002b). Evolutionary studies can point to Darwin's lead (1872), but his interest in emotion was directed more to a rational understanding of its origins and inheritance than to the experience that emotion itself encapsulates. However, the grip of reason on interpreting human behaviour has recently been questioned by neuroscientists (Damasio 2000; Freedberg and Gallese 2007; LeDoux 1998) as well as child psychologists (Gerhardt 2004). They show that our brains are predominantly emotional when it comes to their development and ability to act relationally: a conclusion that confirms what phenomenologists have being saying for a long time (Merleau-Ponty 1962); the experience of being in the world involves more than discursive cognition alone. Emotions, as Wood and Smith (2004, p. 536) point out, infuse the spatialities of everyday life. Moreover, such emotional spaces are relational (ibid. p. 537), based on experience. They entail embodied simulation where understanding others, as revealed through the brain's complex mirror neuron systems, is based on the equivalence of what others do and feel and what we do and feel (Gallese 2006).

This perspective suggests that the appropriate analytical scale to consider a trait such as music is not in the first instance the full sweep of human evolution but rather the daily life of hominins. At this scale, performances are structured but highly variable in scale, as is the case of a classical concert compared with a mother singing to her child, the milkman whistling on his rounds or the sound of drums in the forest. Wood and Smith (2004, p. 537) argue that a performance is 'part of a conversation of practices, it is not therefore a fixed, finite, discrete, or finished thing'. In this sense the similarities with material culture, and in particular art as described by Gell (1998), are striking. Both music and art, viewed from the perspective of an emotional geography, are snippets in the continuum of social life. They are projects that are never finished and which have no precise cognitive moment to mark their beginning. They illustrate James's point (2003, p. 299) that historical connections, mediated through motion, form the basis for social understandings.

Previously (Gamble 1999, pp. 68–85), I have used Goffman's dramaturgical model to interpret Palaeolithic sites/locales, stressing the differences between ephemeral encounters and gatherings and contrasting them with the more permanent social occasions and places. Consequently, I was not concerned to find special places in the Palaeolithic where performances were new or different, as has been the case with the appearance of the first subterranean cave art in France (Lewis-Williams 2003). My aim instead was to reconsider the spaces for interaction and the construction of social life through negotiation. But while reinstating a social life to hominins who had previously been denied one by palaeoanthropologists, I was aware that the emotional content of these gatherings had been ignored, and therefore the role of music obscured. Moreover, not all performances are the same. They differ due to the emotional spectrum they engage with. For example, I expect a different emotional content when a forager looking for a meal encounters a stand of plants and when that same person is part of a trance dance. Since emotions form the bonds that create social ties of variable commitment and duration, not only among hominins but also among other hominids, there must also be an evolutionary trajectory to these questions.

Turner (2000) provides a lead in his micro-sociological approach to the origins of emotions. He proposes that natural selection worked to create more associative or positive emotions in order to build sociality and solidarity. He argues (2000, p. 125) that the brain contains universal structures which provide us with emotional absolutes. These structures allow us to interpret those emotional cues and by so doing regulate our responses to each other. He further speculates that natural selection has created an emotional syntax, comparable to a linguistic one, out of these emotional responses (ibid. p. 127). He then goes on to suggest that during human evolution a brain was created that made our ancestors more social and group-oriented (ibid. p. 153). He emphasizes that positive emotions, such as happiness, sympathy, and love, rather than the negatives of fear and aggression, would be selected for alongside ritual behaviour as the context where participation initiates the desired emotional response (ibid. p. 114).

However, Turner does not specify why such behaviours as dance and music, not to mention language, were also the outcome of selection for a brain better equipped to read other people's emotions. Having specified ritual behaviour in the abstract as the social context where selection occurred he does not examine the material basis of such performance. It is here that archaeologists can contribute.

RATIONAL APPROACHES TO SYMBOLICALLY BASED BEHAVIOURS

The challenge is to link activities such as music, dance, and language to material culture. The problem is that language and music (dance is rarely considered by palaeoanthropologists) are usually described as symbolically based behaviours, and material culture is interrogated for that capacity. But symbolism remains defiantly difficult to type to material culture. In the guise of ideological information it has, together with messages about function and social organization, been seen as one domain of material culture. Hence a crown can simultaneously be a hat (function), a status signifier (social), and a symbol of kingship (ideational). Context determines which of the three voices is heard. However, this domain and information approach to material culture overemphasizes the ability of objects to reflect social position and the wider world of symbols. As a result it presents an essentialist view of material culture; like saying that music is *in* the drum. But as Blacking (1973) stated more than thirty years ago, music has no meaning unless it involves social activity.

However, symbolism remains under-theorized by palaeoanthropologists for any but the most obvious artefacts. For example, in their discussion of Neanderthal symbolism, Chase and Dibble (1987, p. 264) took as their guideline the standard anthropological view that cultural behaviour today is largely symbolic, that individual cultures are reproduced by learning these symbols, and that language is central to the process. They then analysed the bits and pieces of Neanderthal archaeology that twenty years ago were regarded as evidence for symbolism. These rare items were examined 'with an eye toward assessing the degree to which arbitrary categories and symbols structured behaviours (ibid. p. 265)'. However, they only accepted material symbols, such as 'figurines' and 'jewellery' which can trace an unbroken ancestry from contemporary societies. In common with many other studies there was no attempt to explain how symbols acquire meaning and, most importantly for an archaeologist, why they changed.

A more inclusive account of evidence for early symbolism is provided by d'Errico (d'Errico et al. 2003; d'Errico 2003) drawing on data from Africa and Europe. Unlike the direct analogical approach of Chase and

Dibble, d'Errico focuses on Leroi-Gourhan's (1993) proposition that through time symbolic storage increasingly took place outside the body. D'Errico's point is that external storage requires a code (d'Errico et al. 2003, p. 31–2) and the one he favours is based on linguistic symbols, although he accords a place to music as well. But what exactly is meant by the conscious, external storage of symbols? Is it anything more than the standard anthropological account of culture as either an extra-somatic means of adaptation or a shared system of beliefs based on symbols with common meanings derived from the process of socialization?

However, no one disputes that during human evolution there has been an exponential increase in the quantity, diversity, and complexity of material culture. Explaining this increase has been the problem. For Henshilwood and Marean (2003) it was the result of 'substantial amounts of brainpower' manifested in the symbols of external storage now available to organize behaviour. But have our brains during that time so exceeded their file-space that the next incoming message, with the equivalent of six massive files attached, was suddenly turned into a painted cave? To explain the timing, recourse has been made either to the changing architecture of the brain, so that the creative explosion reflects new levels of cognitive fluidity (Mithen 1996), or a mutation for the so-called language gene *FOXP2* (Carroll 2003).

RELATIONAL APPROACHES TO MUSICAL METAPHORS

What needs to be questioned is not only the use of material proxies to determine the appearance of symbolically conceived behaviours such as music, but also the metaphor of an endowment that I discussed earlier. In its place I propose the body as the root metaphor that underpins the study of change in human evolution, and for at least five good reasons:

1. Hominin bodies experienced human evolution.
2. The body is the source of symbolic force that informs materiality (Gamble 2004) by providing a metaphor for the structure of material culture as containers and instruments (Gamble 2007). The body is an acknowledged source of metaphor and

symbolism that is often used to provide understandings of the natural and cultural world (Hamilakis et al. 2002a, p. 11–13).

3. Bodies are the way we appreciate change because of the creative processes, biological and cultural, of growth and alteration.
4. The basic aspects of social cognition, as revealed by neuroimaging, depend on activation of neural structures normally involved in our own personally experienced actions or emotions (Gallese et al. 2004). Conceptual knowledge is therefore embodied; mapped within our sensory motor system (Gallese and Lakoff 2005).
5. As Douglas (1973) has shown, the body's main scope is to express the relation of the individual to the group.

A sixth emerges in the context of music:

6. Many of the basic processes of music derive from the constitution of the human body as well as the interaction of human bodies in society. Music has no meaning without associations between people (Blacking 1973).

In short, the body provides experience of the world and an understanding of how it changes that is neither primarily linguistic, musical, nor symbolic. Furthermore, it is 'our experiences with physical objects (especially our own bodies) [that] provide the basis for an extraordinarily wide variety of ontological metaphors, that is, ways of viewing events, activities, emotions, ideas etc. as entities and substances' (Lakoff and Johnson 1980, p. 25). Such metaphors, whether material or linguistic, derive their force by being grounded in physical experience.

As a result, I can suggest a different significance for those stones and bones that palaeoanthropologists discover and analyse. I do not have to match their form to a scheme of linguistic symbols alone, or suppose that the appearance of elements such as hafted tools, ornaments, art, burials, or architecture represent important gifts to the evolving human endowment and infer from these proxies an aspect such as music.

Instead I prefer a more explicitly metaphorical approach that matches the linguistic with the material. As Chapman (2000; Chapman and Gaydarska 2007) has shown, metaphor and metonymy creatively use material culture to understand and experience personhood, whether it is singular or distributed (Field 2005; LiPuma 1998). For example, a whole pot is often an autobiographical object; made, used, and related to a person (Sterner 1989; Gosselain 1999). At death these pots are often intentionally broken and the shards passed around to those attending the funeral. The shards enter into a distributed network of relationships where part stands for whole as a material example of metonymy (Chapman and Gaydarska 2007).

In this example, a material metaphor (Tilley 1999) expresses embodied concepts of our social lives. We experience the concepts of singularity, where we literally stop-at-the-skin, just as the wall of a pot holds in its contents. We understand through the material experience of part-for-whole (shard for pot) that we are constituted by relationships which are fragmented and distributed through time and space. As a result we do not necessarily need a linguistic account of what we think a distributed person either is, or looks, like. Material culture alone is sufficient to provide the means to create and understand such a category through embodied experiences.

But where exactly does this understanding come from—what I have referred to elsewhere as symbolic force (Gamble 2004)? My answer is that the body provides the source for experience and that knowledge is translated into corporeal culture (rhythms, gestures, body techniques) and material culture (artefacts, locales, landscapes). To trace these connections the body provides the basis for two material proxies: containers and instruments (Gamble 2007). The body forms a material container (trunk and head) as well as a set of instruments (limbs) that inscribe and alter materials. This simple division of the body is the basis of the conceptual metaphors that allow us to understand the structure of material culture despite its huge variety. Some musical examples are given in Table 4.1 and the classification could be extended to consider how sound is produced by plucking and hitting with the limbs (instruments) or blowing and breathing through the lips and nose (containers).

Table 4.1. A metaphorical classification of a selected range of musical instruments that relates the experience of the body to the material production of sound. Further elaboration might be achieved by revisiting the Hornbostel-Sachs system of musical classification that currently is distinctly un-embodied in its approach.

Instruments:	*single forms* drumstick, sticks, hands, and feet; *composites* triangle, bow, strings, z reed, keyboard, electric guitar
Containers:	*single forms* whistle, dijeridu, wooden drums; *composites* woodwind, brass, skin drums, strings (violins, cellos etc.), guitars, balalaikas, sitars, harmonica, rattles, harp, piano, organ, bells, voice-box

MUSIC AND EVOLUTION

What is striking when considering the range of musical artefacts is how many take the form of a container to embody sound. Musical instruments, in my terminology (Table 4.1), are by contrast elements such as drumsticks or strings that serve to produce sound from containers. Imagine a musical world that drew on a greater use of instruments rather than containers as metaphors for the body. It would sound very different indeed, although there would still be music. In the same way a social world without that commonplace container, the house, would provide different metaphorical understandings of how people and objects were associated (Wilson 1988). But even so there would still be container-based metaphors in that world as shown by the embrace of an arm or the seating-circle round the hearth.

Our material worlds today are dominated by the vast array of containers in which we shelter, store memories, travel, consume, love, and protect. Our emotional lives are infused with the commonplace experience of containment, either bodily or in our own constructions. In Lakoff and Johnson's view (1980, p. 58), such container-based concepts emerge from the experience of living through our bodies. That is why we use them so frequently in language and why so much of our material culture, for example masks, clothes, houses, trumpets, and cellos, is also container-based.

What, however, has not been adequately considered is the evolutionary implications of this obsession. Music as one subset of this vast material cornucopia provides a much needed focus for such a perspective. What needs to be understood about the evolutionary history of music is how it too came to be dominated by material metaphors that used containers.

GATHERINGS AND MUSIC AT LOCALES

It is time to return to performance and social life. Cook (1998, p. 84) reminds us that 'to study music is to study your own participation in it—to study yourself'. Rather than stand outside the performance looking in, as a rational approach implies, there is much to be gained by getting inside the performance. The question for archaeologists is how?

The answer is through container metaphors applied to performance rather than seeking proxies for musical production. Performance encompasses many scales of interaction, from the fleeting encounter to the institutionalized political process, as well as the spectrum of emotions they stimulate. What they all have in common, irrespective of complexity, is the use of space. It is this aspect of performance where a contained space is created through attention and association and imbued with an emotional charge.

Here I will concentrate on the gathering as performance and ask what music brings to the party. Gatherings have attracted anthropological interest ever since Durkheim emphasized the emotions that are generated by people coming together (Allen 1998, p. 157). He stressed that social gatherings are creative because they are the context in which ritual and religion originate. From gatherings come forces transcending the individual, what Durkheim called *effervescence* (ibid. 1998, p. 150) and which create social bonds.

I will provide two examples: one based on personal experience but outside my normal ethnography; the second from half a million years ago.

Gathering 1: Makuri Village, August 2001

The village lies a few kilometres from Tsumkwe in the Nyae/Nyae Conservancy, northern Namibia. In a pattern exhaustively described by ethnoarchaeologists (Yellen 1977; Whitelaw 1994; Binford 1983) the Ju'hoansi villagers live in small circular huts around a large central communal area. Dancing is common and has attracted a good deal of attention because of the involvement of shamans and curing ceremonies and the links to bushmen rock art elsewhere in southern Africa (Lewis-Williams 1983).

At the night-time dance-gathering I attended, the women sat round a fire, clapping and singing, while the men, wearing rattles on their legs and striking percussion sticks, danced around them. There were no other musical instruments. The dance ended when the men entered the circle they had made in the sand.

This activity, which involved material culture, symbolic behaviour, music, and language, left behind the next morning a very clearly marked performance space (Figure 4.1): an oval track of about eight metres diameter surrounding a central hearth.

Figure 4.1. Makuri Village performance space. Twelve women and their children sat around the fire while on this occasion four men danced around them. The oval encompasses some $50m^2$.

The Makuri performance space for dancing and singing is ephemeral and will not preserve for future archaeologists. Apart from the rhythmically stamped oval there are no proxies to suggest that dance, music, language, and ritual took place here. But that is because we approach such spaces with a rational rather than a relational approach to material culture.

If, instead, we apply the bodily metaphors that structure material culture, instruments and containers, then the picture brightens for the investigation of music and ritual, those 'machines for the creation of time'. At Makuri, containers abound. The performance space is set within the village with huts all around it. The oval is a container and the actions of the men encircled the women. In their turn the women surrounded the container of the hearth. Material proxies in the form of instruments (Table 4.1) are also present. The men's feet stamped the sand to make the oval shape. The surface within the oval bears the impressions of where the women and children sat while the branches used as fuel still lay within the hearth. Music was made by percussion sticks, hands clapping (instruments), the rattles (containers) surrounding the men's legs (instruments) that were pounding the sand,

and the singing that came from the women (containers). The *effervescence* was palpable, as it always is when people gather and perform, even though that night there was no ritual beyond entertainment.

As a result, the Makuri performance I attended was an example of an architecture without walls; a container without hard-edged material boundaries, although delimited by the actions of bodies and the emotions generated. Using a relational perspective the concept of containment that structured the performance could be traced through material proxies of bodily metaphors.

Gathering 2: The horse dance, Boxgrove, 500,000 years ago

The Boxgrove sand and gravel pit in West Sussex, England (Roberts and Parfitt 1999) is one of the best preserved Lower Palaeolithic locales in Europe associated with the large-brained hominin, *Homo heidelbergensis*. The long sequence of marine sands and silts contains evidence for a changing plains landscape in front of a collapsing chalk sea cliff. The locale was occupied during a warm interval half a million years ago and 250,000m^2 has so far been investigated. At this time the evidence for containers in the form of huts, clothing and jewellery that encircled the body is lacking, even though preservation is exceptional. No hearths have yet been found at Boxgrove, although 400,000 years ago these are well preserved elsewhere in England at Beeches Pit (Gowlett 2005) and at Schöningen in Germany (Thieme 2005).

The interest in Boxgrove is threefold

- to contrast rational and relational accounts of the locale
- to consider it as a contained performance space comparable to Makuri village
- to use it to broaden our understanding of the effervescence of gatherings and the role of music.

Pope (2002; Pope and Roberts 2005) has studied the numerous scatters and patches of flint artefact that are found on these ancient mudflats. In particular his work focuses on explaining why some of these patches were rich in finished stone tools, principally hand-axes, knapped from flint taken from the cliff, while others were very poor. The distances across the Boxgrove landscape are short with flint artefacts found up to 250 metres in front of the cliff. There is a consistent fall-off in hand-axes away from this topographic feature, the source of the raw material (Pope and Roberts 2005, Figure 6.6). But as Pope argues, this decline makes no rational sense in terms of energy expenditure since the distances are so short. This is not a simple case of a distance-decay or least-effort model governing the distribution of artefacts in a landscape.

Instead, Pope argues for a landscape richly structured by intentional and complex hominin actions. To understand these scatters and patches of material it is necessary to adopt a relational perspective between hominins, artefacts, and place.

Within the Boxgrove landscape there were places visited once, such as the single horse butchery locale (site GTP17), and others such as the 'waterhole' in Quarry 1/B where visits were repeated over several years. At the former, short-lived gatherings, few or no hand-axes were found. At the horse butchery locale (Figure 4.2) eight blocks of flint were brought in by, it is inferred, eight individuals. Symmetrical, ovate hand-axes were knapped as revealed by the refitted flakes. But all eight hand-axes were carried away from this locale and deposited elsewhere.

The archaeological evidence at the horse butchery site is more robust than the Makuri dance-oval because it consists of durable stone artefacts excavated from $84m^2$ (Roberts and Parfitt 1999, Figures 279 and 280). The blocks of flint carried from the cliff encircled the horse just as the hunters did when it was killed earlier in the day. The hand-axes and flakes they used were instruments (Table 4.1), material proxies for the limbs, as were the antler hammers used to knap the flint and the spears that the excavators believe they possessed (Pitts and Roberts 1997).

We must also fill this container with the varied smells and sounds that attenuated over the landscape, relating other animals and people to the performance. Smells carried on the wind brought in hyenas and carrion feeders. The sounds of stone ringing on stone as artefacts were shaped caught the attention of other people beyond the immediate performance and focused their curious gaze. Each hammer blow to make tools caused mirror neurons to fire in the brains of the eight people present, and anyone else watching or listening, confirming through such embodied simulation (Gallese 2006) the existence of their closely linked, distributed association. Their neural actions formed a metonymy of the body, linking through embodied experience the individual to the group at this gathering. And the actions

Figure 4.2. A 500,000-year-old performance at the horse butchery locale at Boxgrove, GTP 17. The animal was killed and fragmented on an exposed mudflat forty metres from the cliff. Blocks of flint were brought in, knapped, and then the hand-axes taken away. The excavated area covered $84m^2$ (Roberts and Parfitt 1999: back cover).

that fragmented stone to produce hand-axes and flakes formed a further material metaphor of part-for-whole association.

The container for this particular performance had no walls or fences to delimit it, only the actions of those engaged in the varied activities with flesh, bone, stone, mud, and people. It may have lasted only a few hours before they moved elsewhere to consume and share the results of their labours. Nonetheless, this was a highly charged gathering, not just because of the prospect of feasting on fat, marrow, and meat but because of the association of people in this familiar well-used landscape that held memories of previous encounters and gatherings. Moreover, it was these historical connections that formed the basis for shared understanding and which, as we have seen, result from strategically significant motion (James 2003).

The power of the place was governed by closely observed rules, an aesthetics of mobility and material, in which there was both pleasure and power. Pope (2002, p. 171) concludes that where artefacts were thrown away depended on those places being visited more than once. One such locale was the 'waterhole' in Quarry 1/B about one kilometre away from GTP 17. Here the excavators found not only the richest density of cores and struck flakes but also 321 hand-axes, not all of them showing traces of having been used.

This challenge to a rational account of life half a million years ago emphasizes further the need for a relational perspective based on the concept of performance. Pope (2002, p. 286) argues that these structured patterns of landscape use were one of the mechanisms by which more complicated systems of land use evolved. Group cohesion was assisted under the daily routines of fission and fusion. Artefacts such as hand-axes stood as proxies for individuals (ibid. p. 288). They provided markers on return visits and suggest that 500,000 years ago culture was a hybrid network involving relations between people and objects.

The spatial separation in front of the Boxgrove cliff shows that repeated gatherings were held. These effervescent performances required a dance to bring people, things, and place into relationship with each other. The metaphorical structure of the Boxgrove landscape with its abundant material instruments was therefore created from the choreography of bodies.

CONCLUSION: A DANCE TO THE MUSIC OF TIME

So where does this leave the evolution of music? How do we insert ephemeral performances such as Boxgrove and Makuri into the long-term process of human evolution?

The first shift in perspective has been to add a relational approach to the customary rational assessment of life in the Palaeolithic (Gamble and Gaudzinski 2005). Part of that relational approach recognizes that artefacts are material metaphors grounded out in bodily experience. They are an integral aspect of association and relationships between people, their landscapes, and material worlds. Hand-axes were not musical instruments, although they produced sounds, but they were always much more than just an efficient solution to cutting up a horse. The hand has always been an instrument that experiences and changes the world through touch and manipulation. When a hand-axe was held, it changed the boundaries of the body creating new associations and relationships. At such times it was an embodied concept in action, and such metaphorical extensions to the body added to the *effervescence* generated at gatherings. My emphasis on performance has also shown that spatial metaphors of containment are as significant as those that are framed either linguistically or materially in relating people at such gatherings.

The second change in perspective is to reconsider the breakpoints in human evolution. Instead of following a model of a gradually accrued endowment, where the gift of music was a late addition to the basket of anatomically modern humans, it has been shown by Aiello and Dunbar (1993) that the proxy for our evolving social and cultural complexity is the size of the brain itself. This expensive tissue (Aiello and Wheeler 1995) required strong selection if it was to treble in size from *Australopithecus afarensis* to *Homo sapiens*. The outcome can be measured through the greater group sizes and networks that result from increases in primate and, by extrapolation, hominin brains (Dunbar 2003). These larger social worlds presented challenges to integration and communication as the pattern moved from tactile grooming to verbal address.

Therefore, what archaeologists add to this model are not the artefact proxies for language or music but rather those performance spaces such as Boxgrove where social interaction took place during gatherings. Half a million years ago was a time of rapid increase in brain size. The neocortex doubled in about 200,000 years and group sizes with it (Aiello and Dunbar 1993). However, the archaeological evidence for symbols that might suggest a semiotically-based system of language and music are much later in date (Henshilwood and Marean 2003), hence the need for palaeoanthropologists to delay modernity by creating the holding-category of AMH while we catch up with our evolutionary endowment.

But if we are to achieve a proper evolutionary perspective we need to reject such staged delays. We can do this by acknowledging that gatherings were common to all hominins and that they used emotion to form relationships. The concepts that structured their emotional geographies were embodied, and those experiences were open to elaboration, through the material metaphors that brought people and things into relationship.

Emotion, as Darwin pointed out, was original in hominins and animals. Moreover, as Turner (2000) has suggested, there has been selection for particular emotions—happiness, sympathy, and love—

to reinforce the trend for a more complex social brain. Viewed from this evolutionary perspective, the importance of music is not that it marks the advent of modernity but rather that it has always acted to frame positive social outcomes from gatherings. Turner is probably correct when he argues for a syntax of emotions that preceded those for music, language, and the formal rules of decoration and art.

My conclusion is that since emotions are embodied (Freedberg and Gallese 2007), we can trace the broad outline of changes in syntax through material metaphors that are similarly embodied. Here the archaeological evidence is illuminating (Gamble 2007, Figure 7.1). The artefact record from 2.5 million years ago is dominated by instruments that were brought to gatherings; and where, as we have seen, containers were spun instead from action and association. When Boxgrove was visited by large-brained, socially complex hominins such an instrument-based means to channel the emotions was no longer sufficient. Voices were now recognized as proxies for the containers from which they issued. It is only much later that musical containers in the form of flutes and drums appear, and it is significant that this occurs when the dominant material metaphor changes from instrument to container. This occurs as a shallow gradient rather than a revolution between 100,000 and 20,000 years ago (Gamble 2007, Chapter 8). Music in human evolution becomes intelligible when it is linked to emotion and performance, and translatable when referenced to material metaphors drawing on the bodily proxies of instruments and containers. Music has at all times been the outcome of the creative potential of hominin social life, where there have always been dances with wolves to the music of time.

ACKNOWLEDGEMENTS

I am very grateful to the organizers of the Music, Language, and Human Evolution Research workshop for such a thought-provoking and provocative meeting and to Steve Mithen whose *The Singing Neanderthals* opened up an area for fresh archaeological enquiry; Matt Pope for many stimulating discussions and insights into the Boxgrove data; Karen Till for her thoughts on architecture without walls; Tara Woodyer for her insights on performance and emotional geographies. The British Academy Centenary Project *From Lucy to*

Language: the archaeology of the social brain has provided a further context for the development and discussion of these ideas with colleagues on the project: Margaret Clegg, Wendy James, Robin Dunbar, Garry Runciman, Matt Grove, John Gowlett, James Steele, Simon Buteux, and Fiona Coward.

REFERENCES

Aiello, L. and R. Dunbar (1993). 'Neocortex Size, Group Size and the Evolution of Language', *Current Anthropology*, 34: 184–93.

—— and P. Wheeler (1995). 'The Expensive-Tissue Hypothesis: The Brain and the Digestive System in Human and Primate Evolution', *Current Anthropology*, 36: 199–221.

Allen, N. J. (1998). 'Effervescence and the Origins of Human Society'. In: N. Allen, W. Pickering and W. Watts Miller (eds.), *On Durkheim's Elementary Forms of Religious Life*, pp. 149–61. London: Routledge.

Binford, L. R. (1983). *In Pursuit of the Past*. London: Thames and Hudson.

Blacking, J. (1973). *How Musical Is Man?* Seattle, WA: University of Washington Press.

Carroll, S. B. (2003). 'Genetics and the Making of *Homo Sapiens*', *Nature*, 422: 849–57.

Chapman, J. (2000). *Fragmentation in Archaeology: People, Places and Broken Objects in the Prehistory of South-Eastern Europe*. London: Routledge.

—— and B. Gaydarska (2007). *Parts and Wholes: Fragmentation in Prehistoric Context*. Oxford: Oxbow Books.

Chase, P. and H. L. Dibble (1987). 'Middle Palaeolithic Symbolism: A Review of Current Evidence and Interpretations', *Journal of Anthropological Archaeology*, 6: 263–96.

Cook, N. (1998). *Music: A Very Short Introduction*. Oxford: Oxford University Press.

Crang, P. (1994). 'It's Showtime: On the Workplace Geographies of Display in a Restaurant in Southeast England', *Environment and Planning D: Society and Space*, 12: 675–704.

Cresswell, T. (2006). *On the Move*. London: Routledge.

d'Errico, F. (2003). 'The Invisible Frontier: A Multiple Species Model for the Origin of Behavioural Modernity', *Evolutionary Anthropology*, 12: 188–202.

—— C. S. Henshilwood, G. Lawson et al. (2003). 'Archaeological Evidence for the Emergence of Language, Symbolism, and Music: An Alternative Multidisciplinary Perspective', *Journal of World Prehistory*, 17: 1–70.

Damasio, A. (2000). *The Feeling of What Happens: Body, Emotion and the Making of Consciousness*. London: Vintage.
Darwin, C. (1872). *The Expression of the Emotions in Man and Animals*. London: John Murray.
Douglas, M. (1973). *Natural Symbols*. Harmondsworth: Pelican Books.
Dunbar, R. (2003). 'The Social Brain: Mind, Language, and Society in Evolutionary Perspective', *Annual Review of Anthropology*, 32: 163–81.
Field, A. S. (2005). 'Transformations in Individuality: Personhood and Palaeoliths in the Middle Pleistocene'. In: C. Gamble and M. Porr (eds.), *The Individual Hominid in Context: Archaeological Investigations of Lower and Middle Palaeolithic Landscapes, Locales and Artefacts*. London: Routledge.
Freedberg, D. and V. Gallese (2007). 'Motion, Emotion and Empathy in Esthetic Experience', *Trends in Cognitive Sciences*, 11, 5: 198–202.
Gallese, V. (2006). 'Embodied Simulation: From Mirror Neuron Systems to Interpersonal Relations'. In: Novartis Foundation Symposium 278, *Empathy and Fairness*, pp. 3–19. Chichester: Wiley.
—— C. Keysers, and G. Rizzolatti (2004). 'A Unifying View of the Basis of Social Cognition', *Trends in Cognitive Sciences*, 8, 9: 396–402.
—— and G. Lakoff (2005). 'The Brain's Concepts: The Role of the Sensory-Motor System in Conceptual Knowledge', *Cognitive Neuropsychology*, 22, 3/4: 455–79.
Gamble, C. S. (1999). *The Palaeolithic Societies of Europe*. Cambridge: Cambridge University Press.
—— (2004). 'Materiality and Symbolic Force: A Palaeolithic View of Sedentism'. In: E. DeMarrais, C. Gosden, and C. Renfrew (eds.), *Rethinking Materiality: The Engagement of Mind with the Material World*, pp. 85–95. Cambridge: McDonald Institute of Archaeological Research.
—— (2007). *Origins and Revolutions: Human Identity in Earliest Prehistory*. New York: Cambridge University Press.
—— and S. Gaudzinski (2005). 'Bones and Powerful Individuals: Faunal Case Studies from the Arctic and European Middle Palaeolithic'. In: C. S. Gamble and M. Porr (eds.), *The Individual Hominid in Context: Archaeological Investigations of Lower and Middle Palaeolithic Landscapes, Locales and Artefacts*, pp. 154–75. London: Routledge.
—— and M. Porr (2005a). 'From Empty Spaces to Lived Lives: Exploring the Individual in the Palaeolithic'. In: C. S. Gamble and M. Porr (eds.), *The Individual Hominid in Context: Archaeological Investigations of Lower and Middle Palaeolithic Landscapes, Locales and Artefacts*, pp. 1–12. London: Routledge.
—— and M. Porr, eds. (2005b). *The Individual Hominid in Context: Archaeological Investigations of Lower and Middle Palaeolithic Landscapes, Locales and Artefacts*. London: Routledge.

Gell, A. (1998). *Art and Agency: Towards a New Anthropological Theory.* Oxford: Clarendon Press.

Gerhardt, S. (2004). *Why Love Matters: How Affection Shapes a Baby's Brain.* London: Routledge.

Goffman, E. (1963). *Behaviour in Public Places: Notes on the Social Organization of Gatherings.* New York: Free Press.

Gosselain, O. P. (1999). 'In Pots We Trust: The Processing of Clay and Symbols in Sub-Saharan Africa', *Journal of Material Culture*, 4: 205–31.

Gowlett, J. A. J. (2005). 'Seeking the Palaeolithic Individual in East Africa and Europe during the Lower-Middle Palaeolithic'. In: C. S. Gamble and M. Porr (eds.), *The Individual Hominid in Context: Archaeological Investigations of Lower and Middle Palaeolithic Landscapes, Locales and Artefacts*, pp. 50–67. London: Routledge.

Hamilakis, Y., M. Pluciennik, and S. Tarlow (2002a). 'Introduction: Thinking through the Body'. In: Y. Hamilakis, M. Pluciennik and S. Tarlow (eds.), *Thinking through the Body: Archaeologies of Corporeality*, pp. 1–21. New York: Kluwer Academic/Plenum.

——————eds. (2002b). *Thinking through the Body: Archaeologies of Corporeality.* New York: Kluwer Academic/Plenum.

Henshilwood, C. S. and C. W. Marean (2003). 'The Origin of Modern Human Behaviour: Critique of the Models and Their Test Implications', *Current Anthropology*, 44, 5: 627–51.

Ingold, T. (2000). *The Perception of the Environment: Essays in Livelihood, Dwelling and Skill.* London: Routledge.

James, W. (2003). *The Ceremonial Animal: A New Portrait of Anthropology.* Oxford: Oxford University Press.

Lakoff, G. and M. Johnson (1980). *Metaphors We Live By.* Chicago, IL: University of Chicago Press.

Leach, E. R. (1970). *Lévi-Strauss.* London: Fontana.

LeDoux, J. (1998). *The Emotional Brain.* London: Orion Books.

Leroi-Gourhan, A. (1993). *Gesture and Speech.* Cambridge, MA: MIT Press.

Lewis-Williams, J. D. (2003). *The Mind in the Cave.* London: Thames and Hudson.

—— (1983). *The Rock Art of Southern Africa.* Cambridge: Cambridge University Press.

LiPuma, E. (1998). 'Modernity and Forms of Personhood in Melanesia'. In: M. Lambek and A. Strathern (eds.), *Bodies and Persons: Comparative Perspectives from Africa and Melanesia*, pp. 53–79. Cambridge: Cambridge University Press.

Merleau-Ponty, M. (1962). 'The Spatiality of the Lived Body and Motility'. In: *Phenomenonology of Perception*, trans. C. Smith, pp. 241–71. London: Routledge and Kegan Paul.

Mithen, S. (1996). *The Prehistory of the Mind.* London: Thames and Hudson.

—— (2005). *The Singing Neanderthals: The Origins of Music, Language, Mind and Body*. London: Weidenfeld and Nicolson.

Nash, C. (2000). 'Performativity in Practice', *Progress in Human Geography*, 24: 653–64.

Noble, W. and I. Davidson (1996). *Human Evolution, Language and Mind*. Cambridge: Cambridge University Press.

Pinker, S. (1997). *How the Mind Works*. New York: Norton.

Pitts, M. and M. Roberts (1997). *Fairweather Eden*. London: Century.

Pope, M. (2002). 'The Significance of Biface Rich Assemblages: An Examination of Behavioural Controls on Lithic Assemblage Formation in the Lower Palaeolithic'. PhD dissertation, Department of Archaeology, University of Southampton.

—— and M. Roberts (2005). 'Observations on the Relationship between Palaeolithic Individuals and Artefact Scatters at the Middle Pleistocene Site of Boxgrove, UK'. In: C. S. Gamble and M. Porr (eds.), *The Individual Hominid in Context: Archaeological Investigations of Lower and Middle Palaeolithic Landscapes, Locales and Artefacts*, pp. 81–97. London: Routledge.

Proctor, R. N. (2003). 'Three Roots of Human Recency: Molecular Anthropology, the Refigured Acheulean, and the UNESCO Response to Auschwitz', *Current Anthropology*, 44: 213–39.

Roberts, M. B. and S. A. Parfitt (1999). *Boxgrove: A Middle Pleistocene Hominid Site at Eartham Quarry, Boxgrove, West Sussex*. London: English Heritage.

Sterner, J. (1989). 'Who Is Signalling Whom? Ceramic Style, Ethnicity and Taphonomy among the Sirak Bulahay', *Antiquity*, 63: 451–9.

Tarlow, S. (2000). 'Emotion in Archaeology', *Current Anthropology*, 41: 713–46.

Thieme, H. (2005). 'The Lower Palaeolithic Art of Hunting: The Case of Schöningen 13 II-4, Lower Saxony, Germany' In: C. S. Gamble and M. Porr (eds.), *The Individual Hominid in Context: Archaeological Investigations of Lower and Middle Palaeolithic Landscapes, Locales and Artefacts*, pp. 115–32. London: Routledge.

Thomas, J. (2004). *Archaeology and Modernity*. London: Routledge.

Thrift, N. (2000). 'Performance'. In: R. Johnston, D. Gregory, G. Platt et al. (eds.), *The Dictionary of Human Geography*, p. 577. Oxford: Blackwell.

Tilley, C. (1999). *Metaphor and Material Culture*. Oxford: Blackwell.

Turner, J. H. (2000). *On the Origins of Human Emotions: A Sociological Inquiry into the Evolution of Human Affect*. Stanford, CA: Stanford University Press.

UNESCO (1997). 'Universal Declaration on the Human Genome and Human Rights'. <http://www.unesco.org/new/en/social-and-human-sciences/themes/bioethics/human-genome-and-human-rights/>.

Watkins, T. (2004). 'Building Houses, Framing Concepts, Constructing Worlds', *Paléorient*, 30: 5–24.

Whitelaw, T. M. (1994). 'Order without Architecture: Functional, Social and Symbolic Dimensions in Hunter-Gatherer Settlement Organisation'. In: M. Parker-Pearson and C. Richards (eds.), *Architecture and Order: Approaches to Social Space*, pp. 217–43. London: Routledge.

Wilson, P. (1988). *The Domestication of the Human Species*. New Haven, CT: Yale University Press.

Wood, N. and S. J. Smith (2004). 'Instrumental Routes to Emotional Geographies', *Social and Cultural Geography*, 5: 533–48.

Yellen, J. E. (1977). *Archaeological Approaches to the Present: Models for Reconstructing the Past*. New York: Academic Press.

Part III

Perspectives on the evolutionary prerequisites for musical behaviour

5

Hominin Physiological Evolution and the Emergence of Musical Capacities

Iain Morley

INTRODUCTION

The production and processing of the various elements of musical behaviours relies upon the integration of a number of physiological and neurological capacities. These have evolved into their current form in modern humans from foundations present in our hominin ancestors. Fundamental questions associated with this process concern the original roles of these component capacities, and how and why they developed and integrated in the way that they did, allowing possible musical behaviours of the form with which we are now familiar.

Examination of the fossil record of hominin physiological evolution, of neurological interdependencies, and of primate and developmental studies, can go some way towards addressing these questions, and give some insight into the essential underlying foundations of musical behaviours.

Before the presence of musical instruments in the archaeological record, discussion of the earliest foundations of musical behaviours in the human lineage must make recourse to the features of physiology and neurology that are used in such activities—one is necessarily investigating the origins of the production and processing of complex vocalizations and muscular movements. Vocalization is a fundamental component of musical behaviour, and the production and perception of long sequences of tonally varied and rhythmic sound sequences would be a prerequisite to any behaviour that we would recognize as musical. This chapter focuses on some of the physiological (palaeoanthropological), neurological, developmental, and primatalogical evidence available relevant to the emergence of such capabilities.

PHYSIOLOGICAL EVOLUTION—AN OVERVIEW OF PALAEOANTHROPOLOGICAL EVIDENCE

Although there are clearly difficulties with carrying out reconstructions and analyses of fossil specimens, these techniques are becoming increasingly refined, and the evolution of various elements of physiology related to vocalization can be tracked with reasonable confidence in the fossil record. The following constitutes a very brief summary of the findings of a variety of research related to the evolutionary development of these elements of hominin physiology (for a more detailed treatment the reader is directed to Morley 2003).

The major influences on vocal ability are the position of the larynx in the throat, the size of the oral cavity and the size and control of the tongue, all of which are inextricably linked (Lieberman 1989). We know already that non-human primates (and other mammals) are capable of creating vocalizations, but these are of a limited range, at least in part because of the position of the larynx near the top of the throat (Duchin 1990; Budil 1994). This also restricts the resonance (and thus variety of vowel sounds) produced by the pharyngeal cavity, the area between the epiglottis and soft palate (Lieberman 1984; Budil 1994). In adult humans, the larynx instead occupies a space in the pharynx further down the throat, opening the pharyngeal resonance cavity and allowing the production of the wide range of vowels (such as *a*, *u*, and *i*, the 'vowel triangle') that are universals in all human languages (Lieberman 1984). This capacity to produce a wide variety of sustained vowel sounds is also fundamental to the ability to produce vocal melody.

At birth, the larynx occupies the top of the throat in humans, but as the child grows up, particularly after the first year of life, the larynx descends towards its adult position. This process is only completed during adolescence (Magriples and Laitman 1987). It seems that a similar process has occurred over the course of human evolution.

Detailed analysis of the vocal anatomy of chimpanzees (Laitman and Heimbuch 1982; Duchin 1990) and other non-human mammals, including baboons (Laitman and Reidenberg 1988), shows that non-human primates have a high larynx; in contrast, humans, and only humans (Duchin 1990, p. 688), have a larynx low in the throat (see Figure 4.2, Lieberman 1992, p. 135). Thus, if it is possible to reconstruct the position of the larynx in fossil hominins it should be possible to draw conclusions as to the extent of their vocal abilities, in comparison with modern humans and non-human primates.

Laitman has asserted that there is a direct relationship between laryngeal position and the degree of curvature of the underside of the base of the skull, or *basicranial flexion* (Laitman et al. 1979; Laitman and Reidenberg 1988). Basicranial flexion appears to progressively increase between *Homo ergaster* 1.7–1.9 million years ago and *Homo heidelbergensis*, *c.*300,000 years ago, suggesting a continued lowering of the larynx and thus increase in the supralaryngeal sound-space available for the moderation of laryngeal sounds (Laitman et al. 1979; Laitman and Heimbuch 1982; Laitman 1984; Budil 1994). Not only does laryngeal lowering result in greater supralaryngeal space in which to modify sounds produced by the larynx, it also allows a greater pitch-range of sounds to be produced by the larynx itself. The first flexing of the basicranium is exhibited by *Homo ergaster*, and it seems that this is connected with the shift to fully bipedal posture (Arensberg et al. 1990; Spoor and Zonneveld 1998). As the degree of basicranial flexion increased, this may have continued to add to the supralaryngeal/nasopharyngeal sound-space volume too, though decreases in maxillary width and posterior height at the same time may have limited this increase (Arensberg et al. 1990). As well as allowing a greater variety of sounds to be produced, larger supralaryngeal volume also allows for greater tongue movement, and possibly concomitant control of the tongue (Fitch 2000).

The position of the muscular systems supporting the hyoid bone and larynx are also indicated by the angle of a groove on the inside of the mandibles, called the *mylohyoid groove* (Arensberg et al. 1990). Evidence from Middle Palaeolithic hominins, including Neanderthals and early moderns (e.g. the fossils known as Kebara 2, Amud 1, Qafzeh 9, Skhul V, Tabun I and II, Regourdou, and Krapina J and C) show that this is identical in position and orientation to that of modern humans 'suggesting a suite of muscular relations similar to modern humans must have been present' (Arensberg et al. 1990, p.141).

The dimensions of the neural canal through the cervical and thoracic vertebrae have been taken as indicative of the level of innervation of musculature involved in the control of the larynx and breathing respectively (MacLarnon and Hewitt 1999; Frayer and Nicolay 2000). Thoracic vertebral canal dimensions (indicating fine control of breathing and thus over duration and volume of utterances) in *Homo ergaster* are not increased relative to body size (in comparison with modern primates). However, cervical vertebral dimensions (indicating level of laryngeal muscular control over pitch, intonation, and emphasis, for example) are increased in *Homo ergaster*, equivalent to modern humans (MacLarnon and Hewitt 1999; Frayer and Nicolay 2000). Australopithecines display no increase in thoracic dimensions relative to body size, in comparison to the great apes.

It seems that *Homo ergaster* had some increased control over pitch, intensity, emphasis, and intonation, as indicated by basicranial flexion and cervical vertebral dimensions, but not over duration of utterances—and thus would not have been capable of producing finely controlled long sequences of such tonal vocalizations. However, this latter capability appears to have been fully developed by the time of *Homo neanderthalensis* and early modern humans (MacLarnon and Hewitt 1999), who also have modern thoracic vertebral canal dimensions and hyoid bone morphology, suggesting modern levels of control over extended exhalations, and modern laryngeal anatomy, respectively.

This all suggests the subsequent development of the ability to produce extended sequences of controlled vocalizations out of an initial ability to make discrete vocalizations which were first controlled for pitch and tone, the foundations for which were already in place (as evident in higher primates). An initial increase in control over pitch, intensity, and intonation had probably occurred with *Homo ergaster*, followed by the development of control (and planning) of longer sequences of such vocalizations, probably in place in the common ancestor of *Homo neanderthalensis* and *Homo sapiens*.

In addition to the production of sounds, there are also various indicators of the emergence of the perception of sounds. There exists a set of 'natural auditory categories' (Kuhl 1988) or 'articulatory invariants' (Liberman et al. 1967; Wind 1990) which are universal across all human languages, and these are perceived more effectively than other sounds, even by pre-linguistic infants and non-human mammals. This suggests that the auditory capacity to perceive these sounds predates the ability to produce them, and that the vocal capacity to produce these sounds (a fundamental component of all human vocalization) evolved in response to this perceptual ability.

On the other hand, aspects of audition have certainly been shaped by vocalization abilities. The stapedius inner-ear muscles regulate the intensity of our perception of our own voice, so that external sounds are not drowned-out by our own vocalizations (Borg and Counter 1989). This facility would presumably have become increasingly important as the range and duration of utterances increased, potentially obscuring more environmental sounds.

Modern-like inner-ear morphology is also first exhibited by *Homo ergaster* (Spoor et al. 1994; Spoor 1996; Spoor and Zonneveld 1998). As with laryngeal anatomy, at least some of the changes in the features of the labyrinthine (inner-ear) anatomy appear to be correlated to changes in the shape of the basicranium. The cranial base accommodates the labyrinthine structure, and with the basicranium shortening and flexing with increased brain size, and the change in position of the foramen magnum (where the spinal cord enters the base of the skull) with full bipedalism, it seems this has had distorting effects upon the labyrinth. Again the first evidence of this is with *Homo ergaster*, and this process of flexion of the cranial base continued to some extent in all subsequent species of hominin up to *Homo heidelbergensis*, as elaborated above.

In summary, the foundations for the control of pitch and contour, as controlled by the cervical vertebral innervation, would have been present in *Homo ergaster*, but the duration of such utterances would have been limited until the time of *Homo heidelbergensis*, or possibly the common ancestor of Neanderthals and modern humans. However, it is with *Homo ergaster* that we see the first evidence for an increase in vocal versatility in the form of laryngeal development. This indicates the beginnings of an increase in tonal versatility and control before increased duration of utterances. It would seem likely that the requirement to control the pitch and contour of vocal sequences, and subsequently their duration, would have increased proportionally with the increase in range of pitch tones producible by the vocal tract as it evolved.

As MacLarnon and Hewitt (1999) point out, many primates vocalize in the form of discrete units of sound created with single air movements, but are limited in the duration of these and the order in which certain sounds can be made in the breathing cycle. They are also limited in the diversity of such sounds that they can make. This again suggests the subsequent development of long sequences of controlled vocalizations out of an initial ability to make discrete vocalizations which were first controlled for pitch and tone, the foundations for which are already evident in higher primates.

THE FUNCTIONS AND LOCATIONS OF THE NEUROLOGY FOR MUSIC

The fossil record

Any discussion of the evolution of physiology would be incomplete without consideration of neurology, which is, of course, not only itself physiological, but is also intimately related to the control of other physiological capabilities. There are, however, several difficulties associated with making inferences regarding the development of hominin neuroanatomy from the palaeoanthropological evidence. Not least of these is that there is no *direct* fossil evidence for the nature of early human brains, because brains do not fossilize. Nevertheless, the fossil record can provide some valuable information.

Some light can be shed on the anatomy of early brains by the study of endocasts. These are casts (naturally formed or man-made) of the inside of fossil skulls, which can show the shape and features of the brain which occupied them (e.g. Kochetkova 1978; Falk 1992). However, the extent to which one may extrapolate about brain function from brain physiology is debated. Firstly, an endocast only shows the outer surface of the brain, as it is a cast of the imprint left by the brain on the inner surface of the skull; it can tell us nothing about the internal brain structure and little about connections between different areas (Leakey 1994). Endocasts can, though, show the *shape* of the brain, and this has allowed the identification of development in Broca's area (important in language production) of hominin brains (Holloway 1983). This identification has led to the assertion that *Homo rudolfensis* (skull KNM-ER 1470, from East Turkana) had some language abilities, as it shows evidence of development in that area of the brain (Holloway 1983).

There are some problems, however, with this inference. As discussed further below, language ability is not located exclusively in

Broca's area, but in many other areas too (e.g. Borchgrevink 1982; Benson 1985). Furthermore, areas around Broca's area are also used for processes other than language (Calvin 1996). For example, complex sequences of muscular movements and fine motor muscular coordination of oral and facial muscles are controlled in this region (Calvin 1996). This includes musculature necessary for the production of facial expressions and vocalizations, as well as sequences of movements for rhythm production. Note that *vocalizations* are not the same thing as *language*. Thus development in Broca's area and the surrounding areas may only indicate the development of the ability to produce complex sequences and finely controlled vocalizations, and not have any bearing on language ability in the sense of syntax and grammar. In fact, Peterson et al. (1988) conclude from PET studies of the brain that the classic Broca's area (Brodmann's areas 44 and 45) is not concerned with semantic associations of words at all, but exclusively with motor functions; it seems to be the nearby Brodmann's area 47 that is responsible for semantic associations. Despite the fact that uncertainty about the precise functions of Broca's area may have led to the attribution of language abilities to hominins who probably had none, the identification of Broca's area in endocasts is still very useful to this study, as it gives important information regarding capabilities for vocalization and fine vocal control.

Before about 2 Mya, fossil hominin brains show no development in the area of endocasts corresponding to Broca's area (Falk 1992). The first evidence for any growth in this area comes from KNM-ER 1470, *Homo rudolfensis* (Falk 1992). Between *Homo rudolfensis* and *Homo ergaster* there is a disproportionate level of growth in this area, in comparison to the rest of the brain's growth (Kochetkova 1978). While there is a general increase in brain size of around 50 per cent between *Homo habilis* OH7 (at around 674cc; Johanson and Edgar 1996) and *Homo ergaster* KNM-WT 15000 (estimated adult cranial capacity 909 cc; Johanson and Edgar 1996), there is an 'especially intense' growth of the lateral tuba, corresponding to Broca's area (Kochetkova 1978, p. 200). This suggests that there was a significant development of fine motor control of vocalization as part of the speciation of *Homo ergaster*. By *Homo neanderthalensis* (e.g. Djebel Ihroud I, 1,305 cc; Holloway 1981), this area is 'very protuberant' (Holloway 1981 p. 387).

Neurological foundations for production and perception of speech, melody, and rhythm

Other methods must be used to investigate connections between different areas of the brain, as these are unlikely to show up in endocasts. In addition to primatological and developmental studies, explored further below, clues relating to the evolutionary emergence of vocal complexity can be gained from studies of the interdependencies of neurological systems relating to musical behaviours in modern humans. The following represents a very brief overview of early findings in what is a burgeoning area of research.

Musical and linguistic vocalization capabilities seem to share neurological overlaps at a fundamental level concerned with the instigation and control of vocalization, but are dissociated at higher levels concerned with the specific production and processing of *content* in melody and speech (Marin and Perry 1999). The neurology responsible for complex vocal production is quite distinct from that responsible for lexical linguistic production and comprehension—pathologies show that it is quite possible to maintain full ability to produce complex linguistic sentences, whilst having lost all linguistic comprehension (Geschwind, Quadfasel, and Segarra 1965); that is, the *structure and rhythm of speech* can be perfectly preserved, along with *melodic contour and prosody*, whilst all lexical linguistic ability is lost.

Intonation patterns in speech, even if apparently non-emotional, seem to have their underpinnings in the speaker's emotions, and are mediated by right-hemisphere substrates specialized for emotional experience (Snow 2000). These substrates seem to be used in the comprehension and production of both the prosodic intonation of speech and the melodic elements of music (Mazzucchi et al. 1982; Peretz 1993; Peretz et al. 1994; Patel et al. 1998)—when one is damaged, so is the other, indicating common substrates for those elements of speech and musical functions concerned with prosodic intonation and melodic contour. This ability to process emotional-tonal information in vocalizations seems to be integrated with the ability to process emotional information in other media such as facial expression (Karow et al. 2001), which are evolutionarily ancient sub-cortical systems. They are separate from the ability to process linguistic content, and almost certainly far older. This suggests, then, that whatever the content of vocalization at 'higher' linguistic or melodic levels, its production and perception unavoidably involve some input from neurological systems concerned with affective function.

Brain scanning studies reveal similar patterns of function distribution: lateral biases between hemispheres occur for various auditory processing tasks. Emotional tone analysis, whilst resulting in bilateral activation, involves significant activity in the right inferior frontal lobe, whereas phoneme analysis of words results in activation of the left inferior frontal lobe (Buchanan et al. 2000). Both involve bilateral activation and clearly share some neurology, but neurological specialization is also required. In the production and perception of speech sounds, both prosodic melody and semantic visual images also rely on right-hemisphere activity, whereas semantic meaning and syntactic sequencing and relationships result in left-hemisphere activation. Right hemisphere structures seem to be implicated in the control of pitch and tonality in singing, whilst musical rhythm seems to be controlled by left hemisphere structures (Borchgrevink 1980, 1982, 1991).

Further reinforcement of the association between linguistic and rhythmic function, and the distinction between melodic and rhythmic function, comes from observations by Alcock et al. (2000), who studied a family with an inherited developmental speech and language disorder. The family members were tested on various tasks related to the production and perception of melody and rhythm. It was found that, whilst the affected family members were not deficient in the production or perception of pitch, be it in the form of melodies or single notes, they were deficient in the production and the perception of rhythm, both vocally and manually. Alcock et al. conclude that the oral and praxic defects of the family's condition cannot be at the root of the impairment in timing ability, but that the reverse must be the case; that is, rhythmic capacity forms an important component of oral/praxic ability. It seems that the capacity to perform planned sequences of complex muscular movements of rhythmic behaviour, both orally and manually, predate oral/praxic abilities, and that oral/praxic abilities are built upon muscular timing and sequencing capabilities.

So, rhythmic ability appears to form a foundation both for linguistic ability and for rhythmic planning, execution, and processing; and linguistic, semantic and structural processing seem to be handled largely by left-hemisphere structures. Prosodic elements of speech, melody, pitch, and the emotional content of sound production and perception seem to be handled predominantly by related right-hemisphere structures. It also seems that analysis of sounds with a greater amount of timbral quality (such as organ chords or the human voice, as opposed, for example, to the flute) requires greater right-hemisphere involvement than sounds with more neutral, or less rich, timbre. Human voice recognition also relies on right-hemisphere neurology, and little or no input from left-hemisphere analytical structures; it may be that voice recognition and timbre processing rely on related or overlapping structures (Bogen, 1985). The mechanisms for voice recognition are almost certainly evolutionarily far older than those for linguistic processing; the fact that timbre-rich musical sounds are processed exclusively by these mechanisms could suggest that musical processing also predates linguistic processing, *or at least that the processing of tonal content predates the processing of semantic content*. This conclusion is given extra depth by findings by Watt and Ash (1998) that music seems to have an action on the mind similar to the action of interacting with a person. The above are just a few of many studies from pathologies, fMRI, PET, EEG and other scans to demonstrate overlaps between prosodic, melodic, tonal, and emotional sound content production and perception on the one hand, and structural, rhythmic, syntactic, and sequencing content production and perception on the other (Noffsinger 1985; Marin and Perry 1999; Fujii et al. 1990).

In summary, right hemisphere structures appear to be responsible for processing and production, in both melody and speech vocalization, of prosodic melody, pitch control, tonality of singing, timbre processing, and voice recognition. The left hemisphere appears to be implicated in production and processing of semantic verbal meaning and syntactic sequences, as well as rhythmic and analytical functions. Thus, structures in both hemispheres are involved in the production and processing of both music and language; some of the fundamental elements of music and language production and perception are shared, using the same neurological mechanisms, and some have subsequently become specialized. Musical functions as a whole are less clearly lateralized than language functions, but tasks relating to pitch and pitch discrimination do seem to be right-hemisphere dominated. Linguistic functions seem to be most detrimentally affected by left-hemisphere lesions; most musical functions seem to be impaired in some respect by damage to either hemisphere.

The evidence from pathologies resulting in aphasias and amusias suggests that we have a common neurology for musical and linguistic vocalizations at a 'lower' emotive prosodic level, and contralateral localization of neural processing at 'higher' linguistic, semantic levels. This would suggest that the specialized human neurology dedicated to higher linguistic and musical functions emerged later out of a set of common neurological substrates concerned with (complex) vocal emotional expression and comprehension. Indeed, following their review of a large body of research, Marin and Perry (1999) propose that 'The close correspondence between the networks of regions involved in singing and [linguistic] speaking suggests that [linguistic] speech may have evolved from an already-complex system for the voluntary control of vocalization. Their divergences suggest that the later evolving aspects of these two uniquely human abilities are essentially hemispheric specialisations' (p. 692).

PHYLOGENY AND INNATENESS OF CAPACITIES UNDERLYING MUSICAL BEHAVIOURS

Use of rhythm and of music both seem to have important benefits with regard to interpersonal social bonding and group cohesiveness. Whether or not this is itself selectively beneficial, it betrays a fundamental social-emotional element to music-use. Indeed, human musical vocalizations seem to have far more in common with emotive communicative vocalizations than with territorial and mating song in other animals (Slater 2000).

Primate evidence for neurological structures

Studies of primate vocalization neurology can inform as to the phylogenetic origins of such neurological systems. Neuropathological studies have identified the anterior limbic cortex as being responsible for the production of voluntary emotional vocal expressions in all primates (Jürgens 1992). Destruction of this area in humans does not affect the ability to produce involuntary vocal reactions to external stimuli, but does result in an inability to *voluntarily* produce joyful exclamations, angry curses, or pain outbursts. Without the use of the anterior limbic cortex, voluntary vocal utterances, which would normally be highly emotionally communicative, sound more or less monotonous, with very flat intonation (Jürgens and von Cramon 1982; Jürgens 1992). This is an evolutionarily ancient area of the brain, fulfilling the same function in rhesus monkeys (and probably all primate species in the intervening lineage), with the same effects when destroyed (Aitken 1981; Kirzinger and Jürgens 1982, reported in Jürgens 1992).

On the basis of many years researching the neurological roots of vocalization, Jürgens (1998) considers that 'Squirrel monkey vocalization can be considered as a suitable model for the study in humans of the biological basis of non-verbal emotional vocal utterances, such as laughing, crying and groaning' (p. 376), and much of his research has been carried out on these primates. Critical in the production of non-verbal vocal emotional utterances is the periaqueductal grey matter (PAG) of the mid-brain, which acts as a critical relay station for such utterances (Jürgens and Zwirner 1986). Interestingly, recent findings show that the PAG is only implicated in the origination of the limbic *non-verbal emotional* vocal utterances, and not *learned* utterances with a neocortical origin; a second neurological pathway not involving the PAG seems to be responsible for learned vocal patterns (Jürgens and Zwirner 1986), although these would also be integrated at the nucleus ambiguus where control of the vocal folds occurs (Vanderhorst et al. 2001).

Muscle control for both voiced 'vowel' (laryngeal) and voiceless 'consonant' (orofacial) vocal sounds seems to be generated in the lateral column of the mid-brain PAG (Davis et al. 1996). Laryngeal muscle control is thought to be mediated by a neurological pathway from the PAG via the nucleus retroambiguus to the nucleus ambiguus, where monosynaptic connections exist to the laryngeal motoneurons (Vanderhorst et al. 2001, on the basis of the study of rhesus monkeys). This pathway is critical to the production of vocalizations; as well as being responsible for integrating vocal fold control, the nucleus ambiguus is also crucial for expiratory control, orofacial muscular control, and overall control of the laryngeal system (Jürgens 1998), and is directly adjacent to Broca's area (Carlson 1994). This alone indicates how closely interrelated these functions are. As Davis et al. (1996) propose, 'the PAG is a crucial brain site for mammalian voice production, not only in the production of emotional or involuntary sounds, but also as a generator of specific respiratory and laryngeal motor patterns essential for human speech and song' (p. 34).

So it seems that these foundations for fine laryngeal and orofacial control necessary for vocalization have a long phylogenetic history, and that they have since been built upon during the *Homo* lineage. Jürgens (1992) reports that in contrast to the primates studied, modern humans also have a direct connection to the nucleus ambiguus (site of the laryngeal motoneurons) from the primary motor cortex (responsible for planned motor control) (Jürgens 1976, 1992). This connection is not present in monkeys, and probably not in any non-human mammals (Jürgens 1992), and would seem to be what allows humans to carry out planned, structured vocal utterances. Such an ability is not present in other higher primates. As Lieberman (1994) reiterates, although apes can produce many vowel sounds and phonetic features of human speech, these are 'bound' into stereotyped calls based on affect, and they are limited in their production of novel voluntary motoric sequences. The evidence considered in the previous section concerning cervical and thoracic vertebral innervation suggested that controlled vocalizations of extended duration became possible between *Homo ergaster* and the common ancestor of *Homo neanderthalensis* and *Homo sapiens*. It seems reasonable to suggest that the primary connection between the motor cortex and the nucleus ambiguous for planning such utterances developed at the same time.

It is thus highly likely that the ability to perform voluntary, emotional vocal expression, on the basis of external stimuli and internal affective state, was present in all members of the *Homo* genus, and probably in all primates on the lineage between rhesus monkeys and humans. What separates us from other primates, however, is the vocal behaviour which involves voluntary control over the *structure* and *complexity* of vocal utterances, and the degree of the social functionality of these. Although monkeys have a separate pathway for controlling learned vocalizations, without the connection between the primary motor cortex and the laryngeal motoneurons in the nucleus ambiguus, other primates cannot show the human capacity for learning complex vocal patterns by imitation *and by invention*, and for adapting them to novel situations.

Innate capacities for vocal and rhythmic complexity

So what vocal and perceptual abilities are innate in humans at birth? If a sense of rhythm and a sense of melody exist in neonates and infants, this would indicate that perception of these phenomena (inherent in music) is not simply a product of cultural influences, but instead has some basis in hereditary factors. Further, if prelinguistic (and specifically, pre-lexical) infants can be shown to understand some content from rhythm and tone alone, this would provide some confirmation of the ability of non-verbal (rhythmic and melodic) vocal utterances to communicate important emotional information. Fernald (1989a, 1989b, 1992a) showed that exactly this is the case. Four-month-olds heard utterances of approval and of prohibition in four languages, and displayed significantly different affective (emotional) response to the two types of utterances, regardless of which language, and regardless of whether they had heard that language before. The test was also carried out with nonsense-syllables, with the same result. In all instances, intensity of utterance was controlled for, so that the only variations were in rhythm and tone. This inherited capacity to perceive melody and rhythm and to gain affective information from them does not necessarily mean that these capacities were selected for by music, but it does suggest that the ability to comprehend non-linguistic emotional utterances was selectively important.

It also seems that infants have an innate ability to accompany the emotive-tonal prosodic infant-directed speech (discussed in greater depth in the following section) known as 'motherese' (or more accurately, 'parentese'), or in other circumstances singing, with rhythmic body movements which coordinate with the rhythm of the vocalization (Trevarthen 1999). This can occur even in congenitally blind infants, who cannot possibly be imitating the movement of the parent.

As well as displaying sophisticated vocal-perceptual abilities soon after birth, human infants also undertake prolific vocal production at an early age. Elowson, Snowdon, and Lazaro-Perea (1998a) review and outline seven key features of human infant 'babbling' behaviour (a form of pre-linguistic vocal behaviour):

1. Babbling is universal and frequent, irrespective of the infant's cultural background.
2. Babbling is rhythmic and repetitive.
3. Babbling begins between six and ten months, peaking at seven months of age.
4. Babbling comprises a subset of the phonetic sounds found in adult speech.
5. Babbling has well-formed units with the consonant-vowel structure of adult speech.

6. Babbling lacks apparent meaning with respect to how syllables are used by adults in language.
7. Babbling is instrumental in moulding the caregiver-child bond in the infant's first year (p. 33, box 2).

Likewise, primate infants also display vocalization behaviours from an early age, and these may be valuable in informing us of the foundations for the way in which such behaviours develop, and the roots of their use. In those examples reported, primate infant vocalizations are dominated by social, interpersonal concerns. Following his study of trill vocalizations in white-faced capuchin monkeys *Cebus capucinus*, Gros-Louis (2002) observed that they seem to facilitate social interactions, and that such vocalizations were carried out to the greatest extent by immature and adult females. Infants trilled most when approaching other individuals, and those that did tended to interact with them in an affiliative way afterwards; Gros-Louis proposes that infants' trilling action may have an immediate effect on the listener's behaviour towards them, particularly with regard to such socially important activities as receiving grooming, touching, and inspecting food.

Similar findings were made by Elowson et al. (1998b), observing infant pygmy marmosets; response from a caregiving adult was far more likely to be given to an infant when vocalizing than when it was not. The only known study to date of non-human primate 'babbling' behaviour has identified considerable parallels with human infant babbling behaviour, on a number of counts (Elowson et al. 1998a).

In their long-term study of Amazonian pygmy marmosets *Cebuella pygmaea*, Elowson et al. (1998a) observed that '*(1)* both humans and pygmy marmosets have evolved a family-based social unit of cooperating individuals; *(2)* both humans and marmosets have an open, plastic system of communication where subtle changes of vocal production occur over the lifetime of the individual; and *(3)* both go through a period of babbling as part of their vocal development' (p. 36). In particular, 'the long, complex strings of vocalizations produced by the infant monkeys' (p. 32) have much in common with the babbling of human infants outlined above, such as 'universality, repetition, use of a subset of the adult vocal repertoire, recognizably adult-like vocal structure and lack of a clear vocal referent' (p. 31).

Future research will hopefully show whether this type of vocal development is exclusive to these particular primates, or whether equivalent babbling behaviours are exhibited by other primate infants. It is conspicuous, however, that although repetitive, rhythmicity appears to be lacking from the vocalizations of the monkeys. What is especially interesting about this case is that the pygmy marmoset and other species in the *Callitrichidae* family are unique amongst primates in that they live in groups of extended family which are based on a sophisticated social system. These groups include 'in addition to infant twins, the two parents, older juvenile and subadult siblings, and often unrelated individuals . . . the analogy to human families is striking' (Elowson et al. 1998a, p. 32). In these groups, all individuals contribute to the care of the infants, by carrying and grooming them and by providing close physical contact in the form of huddling together at night and as protection against predators. It is possible that it is this form of social organization that leads to the early development and importance of such socially orientated vocal behaviour.

For Elowson et al. (1998a), the most interesting element of their findings is that it indicates that the processes which underlie (at least some) primate vocal learning are similar to those that underlie human vocal learning, and they accordingly propose that pygmy marmoset babbling has a relevance to understanding the evolutionary processes of vocal development. In the context of the current research, the most interesting implication is a phylogenetic (or evolutionary) one. If this is a form of early vocal behaviour and learning that is phylogenetically ancient and genuinely shared with other non-human primates, then it is interesting in that right alone, indicating that we might expect such behaviours to have been carried out also by intervening ancestral hominin species. If, on the other hand, it is a type of behaviour that has emerged by convergence, it is equally interesting, as it could inform about the circumstances in which such behaviours are selectively useful to the individual and the group. In either case, these similarities indicate that lower primates have the cognitive capacities to develop and employ vocalizations in such a manner; that such vocalizations have an important social function at an early stage; and that there is no issue with the foundations of such behavioural capabilities having an ancient provenance. There is no reason for us to believe that any of our hominin ancestors could not have also exhibited such behaviours. Whether they would have encountered the selective pressure that shaped the ability to do so is a separate issue (see, for example, Morley 2003). If so, differences from species to species would presumably be represented in terms of their vocal range, control, and articulation capabilities, since these ultimately bear upon the complexity of the utterances possible, and would in turn influence those utterances most characteristically used.

Tonal and emotional content of vocalization and melody

In relation to the emotional states that they express, some vocal sounds and frequency changes are fundamental, invariant across cultures and even species (Scherer 1985, 1986; Morton 1977, 1994). Part of the reason for this is that facial expression has a fundamental influence on vocal quality, as orofacial musculature helps determine properties of vocalization such as frequency and vowel duration (Tartter 1980; Tartter and Braun 1994). Given the universality and innateness of certain fundamental facial expressions (e.g. Ekman and Friesen 1971; Ekman 1980; Carlson 1994), and the correspondence between these and characteristics of vocalizations, we can also expect characteristics of particular emotional vocalizations to be universal and innate too.

This correlation between facial expression and vocal quality also has an apparently ancient provenance, being shared by our nearest primate relatives, and similar correlations between vocal sound and emotional expression are also exhibited by several other species (Morton 1977; Falk 2004; Bermejo and Omedes 1999). Facial affect and vocal affect inform each other, and we employ and interpret them as signals inter-dependently in both production and perception (DeGelder and Vroomen 2000).

There is an additional factor that comes into play in the relationship between emotion, vocalization, and facial expression: the production of a particular facial expression whilst producing a particular vocalization results in some degree of feedback which actually affects emotional state (Levenson, Ekman, and Friesen 1990). Furthermore, there is a natural inclination to mimic such expressions (Kraut and Johnston 1979; Field et al. 1982) and consequently to feel such associated emotions. This physiological-emotional feedback may occur not only during production but also during *perception* of such a stimulus (Wild, Erb, and Bartels 2001). This means that in producing, and even to an extent in perceiving, a particular sound, we generate some emotional response in ourselves due to the kinaesthetic feedback from the physiology required to produce that sound.

The universality of the vocal sounds and frequency changes that express particular emotions is especially evident in infant-directed (ID) speech, as the exaggeration of these elements of the vocalization is a characteristic feature of ID speech (e.g. Trainor et al. 2000; Fernald 1992b). This seems to be the case across cultures, and in both tonal and non-tonal languages (Greiser and Kuhl 1988; Fernald 1992b; Fernald 1993; Werker et al. 1994; Kitamura et al. 2002). ID vocalizations can tell us a great deal about the nature and role of the prosodic elements of speech and their relationship to melodic musical behaviour. It should be noted that the characteristic features of ID speech are also characteristics of the tonal (non-linguistic) elements of adult-directed (AD) speech, and they apparently share the same foundations and roles in emotional expression (Trainor et al. 2000).

Many of the properties of ID speech are shared with music. There are numerous parallels, in terms of: variable pitch contour; high rhythmicity; repetitive motifs; and the communication of affect, modulation of arousal, and eliciting of attention and affective response (e.g. Stern 1977; Fernald 1989a, 1989b, 1992b; Fernald and Simon 1984; Papousek et al. 1991; Trainor et al. 1997; Trehub et al. 1993). Vocalizations of pre-school children in particular are often difficult to classify as either linguistic or musical (Mang 2000).

Dissanayake (2000) hypothesizes that this mother-infant behaviour emerged as a consequence of hominin infants being born progressively earlier as brain size increased, and that their thus being more helpless for longer after birth increased the need for emotional conjoinment at this early stage of life. This would, she suggests, have become particularly important at the time of archaic *Homo sapiens* (*Homo heidelbergensis*), which exhibited a large increase in brain size. She proposes that these vocalizations then came to be adopted increasingly into daily life and ritual, eventually becoming the basis for musical behaviour, because they 'were found by evolving human groups to be emotionally affecting and functionally effective . . . to emotionally conjoin and enculturate the participants' (Dissanayake 2000, p. 401).

The benefits of such social-emotive interaction between individuals (emotional conjoinment and enculturation) is well argued and supported by much of the evidence cited above, and the connection between musical behaviours and affective vocalization is an important one (as demonstrated by the neurological evidence considered previously). However, it may appear difficult to envisage hominins beginning to employ a behaviour which originally evolved as a 'baby-talk' in other situations in which they realized it 'promoted affiliation and congruence in adult social life' (Dissanayake 2000, p. 401). As discussed previously, there are precedents for the importance of affective-communicative elements of such vocalizations in adult vocalizations, as well as in primates without language.

Additionally, the fact that the vocal tract of *Homo heidelbergensis* seems to have been effectively fully modern (see above) would suggest that selective pressure for its use for wide-ranged tonal expression had come much earlier than Dissanayake proposes. This form of communication might gradually have evolved into music as Dissanayake suggests, or at least provided shared foundations, but it could also have been the basis for language amongst all of a population (cf. Aiello and Dunbar 1993; Scherer 1991; Brown 2000). It may be that the use of this mode of emotive utterance for communication between (pre-linguistic) infants and adults today is a vestige of this earliest form of social communication; after all, it is potentially still selectively useful, at least until a child gains full linguistic ability, and is still inherent in the tonal elements of *any* emotional vocal communication (Trainor et al. 2000). Thus an alternative, and perhaps more parsimonious, explanation to Dissanayake's suggestion is that this (social-emotive vocalization) was a form of communication that came to be used *throughout* the social group at a much earlier time, without preference, both adult-adult and infant-adult, but is now perpetuated, in this predominantly non-lexical form, in adult-infant interactions and in the prosodic content of adult speech. Furthermore, the shared prosodic pitch- and tempo-related properties of emotional vocalization (ID and AD) and music are not borrowed from one to the other, in either direction, but are, and always have been, a shared fundamental component of both.

Music elicits emotional responses in listeners for a variety of reasons. In eliciting an emotional reaction, music can generate genuine physiological responses such as changes in respiration, heart-rate, and temperature; and tingling (Krumhansl 1997; Sloboda 1991, 1998; Panksepp 1995; Panksepp and Bernatsky 2002). Some emotional responses can be elicited, for example, as a consequence of learned association with particular circumstances; others as a consequence of direct resemblance to ecological phenomena to which we have instinctive or conditioned responses (Sloboda 2001; Sloboda and Juslin 2001; Davies 2001; Cross 2003). A third form of emotional response occurs as a consequence of processing by, and stimulation of, auditory and kinaesthetic mechanisms associated with interpersonal interaction. We can react to and process music in many of the same ways as we do interaction with a person (Kivy 1989; Davies 2001; Lavy 2001); it shares many of the properties of such interaction, and can apparently be attributed with similar traits (Watt and Ash 1998). Consequently, we can (although don't necessarily) emotionally empathize with music, if it elicits the same emotion as is being expressed. The context in which we experience music is also very important in determining the emotion, and intensity of emotion, experienced (Sloboda and Juslin 2001). Especially important in this respect is the social context—the extent to which the experience is shared with others.

Vocalization, corporeal expression, and rhythm

Gesture and speech are interdependent. Both are affected simultaneously in stutterers (Mayberry and Jaques 2000) and gestural and vocal behaviours are interrelated from the earliest babbling in infants (Masataka 2000; Mayberry and Jaques 2000; Locke et al. 1995). They can operate independently, but when the upper limbs are otherwise unoccupied they can be sequestered into speech-related gesture. It seems that there is a cognitive rhythmic motor coordinator which instigates such muscular sequences irrespective of the musculature that is used, and that the complex patterns of muscular movement of gesture (finger, hand, arm, shoulder, and joint musculature) and vocalization (orofacial, laryngeal, and respiratory musculature) are coordinated (Franz et al. 1992; cf. Alcock et al. 2000). The concordance between gesture and speech is instigated early in the vocalization process, by cyclical motor control, with gestural movements being associated with the nuclei of tone groups—prosodic rhythm—rather than the lexical elements of speech (McClave 1994; Mayberry and Jaques 2000). This is also evident in the gestures accompanying infant vocalizations. Greater lateralization of gesture occurs after babbling begins, implying left-hemisphere development at this time in the integration of the two actions (Locke et al. 1995). These earliest gestures are emotive and rhythmic rather than iconic (Trevarthen 1999) (corresponding with the above finding that gesture corresponds with prosodic rhythm rather than lexical content), and only start to be used iconically and in combination with words when lexical behaviour has started to develop (Messinger and Fogel 1998; Butcher and Goldin-Meadow 2000). In the case of speech (as opposed to other, non-linguistic, vocalizations), linguistic meaning and narrative sentence structure are integrated into the gesture-speech system after the integration of gesture with prosody, but before their physical manifestation (Mayberry and Jaques 2000).

Affective content can apparently be interpreted equivalently in visual, auditory and kinaesthetic media, each of which can represent tension, release and particular emotions, underlining the cross-modality of such affective expression and interpretation (Krumhansl and Schenck 1997). As with facial expression, it seems that this can be a two-way process too, with feedback occurring to some extent between the modes of experience (Mitchell and Gallaher 2001). Consequently, our posture and movements can express a great deal about our emotional state, intentionally and unintentionally, and others' body-posture and movements thus provide important cues as to their emotional state. As well as being able to observe such cues, we can empathically experience something of their emotional state in mirroring them with our own bodies. Musicality and rhythmic movement involve deliberate control and sequencing of this system (Trevarthen 1999). There seems to be a close interrelationship between the emotion-controlling elements of the limbic system and the areas responsible for the coordination of motor sequences and posture, driven by the Intrinsic Motor Formation, which consists of the brain stem, basal ganglia and the limbic structures of the Emotional Motor System (Trevarthen 1999). The same systems that select and control movements also cause changes in the emotion-controlling elements of the limbic system. This can result in a self-directed feedback from movement into emotional state and, importantly, feedback and interaction between individuals, in terms of synchrony of movement and of emotional state (Trevarthen 1999; McNeill 1995).

Production of complex vocalization relies on priming of the whole motor system (Blount 1994). Particular non-lexical vocal utterances (and non-linguistic content of speech) are accompanied by quite specific involuntary body-language (Kappas et al. 1991; Wagner 1989; Lavy 2001); they share an affective origin in a system in which vocalization and corporeal expression, or to put it another way, vocal and corporeal gesture, are intimately linked (Blount 1994; cf. Clynes's (1977) 'sentic modulation' and Donald's (1991) 'mimesis'). Note that this is at least equally applicable to body movements accompanying musical vocalization and to vocalization accompanying dance. There are not only shared neurological foundations between the ability to execute vocal and manual gestural sequences, but also a link between vocal and manual *rhythmic* capability. Humans are the only primates to carry out vocalizations with rhythmic content (Geissman 2000), and the capacity to perform rhythms, both manually and verbally, forms an important component of oral/praxic ability—detriment to one results in detriment to the other. This integration occurs whether utterances are linguistic or not (Alcock et al. 2000).

Whilst rhythm and melody are apparently produced and processed in neurologically specialized and distinct areas of the brain which are somewhat independent of each other (as illustrated by the neurological evidence discussed previously), there is also clearly important integration of these systems, with rhythmic muscular movements being coordinated with prosodic elements of vocalization in their production. Specifically, they are interdependent in the planning and execution of sequences of muscular movement associated with instigation of vocalization, rhythmic physiological movement, and expression of emotional state in these media. The production and perception of tonal *content* in vocalizations does not appear to require any input from rhythm-controlling systems in the left hemisphere (see above) but the planning and execution of sequences of such vocalizations does.

The role of this system in evolution

There is evidence that group size increased over the course of hominin evolution, and it has been suggested that complex vocalization and language became increasingly important as a consequence (Aiello and Dunbar 1993; Dunbar 1998; Kudo and Dunbar 2001). The role of complex vocalization of affect, and sensitivity to the prosodic content of vocalizations, has already been elaborated; there is much evidence to suggest that full language also remains important in this role (Dunbar et al. 1997; Emler 1992).

Melody and linguistic speech share common features in the form of intonational phrasing and combinatorial syntax, and share a 'phonological level' of meaning, an 'acoustic mode' involving emotive meaning and interpretation (Brown 2000). Whilst music often does use the full range of the vocal tract, full language does not need to in order to communicate effectively, as linguistic structures disambiguate meaning (e.g. in whispering; or in written text). This would not have been the case for a pre-syntactic, pre-lexical proto-language.

The fossil and neurological evidence discussed in the preceding sections suggests that the ability to produce short, tonally varied utterances emerged before the ability to produce extended sequences of vocalizations. It seems likely that discrete tonally-contoured units expressive of affective state would have become combined into progressively larger globally-contoured units. What must have emerged in the course of the evolution of pitch-contoured vocalizations is an *increased range and control* of pitch contour, allowing greater vocal versatility, expressiveness, and thus efficiency, in proto-linguistic vocal affective communication. This was coupled with the emergence of greater control over the duration of such utterances, and the ability to plan them. This would not necessarily have emerged as a new system of vocalization, however, but may have built upon the limited pitch control already used for emotive-tonal-social vocalization by higher primates. The selective advantages associated with the possession of such capabilities would have resulted in the continual evolution of the vocal tract, and control over it, through the lowering of the larynx and increased innervation of the associated laryngeal and upper vocal tract musculature.

The lowering of the larynx allowed greater versatility of vocal tone and lower formant frequencies to be produced and also allowed greater control and movement of the tongue. It is suggested (Fitch 2000) that the ability to *temporarily* lower the larynx in order to produce more complex vocalizations may have led to a preliminary permanent lowering of the larynx. This would subsequently have continued to lower, with concomitant increasing control over the sound-producing capabilities of the vocal tract and orofacial articulators.

Whilst extremely important in the type of information they express, non-verbal vocal utterances, in meaning, are limited to acting as expressions of personal state and relationships between the self and other. In terms of the emergence of language from such vocalizations, however, it seems likely that such utterances formed the first *vocalized* form of *conceptualizations* relating the self to object/other, and that this ability to vocalize such concepts was subsequently built upon by more complex vocal and gestural sequences (see Armstrong et al. 1994). In fact, it is at that stage that the syntactic potential of the coupling of vocal and *iconic* gestural communication may have become fundamentally important, bridging the gap between non-lexical vocalizations of affect, and multi-part vocalizations with syntactic structure and iconic associations (Armstrong et al. 1994).

The socially important, emotionally communicative elements of such vocalization would likely have remained the dominant element initially, with iconic and then abstract lexical associations gradually increasing in importance. It is the latter—iconic and abstract lexical content—which relies on symbolic and analogical capacity, and which may not have been present in hominins prior to *Homo sapiens*; the former do not rely on symbolic capacity, and, as illustrated, would have been selectively important at a much earlier time. Even with the emergence of full lexical and syntactic language, the social-affective communicative foundations of prosodic contour remained a fundamental element of vocal communication. It is these which formed the foundations of musical behaviours.

In summary, there is a large amount of diverse evidence which suggests that musical behaviours have as their foundation—and constitute a specialized version of—the vocalizations and gestures associated with the expression of emotion and maintenance of social bonds. Accounted for is the production and perception of such prosodically-contoured vocalizations, and the interdependence with them of corporeal and rhythmic expression of the same. Music is built on fundamentally important social and physiological mechanisms and, at an essential level, is processed as such. It uses capacities crucial in situations of social complexity: the vocal, facial, and interactive foundations of these capabilities are evident in other higher primates, and such capacities would have become increasingly important and sophisticated as group size and complexity of social relations increased. Vocal emotional expression, interaction, sensitivity to others' emotional state, and empathic ability would have been selectively important abilities; those individuals in which such capabilities were more developed would have been selectively fitter. Music draws upon these abilities as it uses emotional (vocal and physiological) stimuli as its foundation. Fundamentally integrated into the planning and control of complex sequences of vocalizations, and related to the prosodic rhythm inherent in such sequences, is rhythmic motor coordination. The motor system is primed in the instigation of such vocal behaviours, and corporeal gesture is consequently incorporated into the execution of the vocal behaviour. Musical behaviours constitute a specialized and deliberate use of this rhythmic-tonal vocal-gestural emotional-corporeal system.

REFERENCES

Aiello, L. C. and R. I. M. Dunbar (1993). 'Neocortex Size, Group Size, and the Evolution of Language', *Current Anthropology*, 34: 184–93.

Aitken, P. G. (1981). 'Cortical Control of Conditioned and Spontaneous Vocal Behavior in Rhesus Monkeys', *Brain and Language*, 13: 171–84.

Alcock, K. J., R.E. Passingham, K. Watkins et al. (2000). 'Pitch and Timing Abilities in Inherited Speech and Language Impairment', *Brain and Language*, 75: 34–46.

Arensberg, P., L. A. Schepartz, A. M. Tillier et al. (1990). 'A Reappraisal of the Anatomical Basis for Speech in Middle Palaeolithic Hominids', *American Journal Of Physical Anthropology*, 83: 137–46.

Armstrong, D. F., W. C. Stokoe, and S. E. Wilcox (1994). 'Signs of the Origins of Syntax', *Current Anthropology*, 35: 349–68.

Benson, D. F. (1985). 'Language in the Left Hemisphere'. In: D. F. Benson and E. Zaidel (eds.), *The Dual Brain,* pp. 193–203. New York: Guilford Press.

Bermejo, M. and A. Omedes (1999). 'Preliminary Vocal Repertoire and Vocal Communication of Wild Bonobos (*Pan paniscus*) at Lilungo (Democratic Republic of Congo)', *Folio Primatologica*, 70: 328–57.

Blount, B. G. (1994). 'Comment on Armstrong, Stokoe and Wilcox', *Current Anthropology*, 35: 358–9.

Bogen, J. E. (1985). 'The Dual Brain: Some Historical and Methodological Aspects'. In: D. F. Benson and E. Zaidel (eds.), *The Dual Brain*, pp. 27–43. New York: Guilford Press.

Borchgrevink, H. M. (1980). 'Cerebral Lateralization of Speech and Singing after Intracarotid Amytal Injection'. In: M. Taylor Sarno and O. Hooks (eds.), *Aphasia: Assessment and Treatment*, pp. 186–91. Stockholm: Almkvist & Wiksell.

——(1982). 'Prosody and Musical Rhythm Are Controlled by the Speech Hemisphere'. In: M. Clynes (ed.), *Music, Mind and Brain*, pp. 151–7. New York: Plenum Press.

——(1991). 'Prosody, Musical Rhythm, Tone Pitch and Response Initiation during Amytal Hemisphere Anaesthesia'. In: J. Sundberg, L. Nord and R. Carlson (eds.), *Music, Language, Speech and Brain: Proceedings of an*

International Symposium at the Wenner-Gren Center, Stockholm, 5–8 September 1990, pp. 327–43. Basingstoke: Macmillan.

Borg, E. and S. A. Counter (1989). 'The Middle-Ear Muscles', *Scientific American*, 261, 2: 63–8.

Brown, S. (2000). 'The "Musilanguage" Model of Music Evolution'. In: N. L. Wallin, B. Merker, and S. Brown (eds.), *The Origins of Music*, pp. 271–300. Cambridge, MA: MIT Press.

Buchanan, T. W., K. Lutz, S. Mirzazade et al. (2000). 'Recognition of Emotional Prosody and Verbal Components of Spoken Language: An fMRI Study', *Cognitive Brain Research*, 9: 227–38.

Budil, I. (1994). 'A Functional Reconstruction of the Supralaryngeal Vocal Tract of the Fossil Hominid from Petralona'. In: J. Wind, A. Jonker, R. Allot et al. (eds.), *Studies In Language Origins*, iii, pp. 1–19. Amsterdam: Benjamins.

Butcher, C. and S. Goldin-Meadow (2000). 'Gesture and the Transition from One- to Two-Word Speech: When Hand and Mouth Come Together'. In: D. McNeill (ed.), *Language and Gesture*, pp. 235–57. New York: Cambridge University Press.

Calvin, W. H. (1996). *How Brains Think*. London: Weidenfeld and Nicholson.

Carlson, N. R. (1994). *The Physiology of Behavior*. Boston, MA: Allyn and Bacon.

Clynes, M. (1977). *Sentics: The Touch of Emotions*. New York: Souvenir Press.

Cross, I. (2003). 'Music and Emotion'. <http://www.mus.cam.ac.uk/~ic108/IBMandS/emotion/musicemotion.html>.

Davies, S. (2001). 'Philosophical Perspectives on Music's Expressiveness'. In: P. N. Juslin and J. A. Sloboda (eds.), *Music and Emotion: Theory and Research*, p. 23–44. Oxford: Oxford University Press.

Davis, P. J., S. P. Zhang, A. Winkworth et al. (1996). 'Neural Control of Vocalization: Respiratory and Emotional Influences', *Journal of Voice*, 10: 23–38.

DeGelder, B. and J. Vroomen (2000). 'The Perception of Emotions by Ear and by Eye', *Cognition and Emotion*, 14: 289–311.

Dissanayake, E. (2000). 'Antecedents of the Modern Arts in Early Mother-Infant Interaction'. In: N. L. Wallin, B. Merker, and S. Brown (eds.), *The Origins of Music*, pp. 389–410. Cambridge, MA: MIT Press.

Donald, M. (1991). *Origins of the Modern Mind: Three Stages in the Evolution of Culture and Cognition*. Cambridge, MA: Harvard University Press.

Duchin, L. E. (1990). 'The Evolution of Articulate Speech: Comparative Anatomy of *Pan* and *Homo*', *Journal of Human Evolution*, 19: 687–97.

Dunbar, R. (1998). 'Theory of Mind and the Evolution of Language'. In: J. R. Hurford, M. Studdert-Kennedy, and C. Knight (eds.), *Approaches to the*

Evolution of Language, pp. 92–110. Cambridge: Cambridge University Press.

——N. Duncan, and A. Marriott (1997). 'Human Conversational Behaviour', *Human Nature*, 8: 231–46.

Ekman, P. (1980). *The Face of Man: Expressions of Universal Emotions in a New Guinea Village*. New York: Garland SPTM Press.

——and W. V. Friesen (1971). 'Constants across Cultures in the Face and Emotion', *Journal of Personality and Social Psychology*, 17: 124–9.

Elowson, A. M., C. T. Snowdon, and C. Lazaro-Perea (1998a). '"Babbling" and Social Context in Infant Monkeys: Parallels to Human Infants', *Trends in Cognitive Sciences*, 2: 31–7.

————————(1998b). 'Infant "Babbling" in a Non-Human Primate: Complex Vocal Sequences with Repeated Call-Types', *Behaviour*, 135: 643–64.

Emler, N. (1992). 'The Truth about Gossip', *Social Psychology Newsletter*, 27: 23–37.

Falk, D. (1992). *Braindance: New Discoveries about Human Origins and Brain Evolution*. New York: Henry Holt.

——(2004). 'Prelinguistic Evolution in Early Hominins: Whence Motherese?' *Behavioural and Brain Sciences*, 27: 491–503.

Fernald, A. (1989a). 'Emotion and Meaning in Mothers' Speech to Infants'. Paper presented to *The Society for Research in Child Development*, Kansas City, April 1989.

——(1989b). 'Intonation and Communicative Intent in Mothers' Speech to Infants: Is Melody the Message?' *Child Development*, 60: 1497–510.

——(1992a). 'Meaningful Melodies in Mothers' Speech to Infants'. In: H. Papousek, U. Jürgens, and M. Papousek (eds.), *Nonverbal Vocal Communication: Comparative and Developmental Approaches*, pp. 262–82. Paris and Cambridge: Editions de la Maison des Sciences de l'Homme/ Cambridge University Press.

——(1992b). 'Human Maternal Vocalizations to Infants as Biologically Relevant Signals: An Evolutionary Perspective'. In: J. Barkow, L. Cosmides, and J. Tooby (eds.), *The Adapted Mind: Evolutionary Psychology and the Generation of Culture*, pp. 391–428. Oxford: Oxford University Press.

——(1993). 'Approval and Disapproval—Infant Responsiveness to Vocal Affect in Familiar and Unfamiliar Languages', *Child Development*, 64: 657–74.

——and T. Simon (1984). 'Expanded Intonation Contour in Mothers' Speech to Newborns', *Developmental Psychology*, 20: 104–13.

Field, T., R. Woodson, R. Greenberg et al. (1982). 'Discrimination and Imitation of Facial Expression in Neonates', *Science*, 218: 179–81.

Fitch, W. T. (2000). 'The Phonetic Potential of Nonhuman Vocal Tracts: Comparative Cineradiographic Observations of Vocalising Animals', *Phonetica*, 57: 205–18.

Franz, E. A., H. N. Zelaznik, and A. Smith (1992). 'Evidence of Common Timing Processes in the Control of Manual, Orofacial, and Speech Movements', *Journal of Motor Behavior*, 24: 281–7.

Frayer, D. W. and C. Nicolay (2000). 'Fossil Evidence for the Origin of Speech Sounds'. In: N. L. Wallin, B. Merker, and S. Brown (eds.), *The Origins of Music*, pp. 217–34. Cambridge, MA: MIT Press.

Fujii, T., R. Fukatsu, S. Watabe et al. (1990). 'Auditory Sound Agnosia without Aphasia Following Right Temporal Lobe Lesion', *Cortex*, 26: 263–8.

Geissman, T. (2000). 'Gibbon Songs and Human Music in an Evolutionary Perspective'. In: N. L. Wallin, B. Merker, and S. Brown (eds.), *The Origins of Music*, pp. 102–23. Cambridge, MA: MIT Press.

Geschwind, N., F. Quadfasel, and J. Segarra (1965). 'Isolation of the Speech Area', *Neuropsychologia*, 6: 327–40.

Greiser, D. and P. Kuhl (1988). 'Maternal Speech to Infants in a Tonal Language: Support for Universal Prosodic Features in Motherese', *Developmental Psychology*, 24: 14–20.

Gros-Louis, J. (2002). 'Contexts and Behavioural Correlates of Trill Vocalizations in Wild White-Faced Capuchin Monkeys (*Cebus capucinus*)', *American Journal of Primatology*, 57: 189–202.

Holloway, R. (1981). 'Volumetric and Asymmetry Determinations on Recent Hominid Endocasts: Spy I and II, Djebel Ihroud I, and the Salè *Homo erectus* Specimens, with some Notes on Neanderthal Brain Size', *American Journal Of Physical Anthropology*, 55: 385–93.

——(1983). 'Human Palaeontological Evidence Relevant to Language Behavior', *Human Neurobiology*, 2: 105–14.

Johanson, D. and B. Edgar (1996). *From Lucy to Language*. London: Weidenfeld and Nicolson.

Jürgens, U. (1976). 'Reinforcing Concomitants of Electrically Elicited Vocalizations', *Experimental Brain Research*, 26, 2: 203–14.

——(1992). 'On the Neurobiology of Vocal Communication'. In: H. Papousek, U. Jürgens and M. Papousek (eds.), *Nonverbal Vocal Communication: Comparative and Developmental Approaches*, pp. 31–4. Paris and Cambridge: Editions de la Maison des Sciences de l'Homme/Cambridge University Press.

——(1998). 'Neuronal Control of Mammalian Vocalization, with Special Reference to the Squirrel Monkey', *Naturwissenschaften*, 85: 376–88.

——and D. Von Cramon (1982). 'On the Role of the Anterior Cingulate Cortex in Phonation—A Case Report', *Brain and Language*, 15: 234–48.

——and P. Zwirner (1986). 'The Role of the Periaqueductal Grey in Limbic and Neocortical Vocal Fold Control', *Neuroreport*, 7: 2921–3.

Kappas, A., U. Hess, and K. R. Scherer (1991). 'Voice and Emotion'. In: R. S. Feldman and B. Rimé (eds.), *Fundamentals of Nonverbal Behavior:*

Studies in Emotion and Social Interaction, pp. 200–38. Cambridge: Cambridge University Press.

Karow, C. M., T. P. Marquardt, and R. C. Marshall (2001). 'Affective Processing in Left and Right Hemisphere Brain-Damaged Subjects with and without Subcortical Involvement', *Aphasiology*, 15: 715–29.

Kirzinger, A. and U. Jürgens (1982). 'Cortical Lesion Effects and Vocalization in the Squirrel Monkey', *Brain Research*, 233: 299–315.

Kitamura, C., C. Thanavisthuth, D. Burnham et al. (2002). 'Universality and Specificity in Infant-Directed Speech: Pitch Modifications as a Function of Infant Age and Sex in a Tonal and Non-Tonal Language', *Infant Behaviour and Development*, 24: 372–92.

Kivy, P. (1989). *Sound Sentiment*. Philadelphia, PA: Temple University Press.

Kochetkova, V. I. (1978). *Paleoneurology*. Washington, DC: Winston & Sons.

Kraut, R. E. and R. Johnston (1979). 'Social and Emotional Messages of Smiling: An Ethological Approach', *Journal of Personality and Social Psychology*, 37: 1539–53.

Krumhansl, C. L. (1997). 'An Exploratory Study of Musical Emotions and Psychophysiology', *Canadian Journal of Experimental Psychology*, 51: 336–52.

——and D. L. Schenck (1997). 'Can Dance Reflect the Structural and Expressive Qualities of Music? A Perceptual Experiment on Balanchine's Choreography of Mozart's *Divertimento* No. 15', *Musicae Scientiae*, 1: 63–85.

Kudo, H. and R. I. M. Dunbar (2001). 'Neocortex Size and Social Network in Primates', *Animal Behaviour*, 62: 711–22.

Kuhl, P. (1988). 'Auditory Perception and the Evolution of Speech', *Human Evolution*, 3: 19–43.

Laitman, J. T. (1984). 'The Anatomy of Human Speech', *Natural History*, 93: 20–7.

——R. C. Heimbuch, and E. S. Crelin (1979). 'The Basicranium of Fossil Hominids as an Indicator of their Upper Respiratory Systems', *American Journal of Physical Anthropology*, 51: 13–34.

————(1982). 'The Basicranium of Plio-Pleistocene Hominids as an Indicator of their Upper Respiratory Systems', *American Journal of Physical Anthropology*, 59: 323–44.

——and J. S. Reidenberg (1988). 'Advances in Understanding the Relationship between the Skull Base and Larynx with Comments on the Origins of Speech', *Human Evolution*, 3: 99–109.

Lavy, M. M. (2001). 'Emotion and the Experience of Listening to Music: A Framework for Empirical Research', PhD thesis, University of Cambridge.

Leakey, R. (1994). *The Origin of Humankind*. London: Weidenfeld and Nicolson.

Levenson, R. W., P. Ekman, and W. V. Friesen (1990). 'Voluntary Facial Action Generates Emotion-Specific Autonomic Nervous System Activity', *Psychophysiology*, 27: 363–84.

Liberman, A. M., F. S. Cooper, D. P. Shankweiler et al. (1967). 'Perception of the Speech Code', *Psychological Review*, 74: 431–61.

Lieberman, P. (1984). *The Biology and Evolution of Language*. Cambridge, MA: Harvard University Press.

——(1989). 'The Origins of Some Aspects of Human Language and Cognition'. In: P. Mellars and C. Stringer (eds.), *The Human Revolution: Behavioural and Biological Perspectives in the Origins of Modern Humans*, pp. 391–414. Edinburgh: Edinburgh University Press.

——(1994). 'Comment on Armstrong, Stokoe and Wilson', *Current Anthropology*, 35: 362–3.

Locke, J. L., K. E. Bekken, L. McMinn-Larson et al. (1995). 'Emergent Control of Manual and Vocal Motor Activity in Relation to the Development of Speech', *Brain and Language*, 51: 498–508.

McClave, E. (1994). 'Gestural Beats: The Rhythm Hypothesis', *Journal of Psycholinguistic Research*, 23: 45–66.

MacLarnon, A. M. and G. P. Hewitt (1999). 'The Evolution of Human Speech: The Role of Enhanced Breathing Control', *American Journal of Physical Anthropology*, 109: 341–63.

McNeill, W. H. (1995). *Keeping Together in Time*. Cambridge, MA: Harvard University Press.

Magriples, U. and J. T. Laitman (1987). 'Developmental Change in the Position of the Fetal Human Larynx', *American Journal of Physical Anthropology*, 72: 463–72.

Mang, E. (2000). 'Intermediate Vocalizations: An Investigation of the Boundary between Speech and Songs in Young Children's Vocalizations', *Bulletin of the Council for Research in Music Education*, 147: 116–21.

Marin, O. S. M. and D. W. Perry (1999). 'Neurological Aspects of Music Perception and Performance'. In: D. Deutsch (ed.), *The Psychology of Music*, 2nd edn., pp. 653–724. New York: Academic Press.

Masataka, N. (2000). 'The Role of Modality and Input in the Earliest Stages of Language Acquisition: Studies of Japanese Sign Language'. In: C. Chamberlain, J. Morford and R. I. Mayberry (eds.), *Language Acquisition by Eye*, pp. 3–24. Mahwah, NJ: Erlbaum.

Mayberry, R. I. and J. Jaques (2000). 'Gesture Production during Stuttered Speech: Insights into the Nature of Gesture-Speech Integration'. In: D. McNeill (ed.), *Language and Gesture*, pp. 199–214. Cambridge: Cambridge University Press.

Mazzucchi, A., C. Marchini, R. Budai et al. (1982). 'A Case of Receptive Amusia with Prominent Timbre Perception Defect', *Journal of Neurology, Neurosurgery, and Psychiatry*, 45: 644–7.

Messinger, D. S. and A. Fogel (1998). 'Give and Take: The Development of Conventional Infant Gestures', *Merrill-Palmer Quarterly: Journal of Developmental Psychology*, 44: 566–90.

Mitchell, R. W. and M. C. Gallaher (2001). 'Embodying Music: Matching Music and Dance in Memory', *Music Perception*, 19: 65–85.

Morley, I. (2002). 'Evolution of the Physiological and Neurological Capacities for Music', *Cambridge Archaeological Journal*, 12: 195–216.

——(2003). 'The Evolutionary Origins and Archaeology of Music: An Investigation into the Prehistory of Human Musical Capacities and Behaviours', PhD thesis, Cambridge University.

Morton, E. S. (1977). 'On the Occurrence and Significance of Motivation-Structural Rules in Some Bird and Mammal Sounds', *American Naturalist*, 111: 855–69.

——(1994). 'Sound Symbolism and Its Role in Non-Human Vertebrate Communication'. In: L. Hinton, J. Nichols and J. J. Ohala (eds.), *Sound Symbolism*, pp. 348–65. Cambridge: Cambridge University Press.

Noffsinger, D. (1985). 'Dichotic-Listening Techniques in the Study of Hemispheric Asymmetries'. In: D. F. Benson and E. Zaidel (eds.), *The Dual Brain*, pp. 127–41. New York: Guilford Press.

Panksepp, J. (1995). 'The Emotional Sources of 'Chills' Induced by Music', *Music Perception*, 13: 171–208.

——and G. Bernatsky G (2002). 'Emotional Sounds and the Brain: The Neuro-Affective Foundations of Musical Appreciation', *Behavioural Processes*, 60: 133–55.

Papousek, M., H. Papousek and D. Symmes (1991). 'The Meanings of Melodies in Motherese in Tone and Stress Languages', *Infant Behavior and Development*, 14: 415–40.

Patel, A. D., I. Peretz, M. Tramo et al. (1998). 'Processing Prosodic and Musical Patterns: A Neuropsychological Investigation', *Brain and Language*, 61: 123–44.

Peretz, I. (1993). 'Auditory Atonalia for Melodies', *Cognitive Neuropsychology*, 10: 21–56.

——R. Kolinsky, M. Tramo et al. (1994). 'Functional Dissociations Following Bilateral Lesions of Auditory-Cortex', *Brain*, 117: 1283–301.

Peterson, S. E., P. T. Fox, M. I. Posner et al. (1988). 'Positron Emission Tomographic Studies of the Cortical Anatomy of Single-Word Processing', *Nature*, 331: 585–9.

Scherer K. R. (1985). Vocal affect signalling: a comparative approach. *Advances in the Study of Behavior*, 15: 189–244.

——(1986). 'Vocal Affect Expression: A Review and Model for Future Research', *Psychological Bulletin*, 99: 143–65.

——(1991). 'Emotion Expression in Speech and Music'. In: J. Sundberg, L. Nord and R. Carlson (eds.), *Music, Language, Speech and Brain:*

Proceedings of an International Symposium at the Wenner-Gren Center, Stockholm, 5–8 September 1990, p. 146–56. Basingstoke: Macmillan.

Slater, P. J. B. (2000). 'Birdsong Repertoires: Their Origin and Use'. In: N. L. Wallin, B. Merker, and S. Brown (eds.), *The Origins of Music*, pp. 49–63. Cambridge, MA: MIT Press.

Sloboda, J. A. (1991). 'Music Structure and Emotional Response: Some Empirical Findings', *Psychology Of Music*, 19: 110–20.

——(1998). 'Does Music Mean Anything?' *Musicae Scientiae*, 2: 21–31.

——(2001). 'The Psychology of Music: Affect'. In: L. Macy (ed.) *The New Grove Dictionary of Music Online*. <http://www.grovemusic.com>.

——and P. N. Juslin (2001). 'Psychological Perspectives on Music and Emotion'. In: P. N. Juslin and J. A. Sloboda (eds.), *Music and Emotion: Theory and Research*, pp. 71–104. Oxford: Oxford University Press.

Snow, D. (2000). 'The Emotional Basis of Linguistic and Nonlinguistic Intonation: Implications for Hemispheric Specialization', *Developmental Neuropsychology*, 17: 1–28.

Spoor, F. (1996). 'The Ancestral Morphology of the Hominid Bony Labyrinth: The Evidence from *Dryopithicus*', *American Journal of Physical Anthropology*, 22 (Supplement): 219.

——B. Wood, and F. Zonneveld (1994). 'Implications of Hominid Labyrinthine Morphology for Evolution of Human Bipedal Locomotion', *Nature*, 369: 645–8.

Spoor, F. and F. Zonneveld F (1998). 'Comparative Review of the Human Bony Labyrinth', *Yearbook of Physical Anthropology*, 41: 211–51.

Stern, D. N. (1977). *The First Relationship: Infant and Mother*. Cambridge, MA: Harvard University Press.

Tartter, V. C. (1980). 'Happy Talk: Perceptual and Acoustic Effects of Smiling on Speech', *Perception and Psychophysics*, 27: 24–7.

——and D. Braun (1994). 'Hearing Smiles and Frowns in Normal and Whisper Registers', *Journal of the Acoustical Society of America*, 96: 2101–7.

Trainor, L. J., E. D. Clark, A. Huntley et al. (1997). 'The Acoustic Basis of Infant Preferences for Infant-Directed Singing', *Infant Behavior and Development*, 60: 383–96.

——C. Austin, and R. N. Desjardins (2000). 'Is Infant-Directed Speech Prosody a Result of the Vocal Expression of Emotion?' *Psychological Science*, 11: 188–95.

Trehub, S. E., L. J. Trainor, and A. M. Unyk (1993). 'Music and Speech Processing in the First Year of Life', *Advances in Child Development and Behaviour*, 24: 1–35.

Trevarthen, C. (1999). 'Musicality and the Intrinsic Motive Pulse: Evidence from Human Psychobiology and Infant Communication', *Musicae Scientiae*, Special Issue: 155–215.

Vanderhorst, V. G. J. M., E. Teresawa, and H. J. Ralston (2001). 'Monosynaptic Projections from the Nucleus Retroambiguus Region to Laryngeal Motoneurons in the Rhesus Monkey', *Neuroscience*, 107: 117–25.

Wagner, H. (1989). 'The Physiological Differentiation of Emotions'. In: H. Wagner and A. Manstead (eds.), *Handbook of Social Psychophysiology*, pp. 77–89. Chichester: Wiley.

Watt, R. J. and R. L. Ash (1998). 'A Psychological Investigation of Meaning in Music', *Musicae Scientiae*, 2: 33–54.

Werker, J. F., J. E. Pegg and P. J. McLeod (1994). 'A Cross-Language Investigation of Infant Preference for Infant-Directed Communication', *Infant Behaviour and Development*, 17: 323–33.

Wild, B., M. Erb, and M. Bartels (2001). 'Are Emotions Contagious? Evoked Emotions while Viewing Emotionally Expressive Faces: Quality, Quantity, Time Course and Gender Differences', *Psychiatry Research*, 102: 109–24.

Wind, J. (1990). 'The Evolutionary History of the Human Speech Organs'. In: J. Wind, E. G. Pulleybank, E. de Grolier et al. (eds.), *Studies In Language Origins*, i, pp. 173–97. Amsterdam: Benjamins.

6

Vocal Traditions of the World: Towards an Evolutionary Account of Voice Production in Music

Trân Quang Hai and Nicholas Bannan

INTRODUCTION

While research in the humanities and social sciences has largely embraced the concept of human universals (Brown 1991), it has proved more difficult to establish how a universal, biologically determined human capacity for music-making emerges in the highly distinct and varied practices of different cultures and populations (Blacking 1977; Nettl 1978; Arom 2000; Clayton, Sager, and Will 2004). This poses a considerable challenge, identified by the British singer and voice scientist John Potter:

> Looking backwards to the latter part of the twentieth century and forwards to the twenty-first, it is possible to see a major and continuing change in our perception of what singing is, to the extent that it is no longer possible to come up with any single meaningful definition. (Potter 2000, p.1)

Given that the workings and employment of the human voice are the subject of several disciplines in both the arts and sciences in which research continues to shape the boundaries of the topic, Potter's seemingly pessimistic judgement may well appear realistic. Advances and syntheses in psychology, neurology, linguistics, musicology, semiotics, archaeology, anthropology, zoology, human anatomy, physiology, acoustics, and education have all contributed to the agenda, and the book to which this chapter contributes has set out to reflect and draw together these often distinct and separate strands. In seeking a viewpoint that best accommodates the potential for a consensual debate—one in which there is at least a common understanding of the issues that must be addressed, whether or not this leads to agreed conclusions—a description of voice use that adopts an evolutionary perspective offers a tool for relating the voice of the biological individual to the cultural factors which contextualize its employment that is consistent with ethological methods in related disciplines.

Such a viewpoint is by no means novel. It has its precursors in a long history of attempts to account for the origin of music and language (Levman 2000; see also Thomas 1995). Levman traces this tradition in Western thought back to Plato's *Cratylus*, in which the ideas are examined of communication arising from onomatopoeia and involving a physical relationship between the mode of utterance and objects this indicates. Vico, in eighteenth-century Italy, saw language as arising from the musical nature of vowel sounds differentiated according to the influence of emotion. A generation later, Condillac argued similarly that 'it (language) partook of the nature of music' (Levman 2000, p. 188). Rousseau (1781) continued this line of speculation, defining the kinds of social and collective acts that vocal communication enables while still giving precedence to song over speech. Darwin (1871, 1872) appears to have arrived at similar conclusions through his systematic study of animal communication, though we should recall, too, the extensive role that music played in his own household: his wife was an excellent pianist, his children, of whose infant vocality he made a specific study, were brought up to perform as instrumentalists, and music was employed on several occasions in Darwin's research into animal behaviour (Bannan, in preparation).

Following Darwin, writers in a variety of disciplines reflected similar views. Spencer explored a psychological dimension to the role of rhythm in vocal utterance, differentiating between musical organization and 'emotional speech' (Spencer 1890, p. 460). Jespersen (1922) was perhaps the first linguist to espouse a song-like origin for speech. Slowly but with increasing breadth and depth in the work of the Berlin comparative musicologists Sachs and Hornbostel (see below) at the beginning of the twentieth century and their successors such as Blacking and Lomax, an evolutionary theory of human vocal communication began to develop critical mass, and the need for a comprehensive overview of the varieties and purposes of singing became apparent.

Nevertheless, an obstacle to the widespread understanding of the potential of the human voice, necessary to evolutionary speculation if the field is not to accept the a priori assumption that the voice developed for language, is the lack of clear definitions of the nature of song. In proposing here a taxonomy for singing arising from an evolutionary position, the intention is at least in part to ensure that Eurocentric assumptions about the purpose, repertoire, and style of voice use do not prejudice work achieved in, say, anthropology or archaeology in the absence of a balanced presentation of the vocal achievements of diverse cultures across the globe, including vocal behaviours and repertoires that may be unfamiliar. For this reason, the description of vocal production attempted below takes as its definition of music Cross's (2003) carefully universalist formulation: 'Music embodies, entrains, and transposably intentionalizes time in sound and action', setting out its categories according to how sounds are produced and combined.

A HISTORICAL BACKGROUND TO VOICE THEORY, EVOLUTION, AND CULTURE

The achievements of Charles Darwin (1859, 1871, 1872) provide a significant stimulus to an evolutionary account of human vocalization. Darwin provided the framework by which we see the speech and song of our species as somehow descended from the vocal behaviours of a common ancestor with the chimpanzee. The process by which this must have occurred through the accumulation of small changes over long periods of time is axiomatic, and Darwin not only established the overarching theory that demands this approach, but also provided some glimpses of his own into the possibility that language was founded on existing musical communication in our ancestors (1871). Darwin arrived at this position despite the lack of evidence in his own time to support it, and the weight of Eurocentric and anti-evolutionary prejudices that inhibited effective research. Only seven years after the publication in 1859 of *On the Origin of Species*, which provided a mechanism for speculation on the origin of language, the

Paris Société de Linguistique enacted its infamous ban on speculation of this kind.

In setting out to account for the variety of vocal production reported in different cultures, Darwin was severely misled by his informant, Sir Duncan Gibb, who in 1869 presented a paper to the Anthropological Society of London entitled 'On the Character of the Voice in the Nations of Asia and Africa Contrasted with that in the Nations of Europe'. In the introductory remarks written by an observer, it was recorded that:

> he explained, with the aid of several large coloured diagrams, the different organs of the throat which contribute to the production of speech, pointing out particularly the differences in the position of the ventricles in the throat of the Negro from their position in Europeans.

Gibb, himself, then went on in his own subsequent 'abstract' to assert that:

> The voice of the Chinese and Japanese was of low power, feeble compass, and whining in its tone, possessing at times a sort of metallic twang. Among the natives of Tartary, Thibet (sic), and Mongolia, the voice was stronger, louder, more powerful, yet still partaking of the metallic twang; the female voice was not inferior in power to that of the male sex; the metallic and deafening tones of the voice in those peoples were a well marked and distinctive peculiarity. In India and Birmah (sic), the voice was generally soft and very feminine, not so powerful as shrill; the natives of the hills had a more robust voice than those of the plains—the former possessing a somewhat metallic twang, and the latter, a plaintive and whining tone. In Africa, the Negro was taken as the type, whose larynx was of intermediate proportions between the Chinese and the Tartars, but differed from all other races of mankind in certain peculiarities. . . . The Negro wanted vocal power in whatever part of the world he was placed, but possessed the elements of a bellowing or roaring voice—a deafening noisy sound, without harmony or distinctness. (Gibb 1869, p. lxii)

Unsurprisingly, subsequent to this catalogue of vocal aberration revealed by supposedly scientific investigation, Gibb adds by way of pointed contrast:

> Considered generally, the various nations of Europe possessed strong, powerful, sonorous, and clear voices; variations as to character and tone might and did exist, but, as a rule, they all agreed in power, full compass, range, clearness, and loudness of sound. (ibid.)

Darwin's adoption of these observations in *The Descent of Man* led him to conclude:

> the voice and the form of the larynx differs in the different races of mankind: but with the Tartars, Chinese, &c., the voice of the male is said not to differ so much from that of the female, as in most other races. (Darwin 1871, p. 862)

Echoes of this misrepresentation of the relationship between anatomy and vocal utterance remained current for almost a century after Darwin's death, such as in this entry by the American musicologist Willi Apel in the widely consulted *Harvard Dictionary of Music*:

> Singing is no doubt the most widespread kind of music-making, being the only one (besides whistling) that does not depend on an instrument. The human vocal apparatus can be used in very different ways, depending in part on anatomical differences among various peoples (European, Asian, African) and even more important, on differences in training and taste. (Apel 1973, p. 776)

In the mid twentieth century, the tide began to turn against such uninformed opinion in the systematic and objective observations of ethnomusicologists such as Curt Sachs:

> many tribal groups have several voice mannerisms according to sex, geographic situation, and the singing occasion. No attempt must be made to ascribe the individual styles to 'races', whatever this word may mean. They are not biologically conditioned and hence not inherited but purely environmental, that is, cultural. (Sachs 1962, p. 88)

Objective scientific investigations, unbiased by prejudices we would now view as racist, have more recently come to illustrate that, just as any child will learn the language of the environment in which it is brought up, the vocal styles of all cultures can be acquired, and anatomical features are not a factor in the differences between them (Shehan Campbell 1991; Adkins 2000). Indeed, the apparatus of the voice can properly be viewed in engineering terms as an available sound source capable of description like any instrument. Sundberg's account (2001) presents a dispassionate view of the voice as a wind instrument dependent on a power source, oscillator, filters, and resonator. The physiological, cultural, and educational assumptions of our own time thus allow us to view the voice in both speech and song as employing a universal capacity that has evolved as a defining feature of the human species which is inherited at birth as an inseparable feature of our anatomy. The repertoire of linguistic priming and cultural practice to which each of us is subsequently exposed in turn defines the nature of our lifelong vocal development.

METHODS OF VOICE RESEARCH

The richness of musical expression encountered in the widespread distribution of human cultures through history and across the globe has suggested myriad possibilities of classification: for instance, by continents, by countries, by ethnic groups, or by function (e.g. lullabies, children's songs, festival songs, work songs, love songs, dance songs, funeral songs). These methods are, in particular, the ones typically to be found in music dictionaries and encyclopaedias.

Among the methods adopted to investigate the role of song in human culture has been literary analysis, in which a key work is Bowra's (1962) *Primitive Song*. This arose from the attempt to seek a comparison between the song traditions of surviving hunter-gatherer communities in Australia and southern Africa and the earliest poetic writings of classical Greece, in which Bowra sought evidence of an oral tradition that might represent the foundations of European literature. Inevitably, classification and analysis are applied almost exclusively to verbal texts, themselves translated into English. This provides little trace of the sounds of the originals, nor of the vocal production or acoustic environment within which they were performed. A fictional work influenced by this tradition, however, Bruce Chatwin's *The Songlines* (1987), provides considerably greater insight into the musical properties of Australian aboriginal singing.

A different approach to the nature of singing that explicitly sets out to deal with the bodily origins and sound of the voice is that in the field of semiotics provided by Barthes (1985). Barthes vividly argues the need to account for the physiological processes responsible for 'the grain of the voice', but his methods are themselves largely literary and metaphorical.

The Berlin school of comparative musicology initiated a programme of song analysis and categorization at the beginning of the twentieth century that related musical form to literary and ritual functions. Transcriptions of song types from various cultures permitted comparisons to be made of such features as melodic contour, overall range, and structural organization (Hornbostel 1905; Sachs 1956, 1962). Sachs (1943, p. 41) proposed a threefold taxonomy of song generation:

pathogenic - logogenic - melogenic

Here, *pathogenic* represents vocal calls closest to the instinctive expression of emotion, and is characterized by high, sustained phonation which subsides as breath runs out in a 'tumbling strain'. *Logogenic* denotes vocalization which carries words: chants, prayers, and incantations. *Melogenic* music 'represents the wide middle area between the extremes of logogenic and pathogenic music' (ibid. p. 42), and is characterized by '*structural intervals* as a second principle of organisation'.

This approach was taken up and developed by Révész (1924, 1941), who applied it to the attempt to discern the chronology by which musical culture evolved, analysing the song repertoires of surviving hunter-gatherers as if these represented 'primitive' levels of musical achievement:

> With the aid of ethnological, archaeological, and historical research, an attempt is made to establish those musical configurations and forms that we assume to be the simplest, earliest revelations of musical experiences and activities, and which in consequence may be taken as the basis of our music. (Révész 1924, p. 219)

There is much common ground between the approaches of Sachs and Révész: both took an interest in the relationship between music and language whereby vocal music is taken to be the basis of subsequent instrumental developments. Révész made the assumption that music itself developed biologically, like an organism which grows from the single cell to the elaborate final form. But these attempts to categorize by complexity and determine the perceived level of sophistication of which performers were capable proved ineffective:

> Both horizontal and tumbling strains are often found within the same tribe and indeed within the same piece . . . The ways in which the two species mingle . . . bares two different roots of singing, one derived from the violent howl, and the other from recitation. (Sachs 1962, p. 72)

Other strands in twentieth-century folk song study and comparative musicology have engaged with the problems of song analysis without exploring them in an explicitly evolutionary context. Brailoiu (1931/

1984) laid out categories for the conditions that gave rise to different kinds of song according to a typology of melodic generation. Grainger (1934) similarly devised theories of musical archetypes (later explored in Blacking 1987), and also set high standards in his transcriptions of folk song, testing to the limit the capacity of conventional music notation to represent the expressive idiosyncrasies he encountered in the singing of his informants (Grainger 1908). Metfessel (1928) experimented with mechanical transcription, or what he termed 'phonophotography', in order to capture the character and performance style of the American Negro voice. Olwage (2004) reviewed this and other attempts to discriminate between the timbral character of voice types associated with race and culture, quoting Fales's observation that with 'no domain-specific adjectives, timbre must be described in metaphor or by analogy to other senses' (Fales 2002, pp. 56–7).

An entirely independent yet complementary approach to voice analysis is provided by the 'grounded' taxonomy devised by the British composer and music technologist Trevor Wishart (1979, 1996). Wishart's (1979) initial analysis of the physiological means by which his infant son produced an enormous range of sounds, by a variety of means, represents a top-down approach. He saw the eventual acoustic convergence of the diversity of sounds he recorded onto the target categories of environmental speech as a paring down of the fluidity and breadth of infants' instinctive vocalization. In later work (1996), Wishart modelled strategies in compositions for vocal ensemble that applied to the voice operations then central to work with synthesizers and then in electronic music studio: oscillation, filtering, envelope-shaping, intermodulation, and the relationship between precisely-pitched sounds and 'white' noise. He explored the transitions that can be achieved between the distinctive properties of the conventional phonatory and articulatory mechanisms employed in speech and song—as well as some that may not normally be—viewing the voice as a biologically determined synthesizer. In maintaining a focus on the productive capacity of the instrument, achieved without specific reference to culture or existing repertoire and embracing a wide variety of sounds foreign to what he termed WAM (Western art music), Wishart's taxonomy is lent an objectivity that transcends the limitations of the semiotic-metaphorical tradition. A particular strength of this approach is to view vocal production from the child's point of view, as something that happens through chance and experiment. This offers a model of child vocality in musical terms consistent with the work of linguists such as Locke (1996), who draws attention to a similar quality of 'unintended consequence' in prelinguistic vocalization, and Halliday (1979), whose analysis of the development of prelinguistic semanticity in his own son's infant vocal behaviour parallel's Wishart's musical interpretations. Trevarthen (1994) provides a further insight into how infant spontaneous vocalizations can be perceived as meaningful.

Bearing in mind the emotional motivation associated with the production and perception of infant communication, the field of child development provides a useful reminder that how we perceive timbre is primed by our earliest experiences. The aural mechanism is the first of the senses to lay down retrievable memories (Shetler 1990; Woodward 1992; Lecanuet 1996; Bannan and Woodward 2009) from some three months before birth. Ragozzine and Deutsch (1994) illustrate that early priming provided by environmental exposure to language creates a preferential pattern for the perception of pitched sounds. Kessen et al. (1979) examined the early window for precise pitch-matching with the mother's voice that is associated with the pre-descended larynx in neonates (see also Minami and Nito 1998).

Developmental approaches have provided a fruitful field for modelling evolutionary features of musical behaviour (Bannan 2000). Nelson's (1997) cross-cultural comparison of lullabies reveals some universal features of this intrinsically functional and yet formative musical behaviour. The work of Trehub (2001) and Trevarthen (1997) illustrates the social, multi-modal, and emotional context in which vocal intentionality and perception evolve in the young individual. While much work of this kind has been carried out in a Western milieu, Romet (1992) analysed the means by which children undergo the transition in Javanese Sunda between engagement with the style of maternal song provided for them as infants and the quite different musical world of children's game songs through which they are initiated into groups of peers (see also Bannan, this volume ch. 12).

In the field of ethnomusicology, a relatively new method has proved effective in refuting the prejudicial assumptions we have encountered in the work of Gibb, Apel, and Révész. The anthropological practice of immersion in a culture, mirrored by the methods of experimental archaeology and by research in bilingualism, has focused on those performers who can be viewed as bicultural in their capacity to present authentically the styles of two (or more) independent traditions. Hood (1960) defined the *bimusical* agenda for work on these lines. The approach adopted in all of his work by Trân Quang Hai (Trân Quang Hai 1975, 1990, 1991a and b, 1995, 1997a and b, 1999, 2000, 2002, 2003; Trân Quang Hai and Guillou 1980; Tisato and Cosi 2003; Trân Quang Hai and Zemp 1991; Tokumaru 1994), who has sought to master the song styles and vocal production of several cultures both in performance and in order to develop a scientific understanding of these phenomena, represents this form of practical ethnomusicology. Further examples of bimusicality include, amongst others, Adkins (2000), who has added the difficult genre of Japanese traditional song to her existing capacity for performance in the Western *bel canto* style; Shehan Campbell (1991), who has similarly studied Vietnamese, Slavic, and Native American traditions as both performer and theorist; and Goetze (2000), who has developed with an internationally recruited choir a carefully devised programme for the traditional teaching and learning of choral repertoire from a variety of cultures.

The imitation and mastery of a non-native style permits a form of musical and physiological 'reverse-engineering' (Dawkins 1986; Dennett 1995; see also Bannan 2003), where an achieved performance competency is analysed for its anatomical and acoustic components. The tools for carrying out the latter have provided further, objective data about the quality of human vocal capacity able to relate the acoustic properties of perception (including self-perception) to the physiological processes involved in production. Sundberg's work (1977, 1999, 2000, 2001) in developing laboratory techniques for voice analysis has proved influential both in Western art music and as a basis for investigation into folk and non-Western styles (Sundberg 2003). Johnson (1984) applied Sundberg's methods to the study of the calls to animals of northern Swedish herdswomen.

Acoustic analysis was applied by Classe (1957) to Silbo, the whistle-language of the mountains of Gomera, and by Luchsinger (1942) in determining the nature of Alpine yodelling. More recently, Zemp (1978, 1988), Abitbol (2005) and Trân Quang Hai (1995, 2002, 2003) have employed it as a means of analysing and presenting research in a variety of styles and repertoires. Ternstrom, an associate of Sundberg, has focused on the difficult challenge of analysing simultaneous performance by groups of singers, and the phenomena that affect this (Ternstrom and Sundberg 1989; Ternstrom 1993, 1999). Daugherty (2000) has employed complementary methods drawing on the social sciences to determine choral singers' perceptions of their acoustic and affective interaction with one another.

A sophisticated taxonomy of vocal performance drawing on a variety of methods and principles was devised by Lomax (1968, 1982) with the name *cantometrics*. Lomax's intention was to apply a battery of quantitative analytical descriptors to discriminate the features of sample repertoires so as to devise correspondences between song types and the cultures that gave rise to them. Thirty-seven distinct questions are posed in relation to a given recording or performance, embracing the rating of such features as the social organization of the singing group, the nature of information carried by the text, the complexity of metrical and melodic structures, employment of ornamentation, specific 'noise' categories in the vocal production, and the precision with which language is articulated (Titon 1982). Taken together, these data would 'capture the constant features in a cultural style' (Lomax 1972, p. 143), permitting geographical and social distribution to be mapped since:

> [a] sizeable number of strong relations were found, indicating that song performance symbolizes and reinforces aspects of social structure such as the level of economic development and political organization, the degree of community solidarity, inter-sex complementarity, and the differential socialization of males and females. (ibid.)

The deterministic framework of cantometrics has been criticised as leading to the categorization of cultures on the basis of samples of the most distinct features of only some of their repertoires. For instance, Henry (1976) found that there was a mismatch between the song styles predicted by cantometrics on social and economic grounds for the North Indian village whose music he studied and the breadth and variety of repertoire actually existing. Similarly, Johnson (1984) reported that while the performers she recorded in rural Sweden had a mastery of the unique vocal style of *kulning*, which involves high intensity production of notes at the top of the soprano range, they also sang folk songs of their region in a normal manner. While Lomax's evolutionary stance provides useful insights into the circumstances that may have given rise to particular features of song, and that may illustrate the means by which songs styles have become distributed through population movement and cultural contact, this focus deals relatively poorly with the use of the vocal instrument itself. The typological problem of the relationship between repertoires and the specific vocal techniques employed by singers still remained to be solved.

DEVELOPING A SYNTHETIC AND PRACTICAL MEANS OF VOICE CLASSIFICATION

The classification of musical instruments proposed by Curt Sachs and Erich von Hornbostel was established at the beginning of the twentieth century in Berlin (von Hornbostel and Sachs 1914). This set out to place each instrument in a category according to how the sound was produced, rather than according to the style or culture with which it was associated. So, all stringed instruments entered the category *chordophone*; those activated by the use of wind, *aerophone*; those employing resonating skins or their equivalent as the resonating agency, *membranophone*; and those that 'play themselves' through being struck, blown, shaken or otherwise activated were termed *idiophone*. These classifications played a key role in establishing comparative musicology as a discipline, placing the emphasis on the means of production rather than the perceived conventions (melody, harmony, tuning, dissonance, rhythmic organisation, etc.) of a given musical tradition. However, a systematic yet easily usable general typology of the human voice able to complement the Sachs/Hornbostel organology had yet to be devised.

The system of voice classification here proposed is based on the specific ways in which the phonatory/articulatory apparatus is employed, with the following different items:

1. calls, cries, and clamours
2. voice and breath
3. spoken, declaimed, sung
4. compass and register
5. colours and timbres
6. disguised voices
7. ornamentation
8. voices imitating musical instruments
9. use of harmonics
10. polyphony
    a. heterophony
    b. overlapping
    c. drones and ostinati
    d. parallel, oblique, and contrary motion of two voices
    e. chords
    f. counterpoint and combined techniques.

The system allows considerable fluidity: while categories 1–9 may involve solo voices, there are in most cases also examples of co-ordinated vocalization that embrace these techniques. Only in category 10 is it assumed that more than one singer is strictly necessary in order to combine independent contributions in achieving each of the sub-classifications (a)–(f). While the character of examples of any of these categories may be strongly associated with a specific culture or environment, the exemplification given below (supported by the Discography and links provided on the accompanying CD) should not be taken as exclusive. An important point is that these vocalizations represent what performers *can* do, not, as cantometrics tends to suggest, what they only and always do: there may well be more than one style or means of production, often a great many, in any individual culture, and occasionally even within a single song.

SYSTEM OF VOICE CLASSIFICATION

[*D* numbers correspond to the Discography that follows the references for this chapter]

Calls, cries, and clamours

For this category, the voice is intensified and projected towards the audience. Such calls are directed at potential listeners (divinity, human being, or animal), and thus have the property of *intentionality*. The cries may express pain (funeral songs, D1), joy (ululations of the Arabo-Berber world, D2) or astonishment (the '*olé*' of Flamenco, D3). Cries become clamours when they are made by a group (the slogans of participants in political demonstrations, D4, and sporting events, D5).

Calls, cries, and clamours may have an intelligible text (calls to cattle, D6; funeral lamentations, D7), unintelligible text (cries of a woman bystander while a man plays an African sanza, D8), be interjections (Japanese drummers in a *Noh* play, D9), a signal (Pygmy hunting calls, D10), or imitation of animals (Balinese *Kecak* performance with the coordinated imitation of the cries of monkeys as rhythmical accompaniment, D11).

Voice and breath

The sound of the breath can be exploited for aesthetic purposes in search of a special timbre (whispered voice of the Burundi zither player, D12), a rhythmical component of a musical language (voice of the pearl divers of Bahrain, Persian Gulf, D13; throat game of the Inuit in Canada, D14). Breathing can be used as hyperventilation to induce trance (healer's song of Madagascar, D15; praise of Allah in Kenya, D16).

Spoken, declaimed, sung

The technique of '*sprechgesang*' (spoken-sung) can be found in public speeches, prayers, and incantations (modified pitches). Recitations may be limited to a single pitch (*recto-tono*) as in the Koranic recitation (D17), or with a few successive pitches (Tibetan Buddhist chanting, D18; Indian Vedan prayers, D19). The Polynesian singing game and dance song from Tuvalu (Western Polynesia, D20) employ both monotonous recitation (*recto-tono*) and 'spoken-sung' (*sprech gesang*) of undetermined pitches. The Romanian ballad exploits three possibilities: singing, monotonous recitation (*recto-tono*) and scanned speech (D21).

Compass and register

In Western classical voice taught at conservatories of music, and utilised in recruiting choral singers by range, the register can be classified in different categories. For men's voices, there are tenor, baritone, and bass. For women's voices, there are soprano, mezzo soprano, alto, and contralto (Kersalé 2006). There is another category: counter tenor, a *falsetto* male voice that maps onto the range of the female contralto or mezzo soprano (Giles 1996); as well as the now extinct voice of the castrated man (Brodnitz 1975).

For traditional performance in European folk music and non-Western traditions, we can find four categories of voice: strohbass (Fuchs, 1963) or 'fry' voice of very low pitch; chest voice; head voice; and whistling voice.

The voice of the Abelam man from New Guinea uses the register of 'strohbass' (D22). The Tibetan monks of the Gyutö monastery pray with a deep bass voice in the chest register (D23). The head voice with a high pitch can be heard from the Peking Opera singers (D24). The whistling voice is practised by women in Bolivia (D25).

Korean Buddhist monks use the hoarse chest voice with some parts of prayers in falsetto voice (D26). The alternation between chest and head voices is found in Swiss and Austrian yodel (D27); and also in Pygmy yodel from some countries in Africa (D28).

Colours and timbres

It is difficult to classify voices within this category, since the anatomical fluidity of the voice permits such fine gradations of variation (see, for instance, Kayes 2000; Thurman and Welch 1997). One might characterize perception of timbre as referring to the method of production. For instance, voices can be 'thin', 'thick', or 'coloured', depending on the presence of overtones and formants. A 'thin' voice of Banda girls from Central Africa (D29), a 'thick' voice of a Flamenco singer from Spain (D30), and a 'coloured' voice of a Sardinian singer from Sardegna, Italy (D31) can be considered as three categories of timbre.

Other timbres can be found, such as 'guttural', 'nasal', 'hoarse'. In Vietnam, the voice is classified after different parts of the body: *giong ruôt* (intestine voice, D32), *giong gan* (liver voice, D33), *giong nguc* (chest voice, D34), *giong hâu* (throat voice, D35), *giong mui* (nose voice, D36), *giong oc* (brain voice, D37).

Disguised voices

To some extent, we could say that all or any sung vocal technique is nothing but a disguise in comparison with the normal speaking voice. The word 'disguise' is used for those particular modifications of the voice for symbolic ends. This voice appears as singular; or as unidentifiable in the normal context. Such voices may relate to ritual contexts, the supernatural, or traditions of theatrical presentation.

In the Peking opera, a male singer sings in falsetto to impersonate a female. The disguised voice of the Dan from Ivory Coast (D38), as well as the Mitsogho voice from Gabon (D39) using a mask, are for supernatural purpose. The voice may also be distorted by resonator devices in front of the mouth such as the *didjeridu* of Australia, by the bamboo tubes of the Iatmul of Papua New Guinea, or by clay pots in Rajasthan (India) and Ivory Coast. The European children's comb-and-paper buzzer sounds as if the voice is hummed, with an additional reedy timbre: this playful practice was referred to in the mid-seventeenth century by Mersenne (1957) and the crafted instruments that created and amplified such sounds were known as *myrmidons*, and are still commercially available as *kazoos*.

Ornamentation

Ornamentation is a decoration of the principal melodic line. The Sioux vocal technique is characterized by a forced vibrato, concentrated in the high register (D40). In Asian singing, vibrato is controlled for expressive purposes. The Mongolian vibrato alternates with trills upon two degrees (D41). In the Indian *dhrupad* style, the tongue or the glottis is used to interrupt the emitted sounds for very short moments (D42). The Lebanese voice uses vibratos with trills and extended melismas (D43). We can hear the short and fast vibrato in Berber ululations (D44), and the long vibrato in Korean traditional songs (D45). The technique *dô hôt* (dropping pearls on the table) is particular in the *Ca Tru* singing in North Vietnam (D46).

Voices imitating musical instruments

The imitation of musical instruments is found in some areas of the world. In Mongolia, a singer can imitate the flute *limbe* (D47). The Tuareg singer from Mali imitates the sound of the flute when she rocks her baby (D48). The Indian drummer can reproduce the sound of a *mridangam* drum with rhythmic formulae by means of the conventional syllables called *jati* in the South Indian tradition (D49).

Use of harmonics

As we know, a periodic sound is made of a fundamental (or harmonic 1) and a series of upper harmonics (whole multiples of the fundamental). Aside from these properties influencing musical practice and structure, they relate to the mechanism for the differentiation of vowel sound in the production and perception of language (see Bannan, chapter 12). Some of these upper harmonics can be used for melodic purposes by a technique which modifies the shape of the mouth by changing the position of the tongue and the shape of the lips. The selected harmonics are consequently perceived as variable pitches. In this *overtone* singing (Van Tongeren 2004; Dargie 1988, 1991, 1993), the exciter is internal because it is achieved with the vocal folds. Further examples have been recorded from Mongolia (D50), Tuva (D51), South Africa (D52) and the Dani tribe from Arian Jaya (the Indonesian part of New Guinea, D53).

Polyphony

Classical musicology maintained for a long time that polyphony was, along with musical notation, a major and determining discovery of Western art music (see Schenker 1906/1954, p. 3). Nowadays, we know that polyphony in various traditional cultures is widespread, and is found in many regions of the world. Polyphony is significantly represented (outside Europe) in areas where tribal cultures survive in sub-Saharan Africa, Oceania, and Asia. Voices combine in a variety of ways to achieve polyphonic textures.

Heterophony indicates a musical procedure where several performers sing in a sort of unison in which they coordinate rhythmically, but pitch-contour may not be uniform (Romanian lamentations, D54; the Jivaro song of Ecuador, D55).

Overlapping is the technique used where two choirs alternate, the the entry of the second choir coinciding in unison with the last note sung by the first choir (Kaluli women in Papua New Guinea (D56), in Senegal (D57), in East Timor (D58), and also in the '*Pasi but but*' ritual song by Bunun tribe of Taiwan (D59)).

The *drone* is a continuous sound. It creates an interrupted sound on the same pitch, usually low in range, over which a melodic line is performed (examples are the Toraja song from Sulawesi, Indonesia (D60); the Albanian festive song (D61); the Bororo Fulber's song of Niger (D62); and one from Bulgaria that shows the use of drone as a foundation for polyphony (D63)).

The *ostinato* has a character of permanence based on the cyclic repetition of a limited set of pitches and/or durations to provide a kind of 'animated' drone over which performance develops. Examples are widespread, such as in Georgian vocal music (D64) and from the Dorze of Ethiopia (D65).

Two voices may superimpose in *parallel* motion, with one interval or another between them: the fifth, the fourth, the third, or the second in parallel can be found in different parts of the world. Examples can be found in Ethiopian music (D66), in the Pacific Islands (D67), and in Serbia (D68).

When *chords* are performed, they can be sung in either a 'horizontal' or 'vertical' manner. Chords often distribute pitches derived from the harmonic series (see 'Use of harmonics' above). Sometimes different chords are systematically linked as in the Corsican *paghiella* (D69). The religious polyphony of Sardinia is characterized by triads. It uses a technique known as *falsobordone* in which the different parts double the principal voice at the fifth, at the fourth, and at the third, and are strictly related between themselves by rules of consonance (D70). Singing in parallel harmony and with vibrant relish for acoustic effect is evident especially in southern Africa and in the islands of the Pacific and its continental Rim. It is central to the Xhosa culture of South Africa (D71), the Bantu of Ruanda (D72), and to the islanders of Rapa Iti, Polynesia (D73), Taiwan (D74) and New Ireland (D75).

In *counterpoint*, the parts are differentiated both melodically and rhythmically, whether each part is sung by an individual or a group performing in collective unison. Examples are the 'Aré'Aré of the Solomon Islands (D76), the Ami tribe of Taiwan (D77), and the Gurians of Georgia (D78).

Combined techniques represent the complex forms of polyphony, for example in Albania, where *drone* and *ostinato* alternate with dissonant and consonant chords (D79); in the Banda song of Central Africa (D80); and in the Dorze song of Ethiopia (D81).

CONCLUSIONS

This panoramic view of different aspects of vocal performance from various traditional cultures throughout the world is the first attempt at voice classification on these lines. Related to the Discography provided, and to audio and video files on the accompanying CD-ROM that permit listening to selected musical examples, this description of voice types and purposes offers a concise yet rich inventory of the processes employed in human vocal music.

As the case for considering the role in human evolution of the musical voice gathers momentum (Mithen, 2005; Gamble, Morley, Dunbar, Bannan, Merker, this volume), a considerable barrier to understanding has remained the limited perspective on singing that has characterized contributions to this debate not only of linguists and psychologists but also of musicians themselves. No universalist account of the capacity for singing can afford to ignore the wide variety of styles of vocal production reported in this chapter. Nor can anatomical speculation about the adaptations necessary for language fail to take into consideration those abilities that may have been shaped by musical behaviours: by high energy vocalization, and its effect on both participants and listeners; by summative group singing in unison, harmony and heterophony; and by those developmental features that determine that, while male and female infants can sing in unison with their mothers, the adult male and female voices are eventually pitched about an octave apart—a biologically determined feature that is employed in the music of many cultures that are widely separated geographically, and that is itself open to explanation on evolutionary lines.

Of one thing we can be certain: were we by some miracle or advanced technology to obtain a recording of our ancestors singing some 500,000 years ago, we might be as surprised by the similarity of their performance to one of the many examples of current human singing referred to above as we would be to the extent to which their music-making differed from the styles we are most familiar with ourselves. In celebrating and remaining aware of the great diversity of music that can be, and is, made with anatomically identical vocal mechanisms in the varied environments in which humans live today, we stand a greater chance of gaining an insight into the lost world of song that accompanied the lives of our forebears.

REFERENCES

Abitbol, J. (2005). *L'Odyssé de la voix*. Paris: Robert Laffont.

Adkins, M. (2000). 'Bi-Musicality in Japanese Folk Song and Italian Art Song: A Study on Vocal Timbre'. In: B. A. Roberts and A. Rose (eds.),

The Phenomenon of Singing 2, pp. 1–6. St. John's, Newfoundland: Memorial University of Newfoundland.

Apel, W. (1973). *The Harvard Dictionary of Music,* 2nd edn. London: Heinemann.

Arom, S. (2000). 'Prolegomena to a Biomusicology'. In: N. Wallin, B. Merker, and S. Brown (eds.), *The Origins of Music*, pp. 27–9. Cambridge, MA: MIT Press.

Bannan, N. (2003). '"Reverse-Engineering" the Human Voice: Examining the Adaptive Prerequisites for Song and Language'. In: R. Kopiez, A. Lehmann, and I. Wolther (eds.), *Proceedings of the Fifth Triennial Conference of the European Society for the Cognitive Sciences of Music.* [CD-ROM]. Hanover: Hochschule für Musik und Theater.

—— (2000). 'Instinctive Singing: Lifelong Development of "the Child within"', *British Journal of Music Education*, 17, 3: 295–301.

—— (in preparation). 'Darwin as Musicologist: The Role of Music in His Life and Theory'.

——and S. Woodward (2009). 'Spontaneity in the Musicality and Music Learning of Children'. In: S. Malloch and C. Trevarthen (eds.), *Communicative Musicality*, pp. 465–94. Oxford: Oxford University Press.

Barthes, R. (1985). *The Grain of the Voice*, trans. L. Coverdale. London: Jonathan Cape.

Blacking, J. (1987). *'A Commonsense View of All Music'*. Cambridge: Cambridge University Press.

—— (1977). 'Some Problems of Theory and Method in the Study of Musical Change', *Yearbook of the International Folk Music Council*, 9: 1–26.

Bowra, M. (1962). *Primitive Song*. London: Weidenfeld and Nicolson.

Brailoiu, C. (1931/1984). *Problems of Ethnomusicology*, ed. and trans. A. L. Lloyd. Cambridge: Cambridge University Press.

Brodnitz, F. (1975). 'The Age of the Castrato Voice', *Journal of Speech and Hearing Disorders*, 40, 3: 291–5.

Brown, D. E. (1991). *Human Universals*. New York: McGraw-Hill.

Chatwin, B. (1987). *The Songlines*. London: Jonathan Cape.

Classe, A. (1957). 'Phonetics of the Silbo Gomero', *Archivum Linguisticum*, ix: 43–61.

Clayton, M., R. Sager, and U. Will (2004). 'In Time with Music: The Concept of Entrainment and Its Significance for Ethnomusicology', *ESEM Counterpoint*, 1: 1–82.

Cross, I. (2003). 'Music and Biocultural Evolution'. In: M. Clayton, T. Herbert, and R. Middleton (eds.), *The Cultural Study of Music: A Critical Introduction*, pp. 19–30. London: Routledge.

Dargie, D. (1993). 'Thembu Xhosa Umngqokolo Overtone Singing: The Use of the Human Voice as a Type of "Musical Bow"'. Self-published presentation for the 1993 Berlin conference of the International Council for Traditional Music.

—— (1991). 'Umngqokolo Overtone Singing and the Song Nondel'ekhaya', *South African* (ILAM), 7, 1: 32–47.

—— (1988). *Xhosa Music: Its Techniques and Instruments, with a Collection of Songs.* Cape Town: David Philip.

Darwin, C. (1859/1968). *The Origin of Species.* London: Penguin.

—— (1872). *The Expression of the Emotions in Man and Animals.* London: John Murray.

—— (1871). *The Descent of Man and Selection in Relation to Sex.* London: John Murray.

Daugherty, J. (2000). 'Choir Spacing and Choral Sound: Physical, Pedagogical and Philosophical Dimensions'. In: B. A. Roberts and A. Rose (eds.), *The Phenomenon of Singing 2,* pp. 77–88. St John's Newfoundland: Memorial University of Newfoundland.

Dawkins, R. (1986). *The Blind Watchmaker.* London: Longman.

Dennett, D. (1995). *Darwin's Dangerous Idea.* London: Allen Lane.

Fales, F. (2002). 'The Paradox of Timbre', *Ethnomusicology,* 46, 1: 56–95.

Fuchs, V. (1963). *The Art of Singing and Voice Technique: A Handbook for Voice Teachers, for Professional and Amateur Singers.* London: Calder and Boyars.

Gibb, D. (1869). 'On the Character of the Voice in the Nations of Asia and Africa Contrasted with that in the Nations of Europe', *Journal of the Anthropological Society of London,* 7: lxii–lxvi.

Giles, P. (1996). *The History and Technique of the Counter-Tenor.* Aldershot: Ashgate.

Goetze, M. (2000). 'Challenges of Performing Diverse Cultural Music', *Music Educators Journal,* 87, 1: 23–5, 48.

Grainger, P. (1934). 'A Commonsense View of All Music'. Synopses of 12 illustrated lectures published subsequent to their radio broadcast from their Melbourne studio by the Australian Broadcasting Commission. Reprinted as Appendix A of J. Blacking (1987), *'A Commonsense View of All Music'.* Cambridge: Cambridge University Press.

Grainger, P. (1908). 'Collecting with the Phonograph', *Journal of the Folk-Song Society,* III/iii (12); repr. in T. Balough, ed. (1982), *Percy Aldridge Grainger: A Musical Genius from Australia.* Perth, WA: CIRCME.

Halliday, M. (1979). 'One Child's Protolanguage'. In: M. Bullowa (ed.), *Before Speech,* pp. 171–90. Cambridge: Cambridge University Press.

Henry, E. (1976). 'The Variety of Music in a North Indian Village: Reassessing Cantometrics', *Ethnomusicology,* 20, 1: 49–66.

Hood, M. (1960). 'The Challenge of "Bimusicality"', *Ethnomusicology,* 4, 2: 55–9.

Hornbostel, E. M. v (1905/1975). 'Die Probleme der vergleichenden Musikwissenschaft'; repr. as 'The Problems of Comparative Musicology'. In:

K. P. Wachsmann, D. Christensen, and H.-P. Reineckes (eds.), *Hornbostel Opera Omnia*, pp. 249–70. The Hague: Nijhoff.

Hornbostel, E. v and C. Sachs (1914). 'Systematik der Musikinstrumente: Ein Versuch', *Zeitschrift für Ethnologie*, 46, 4–5: 553–90.

Jespersen, O. (1922). *Language:* Its Nature, Development and Origin. London: Allen & Unwin.

Johnson, A. (1984). 'Voice Physiology and Ethnomusicology: Physiological and Acoustical Studies of the Swedish Herding Song', *Yearbook for Traditional Music*, 16: 42–66.

Kayes, G. (2000). *Singing and the Actor*. London: A & C Black.

Kersalé, P. (2006). *Them'Axe 2: La Voix*. [DVD]. Lyons: Éditions Musicales Lugdivine.

Kessen, W., J. Levine, and K. Wendrich (1979). 'The Imitation of Pitch in Infants', *Infant Behaviour and Development*, 2: 93–9.

Lecanuet, J.-P. (1996). 'Prenatal Auditory Experience'. In: I. Deliège and J. Sloboda (eds.), *Musical Beginnings: Origins and Development of Musical Competence*, pp. 3–34. Oxford: Oxford University Press.

Levman, B. (2000). 'Western Theories of Music Origin, Historical and Modern', *Musica Scientiae IV*, 2: 185–208.

Locke, J. L. (1996). 'Why Do Infants Begin to Talk? Language as Unintended Consequence', *Journal of Child Language*, 23, 2: 251–68.

Lomax, A. (1968). *Folk Song Style and Culture*. Washington, DC: American Association for the Advancement of Science.

—— (1972). 'Brief Progress Report: Cantometrics-Choreometrics Projects', *Yearbook of the International Folk Music Council*, 4: 142–5.

—— (1982). 'Singing'. In: *Encyclopaedia Britannica, Macropaedia*, xvi: 789. Chicago, IL: Encyclopaedia Britannica.

Luchsinger, R. (1942). 'Untersuchungen über die Klangfarbe der menschlichen Stimme', *Archiv für Sprach- und Stimmphysiologie und Sprach- und Stimmheilkunde*, 6: 1–39.

Mersenne, M. (1957). *Harmonie Universelle: The Books on Instruments*, trans. R. E. Chapman. The Hague: Martinus Nijhoff.

Metfessel, M. (1928). *Phonophotography in Folk Music: American Negro Songs in New Notation*. Chapel Hill, NC: University of North Carolina Press.

Minami, Y. and H. Nito (1998). 'Vocal Pitch-Matching in Infants'. In S. Woodward (ed.), *Respecting the Child in Early Childhood Learning: Proceedings of the 8th International Seminar of the Early Childhood Commission of the International Society for Music Education, Stellenbosch, South Africa*. Cape Town: International Society for Music Education.

Mithen, S. (2005). *The Singing Neanderthals*. London: Weidenfeld & Nicolson.

Nelson, J. (1997). 'Lullabies as Human Adaptation: a Cross-Cultural Analysis of Children's Bedtime Songs'. In: D. Schneck and J. Schneck (eds.), *Music in Human Adaptation*, pp. 61–78. Blacksburg, VA: Virginia Polytechnic Institute and State University Press.

Nettl, B. (1978). 'Some Aspects of the History of World Music in the Twentieth Century: Questions, Problems, and Concepts', *Ethnomusicology*, 22, 1: 123–36.

Olwage, G. (2004). 'The Class and Colour of Tone: An Essay on the Social History of Vocal Timbre', *Ethnomusicology Forum*, 13, 2: 203–26.

Potter, J. (2000). *The Cambridge Companion to Singing.* Cambridge: Cambridge University Press.

Ragozzine, F. and D. Deutsch (1994). 'A Regional Difference in Perception of the Tritone Paradox within the United States', *Music Perception*, 12, 2: 213–25.

Révész, G. (1924). *Introduction to the Psychology of Music*, trans. G. de Courcy (1953). pp. 70–5, London: Longman.

—— (1941). 'Der Ursprung der Musik', *International Archiv für Ethnographie*, 40.

Romet, C. (1992). 'Song Acquisition in Culture: A West Javanese Study in Children's Song Development'. In: H. Lees (ed.), *Music Education: Sharing Musics of the World. Proceedings of the 20th World Conference of the International Society for Music Education, Seoul, Korea*, pp. 164–73. Christchurch, NZ: ISME/University of Canterbury.

Rousseau, J.-J. (1781/1817). *Essai sur l'origine des langues. Où il est parlé de la mélodie et de l'imitation musicale.* Paris: A. Belin.

Sachs, C. (1962). *The Wellsprings of Music.* The Hague: Martinus Nijhoff.

—— (1956). *A Short History of World Music.* London: Dobson.

—— (1943). *The Rise of Music in the Ancient World East and West.* London: J. M. Dent.

Schenker, H. (1906/1954). *Harmonie.* O. Jonas (ed.) and E. M. Borgese (trans.). Chicago, IL: University of Chicago Press.

Shehan Campbell, P. (1991). *Lessons from the World.* New York: Schirmer.

Shetler, D. J. (1990). 'The Inquiry into Prenatal Musical Experience'. In: F. R. Wilson and F. L. Roehamann (eds.), *Music and Child Development*, pp. 44–62. St Louis, MI: MMB Music.

Spencer, H. (1890). 'The Origin and Function of Music', *Mind*, 60: 449–68.

Sundberg, J. (2003). 'Research on the Singing Voice in Retrospect'. *TMH-QPSR, 45.* Stockholm: Speech, Music and Hearing, KTH, 11–22.

—— (2001). 'Acoustics, vi: The Voice, 1. Introduction', ed. L. Macy. <http//www.grovemusic.com>.

—— (2000). 'Where Does the Sound Come from?' In: J. Potter (ed.), *The Cambridge Companion to Singing*, pp. 231–47. Cambridge: Cambridge University Press.

—— (1999). 'The Perception of Singing'. In: D. Deutsch (ed.), *The Psychology of Music*, pp. 171–214. London: Academic Press.

—— (1977). 'The Acoustics of the Singing Voice', *Scientific American*, 236, 3: 82–91.

Ternstrom, S. (1999). 'Preferred Self-to-Other Ratios in Choir Singing', *Journal of the Acoustical Society of America*, 105, 6: 3563–74.

—— (1993). 'Perceptual Evaluations of Voice Scatter in Unison Choir Sounds', *Journal of Voice*, 7, 2: 129–35.

—— and J. Sundberg (1989). 'Formant Frequencies of Choir Singers', *Journal of the Acoustical Society of America*, 86: 517–22.

Thomas, D. A. (1995). *Music and the Origins of Language: Theories from the French Enlightenment*. Cambridge: Cambridge University Press.

Thurman, L. and G. Welch (1997). *Bodymind and Voice: Foundations of Voice Education*. Minneapolis, MN: The VoiceCare Network.

Tisato, G. and P. Cosi (2003). 'On the Magic of Overtone Singing', *Voce, Canto, Parlato, Studi in Onore di Franco Ferrero*, pp. 83–100. Padova: Istituto di Scienze e Tecnologie della Cognizione.

Titon, J. (1982). Review of 'Cantometrics', *Journal of American Folklore*, 95, 377: 370–4.

Tokumaru, Y. (1994). Review of Hugo Zemp and Trân Quang Hai, *The Song of Harmonics*, in *Ethnomusicology*, 38, 2: 387–9.

Trân Quang Hai (1975). 'Technique de la voix chantée mongole: Xöömij', *Bulletin du CEMO*, 14/15: 32–6.

—— (1990). 'Les Musiques vocales'. In: C. Alès (ed.), *L'Esprit des Voix*, pp. 43–52. Grenoble: La Pensée Sauvage.

—— (1991a). 'Nouvelles Expérimentations sur le chant diphonique: Nouvelles Voies de la voix, 1ère partie', *Bulletin d'audiophonologie*, 7, 5/6: 607–18.

—— (1991b). 'La Voix dans les cultures extra-occidentales'. In: R. Cross (ed.), *La Voix dévoilée*, pp. 164–86. Paris: Romillat.

—— (1995). 'Le Chant diphonique: description, historique, styles, aspect acoustique et spectra', *Annuario degli Archivi di Etnomusicologia dell'Accademia Nazionale di Santa Cecilia*, 2: 123–50.

—— (1997a). 'Recherches introspectives sur le chant diphonique et leurs applications'. In: G. Dessons (ed.), *Penser la voix*, 195–210. Poitiers: La Licorne.

—— (1997b). 'Overtones in Central Asia and in South Africa (Xhosa Vocal Styles)'. *Confluences: Cross-Cultural Fusions in Music and Dance: Proceedings of the First South African Music and Dance Conference and 15th Symposium on Ethnomusicology*, pp. 422–3. Cape Town: University of Cape Town.

—— (1999). 'Overtones Used in Tibetan Buddhist Chanting and in Tuvan Shamanism'. In: R. Astrauskas (ed.), *Ritual and Music*, pp. 129–36. Vilnius: Department of Ethnomusicology, Lithuanian Academy of Music.

—— (2000). 'Some Experimental and Introspective Researches on Xöömij Overtone Singing', *Proceedings of WESTRPRAC VII*, pp. 593–8. Japan: University of Kumamoto.

—— (2002). 'A la découverte de chant diphonique'. In: G. Cornut (ed.), *Moyens d'investigation et pédagogique de la voix chantée,* pp 117–32. Lyons: Symétrie.

—— (2003). 'Polyphony in one throat'. In: J. Jordania (ed.), *Proceedings of the First International Symposium of Traditional Polyphony,* pp. 274–87. Tbilisi: International Research Center for Traditional Polyphony.

—— and D. Guillou (1980). 'Original Research and Acoustical Analysis in Connection with the Xöömij Style of Biphonic Singing'. In: Y. Minegishi and R. Emmert (eds.), *Musical Voices of Asia*, pp. 162–73. Tokyo: The Japan Foundation, Heibonsha.

—— and H. Zemp (1991). 'Recherches expérimentales sur le chant diphonique', *Cahiers de Musiques traditionnelles: No. 4 VOIX*. Ateliers d'ethnomusicologie/AIMP, Genève, 27–68.

Trehub, S. E. (2001). 'Musical Predispositions in Infancy'. In: R. J. Zatorre and I. Peretz (eds.), *The Biological Foundations of Music*, pp. 11–16. New York: Annals of the New York Academy of Sciences, 930.

Trevarthen, C. (1997). 'Fetal and Neonatal Psychology: Intrinsic Motives and Learning Behaviour'. In: F. Cockburn (ed.), *Advances in Perinatal Medicine: The Proceedings of the XV European Congress of Perinatal Medicine, Glasgow, 1996,* pp. 181–90. London: Parthenon.

——(1994). 'Infant Semiosis'. In: W. Nöth (ed.), *Origins of Semiosis*, pp. 219–52. Berlin: Mouton de Gruyter.

Van Tongeren, M. (2004). *Overtone Singing: Physics and Metaphysics in East and West*, 2nd rev. edn. Amsterdam: Fusica.

Wishart, T. (1996). *On Sonic Art.* Amsterdam: Harwood Academic Press.

—— (1979). *The Book of Lost Voices.* York: Department of Music, University of York.

Woodward, S. (1992). 'The Transmission of Music into the Human Uterus and the Response to Music of the Human Fetus and Neonate'. PhD dissertation, University of Cape Town.

Zemp, H. (1988). 'Filming Music and Looking at Music Films', *Ethnomusicology*, 32, 3: 393–427.

—— (1978). ''Are'Are Classification of Musical Types and Instruments', *Ethnomusicology*, 22, 1: 37–67.

Discography

Calls, cries, and clamours

1. *Musique Mnong Gar du Viet Nam*, recordings (1958) by Georges Condominas, Paris, 1974 (funeral song). (Ocora OCR 80) [LP].
2. *Fantaisies marocaines*, recordings by Maurice Benitah, Paris (ululations of the Arabo-Berber world). (Le Chant du Monde LDX-A 4297) [LP].

3. *Noche Flamenca*, recordings in studio, Paris (Olé in Flamenco). (BAM LD 6008) [LP].
4. *La Voix*, collection Thèm'Axe, video by Patrick Kersalé, Lyons, 2006 (political demonstration, France). (Lugdivine 7347) [DVD].
5. *La Voix*, collection Thèm'Axe, video by Patrick Kersalé, Lyons, 2006 (sporting event, France). (Lugdivine 7347) [DVD].
6. *Voices of the World*, recordings (1986) by Michel De Lannoy, Paris, 1996 (call to cattle, France). (Le Chant du Monde CMX 374 1010) [CD].
7. *Voices of the World*, recordings (1991) by Jacques Bouët, Bernard Lortat-Jacob and Speranta Radulescu, Paris, 1996 (funeral lamentations, Romania). (Le Chant du Monde CMX 374 1010) [CD].
8. *Central Africa: Sanza/Music of the Land of Gbaya*, recordings (1977) by Vincent Dehoux, Geneva (cries of a woman, Central African Republic). (AIMP XXVII, VDE-755) [CD].
9. *Japon—Musique du Nô—Shakkyo-Pont en pierre*, recordings in studio (1983), Paris (vocal interjections, Japan). (Ocora C559 005) [CD].
10. *Centrafrique: Anthologie de la musique des Pygmées Aka*, recordings (1971) by Simha Arom, Paris, 1987 (hunting calls, Central African Republic). (Ocora C 559 012/13) [CD].
11. *Kecak and Sanghyang of Bali*, Tokyo (rhythmical choral chant based on monkey chattering). (King Records KICC 5128) [CD].

Voice and breath

12. *Burundi—Musiques traditionnelles*, recordings (1967) by Michel Vuylsteke, Paris, 1988 (whispered voice, Burundi). (Ocora C 559 003) [CD].
13. *Pêcheurs de perles et musiciens du Golfe Persique*, recordings (1962) by Poul Rovsing-Olsen, Paris, 1966 (voice of the pearl divers, Bahrain). (Ocora OCR 42) [LP].
14. *Canada—Inuit Games and Songs*, recordings (1974–76) by Nicole Beaudry, Paris, 1991 (throat games, Canada). (Auvidis—UNESCO D 8032) [CD].
15. *Musique malgache*, recordings (1963) by Charles Duvelle, Paris, 1965 (healer's song, Madagascar). (Ocora OCR 24) [LP].
16. *The Sound of Africa Series*, recordings by Hugh Tracey, Grahamstown, South Africa (praise of Allah, South Africa). (International Library of African Music TR 171) [LP].

Spoken, declaimed, sung

17. *Morocco 1—The Music of Islam and Sufism in Morocco*, Kassel, Germany (cantillation of the Koran, Morroco). (Barenreiter Musicaphon BM 30 SL 2027) [LP].
18. *Ladakh—Monastic and Village Music*, recordings (1976) by Mireille Helffer, Paris, 1989 (Buddhist psalmody, Tibet). (Le Chant du Monde LDX 274 662) [CD].

19. *South India—Ritual Music and Theatre of Kerala*, recordings (1983) by Pribislav Pitoeff, Paris, 1990 (Vedic chanting, India). (Le Chant du Monde LDX 274910) [CD].
20. *Kanak Song: Feasts and Lullabies*, recordings (1984) by Jean Michel Beaudet, Paris, 1990 (recto-tono, spoken-sung, New Caledonia). (Le Chant du Monde LDX 274909) [CD].
21. *Ballads and Festivals in Romania*, recordings (1981) by Jacques Bouët and Bernard Lortat-Jacob, Paris, 1985 (monotonous recitation, Romania). (Le Chant du Monde LDX 74846.47) [LP].

Compass and register

22. *Musique of the Abelam: Papua Niugini*, recordings (1979) by Brigitta Hauser-Schäublin, Kassel, Germany (strohbass, Papua New Guinea). (Barenreiter Musicaphon BM 30 SL 2704) [LP].
23. *The Gyutö Monks: Tibetan Tantric Choir*, recordings at a concert in the USA, Stanford, CA, 1987 (deep bass voice, Tibet). (Windham Hill Records WH 2001) [LP].
24. *Chinese Opera/Famous Songs Sung by Zhang Junqiu*, recordings (1962) by the Central People's Broadcasting Station of China (high pitched voice/head voice, China). (Shenzen Lu Soft Publishing Company SC-91-004) [CD].
25. *Bolivia: Calendar Music in the Central Valley*, recordings (1990) by Rosalia Martinez, Paris, 1996 (high pitched voice, Bolivia). (Le Chant du Monde LDX 274 938) [CD].
26. *Musique bouddhique de Corée*, recordings (1964) by John Levy, Paris, 1969 (hoarse chest voice, Korea). (Vogue LVLX 253) [LP].
27. *Switzerland/Yodel of Appenzell*, recordings (1979) by Hugo Zemp, Paris, 1990 (alternation between head/chest voices, Switzerland). (Auvidis D 8026) [CD].
28. *Opeka Nyimbo—Musiker—Komponisten us dem Südlichen Malawi*, recordings (1967) by Gerhard Kubik, Berlin (yodel, Malawi). (Museum Collection Berlin MC 15) [LP].

Colours and timbres

29. *Musiques Banda*, recordings (1964–67) by Simha Arom and Geneviève Dournon-Taurelle, Paris, 1971 ('thin voice', Central African Republic). (Vogue LD 765) [LP].
30. *Pepe de la Matrona*, recordings (1957), Paris, 1993 ('thick voice', Spain). (Le Chant du Monde LDX 274 829) [CD].
31. *Polyphonies of Sardinia*, recordings (1992) by Bernard Lortat-Jacob, Paris, 1993 (coloured voice, Sardinia). (Le Chant du Monde LDX 274 760) [CD].

32–37. *Vietnam II*, recordings (1976) by Trân Van Khê, coll. UNESCO/ Musical Atlas, Venice, 1985 ('intestine', 'lever', 'chest', 'throat', 'nose', and 'brain' voices, Vietnam). (EMI Odeon) [LP].
38. *Ivory Coast—Dan Masks*, recordings (1965) by Hugo Zemp, Paris, 1990 (masked voice, Ivory Coast). (Ocora C 580 048) [CD].
39. *Gabon—Musique des Mitsogho et des Batéké*, recordings (1968) by Pierre Sallée, Paris, 1974 (masked voice, Gabon). (Ocora OCR 84) [LP].

Ornamentation

40. *Takini—Musique et chants des Lakota Sioux d'Amérique du Nord*, recordings (1994) by Milt and Jayme, Paris (forced vibrato, USA). (Le Chant du Monde CM 1274–1000) [CD].
41. *Musical Voices of Asia*, recordings (1978) at a festival, The Japan Foundation, Tokyo, 1979 (vibrato/trills, Mongolia). (Victor SJL 209–211) [LP].
42. *India III—Dhrupad*, recordings (1964) by Alain Danielou, Kassel, Germany, 1966 (ornamentation, India). (Barenreiter Musicaphon BM 30 L 2018) [LP].
43. *Folk Music of the Mediterranean*, recordings (1940), New York (ornamentation, Lebanon). (Folkways Ethnic Library FE 4051) [LP].
44. *Fantasias marocaines*, recordings by Maurice Benitah, Paris, 1969 (ululations, Morocco). (Le Chant du Monde LDX A 4297) [LP].
45. *Pansori/Simchong-ga/The Ideal of Korean Epic Vocal Art*, studio recordings, CD Ethnic Sound Series 24, Tokyo, 1988 (ornamentation, Korea). (JVC VID-2519) [CD].
46. *Vietnam—Ca Tru/Tradition of the North*, studio recordings (1995), Paris, 1996 (ornamentation, Vietnam). (Inedit W 260 070) [CD].

Voice imitating musical instruments

47. *Mongol and Buriat Songs*, recordings (1968) by Roberte Hamayon, Paris, 1973 (imitation of the flute, Mongolia). (Vogue LDM 30138) [LP].
48. *Voix du Monde*, recordings (1988) by Bernard Lortat-Jacob, Paris, 1996 (imitation of the flute, Mali). (Le Chant du Monde CMX 374 1011) [CD].
49. *Rythmes de l'Inde—Mahapurush Misra*, studio recordings (1968), Paris, 1969 (imitation of the drum, India). (Philips 8440534 PY) [LP].

Use of harmonics

50. *Mongolie/Musique et Chants de tradition orale*, recordings (1985) by Xavier Bellenger, Paris, 1986 (overtone singing, Mongolia). (GREM G 7511) [CD].

51. *Tuva/Voices from the Land of the Eagles*, studio recordings (1991), Leiden, the Netherlands, 1992 (overtone singing, Tuva). (Pan Records PAN 2005 CD) [CD].
52. *Afrique du Sud—le chant des femmes Xhosa*, studio recordings (1995), Geneva (overtone singing, South Africa). (AIMP & VDE-GALLO VDE-879) [CD].
53. *Les Dani de Nouvelle Guinée*, ii, recordings (1984–1997) by Anne Marie Petrequin, Pierre Petrequin, and Olivier Weller, Paris, 2001 (overtone singing, Arian Jaya, Indonesia). (Nord Sud Music CONS 1102/2) [CD].

Heterophony

54. *Voix du Monde*, recordings (1991) by Jacques Bouët, Bernard Lortat-Jacob, Speranta Radulescu, Paris, 1996 (funeral lamentations, Romania). (Le Chant du Monde CMX 374 1010) [CD].
55. *Jivaro*, recordings by Philippe Luzuy, Paris, 1960 (heterophony, Ecuador). (Vogue Contrepoint MC.20161) [LP].

Overlapping

56. *Music of the Kaluli*, recordings (1977) by Steven Feld, Papua New Guinea, 1981 (overlapping, Papua New Guinea). (Institute for Papua New Guinea Studies IPNGS 001 C) [LP].
57. *Senegal: Peul and Tenda Music*, recordings (1983) by Vincent Dehoux, Paris, 1992 (overlapping, Senegal). (Ocora C 560 043) [CD].
58. *Timor/Songs of the Ema*, recordings (1966) by Brigitte Clamagirand, Paris, 1979 (overlapping, East Timor). (Le Chant du Monde LDX 74693) [LP].
59. *The Songs of the Bunun Tribe*, recordings by Wu Rung Shun, Taipei, 1992 (overlapping, Taiwan). (Wind Records TCD-1501) [CD].

Drone

60. *Indonesia: Toraja/Funerals and Fertility Feasts*, recordings (1993) by Dana Rappoport, Paris, 1995 (drone, Sulawesi). (Le Chant du Monde CNR 274 104) [CD].
61. *Albania: Vocal and Instrumental Polyphony*, recordings (1983) by Bernard Lortat-Jacob, Paris, 1989 (drone, Albania). (Le Chant du Monde LDX 274 897) [CD].
62. *Nomades du Niger/Musique des Touareg, musique des Bororo*, recordings (1963) by Tolia Nikiprowetzky, Paris, 1966 (drone, Niger). (Ocora OCR 29) [LP].
63. *In the Shadow of the Mountains/Bulgarian Folk Music*, recordings (1967) by David B. Jones, New York (drone, Bulgaria). (Nonesuch Records H-72038) [LP].

Ostinato

64. *Georgia 1*, recordings by the Conservatory of Tbilisi for the UNESCO collection: *A Musical Anthology of the Orient*, Kassel, Germany, 1971 (ostinato, Georgia). (Barenreiter Musicphon BM 30 L 2025) [LP].
65. *Ethiopia/Polyphony of the Dorze*, recordings (1974–1975) by Bernard Lortat-Jacob, Paris, 1994 (ostinato, Ethiopia). (Le Chant du Monde CNR 274 646) [CD].

Parallel Voices

66. *Ethiopie: Polyphonies et Techniques Vocales*, recordings (1967) by Jean Jenkins, Paris, 1969 (parallel singing, Ethiopia). (Ocora OCR 44) [LP].
67. *Iles Salomon/Musique de Guadalcanal*, recordings (1970) by Hugo Zemp, Paris, 1994 (parallel singing, Solomon Islands). (Ocora C 580 049) [CD].
68. *Folk Music of the Mediterranean*, recordings selected by Henry Cowell, New York, 1953 (parallel singing, Serbia). (Ethnic Folkways Library FE 4501C) [LP].

Chords

69. *Musique corse de tradition orale*, recordings by Felix Quilici, Paris, 1983 (chord singing, Corsica). (Bibliothèque Nationale APN 82.1/2/3) [LP].
70. *Sardinia: Polyphony for Holy Week*, recordings (1995) by Bernard Lortat-Jacob, Paris, 1996 (chord singing, Sardinia). (Le Chant du Monde LDX 274 936) [CD].
71. *African Sunday*, recording by David Dargie, 1981–4 (harmony singing, Western Cape, South Africa). (Lumko Music Department, tape no. 100, Delmenville, South Africa).
72. *Bantu*, recordings (1982) selected by Charles Duvelle and Mwesa Mapoma, Paris, 1984 (chords, Rwanda). (Ciciba 8401/8402) [LP].
73. *Spirit of Polynesia*, recording by David Fanshawe, 1993 (harmony singing, Rapa Iti, French Polynesia). (Saydisc CD-SDL- 403) [CD].
74. *Voice of the Forgotten World*, recordings selected by Ted Levin, New York, 1993 (chord singing, Taiwan). (Ellipsis Arts CD 3252) [CD].
75. *New Ireland/Papua New Guinea*, recordings (1974) by Charles Duvelle, Paris, 2002 (chord singing, New Ireland). (Philips 458447/2, Prophet 21) [CD].

Counterpoint

76. *Solomon Islands/'Are'Are Intimate and Ritual Music*, recordings (1977) by Hugo Zemp, Paris, 1992 (counterpoint, Solomon Islands). (Le Chant du Monde CNR 274 963) [CD].

77. *Polyphonic Music of the Amis Tribe*, recordings (1992) by Wu Rung Shun, Taipei, 1993 (counterpoint, Taiwan). (Wind Records TCD 1502) [CD].
78. *Georgia 1*, recordings by the Conservatory of Tbilisi for the UNESCO collection: *A Musical Anthology of the Orient*, Kassel, Germany, 1971 (counterpoint, Georgia). (Barenreiter Musicphon BM 30 L 2025) [LP].

Combined techniques

79. *Albania/Vocal and Instrumental Polyphony*, recordings (1983) by Bernard Lortat-Jacob, Paris, 1989 (combined techniques, Albania). (Le Chant du Monde LDX 274 897) [CD].
80. *Centrafrique/Anthologie de la musique des Pygmées Aka*, recordings (1971) by Simha Arom, Paris, 1987 (combined techniques, Central African Republic). (Ocora C 559 012/13) [CD].
81. *Ethiopia—Polyphony of the Dorze*, recordings (1975) by Bernard Lortat-Jacob, Paris, 1994 (combined techniques, Ethiopia). (Le Chant du Monde CNR 274 646) [CD].

7

Found Objects in the Musical Practices of Hunter-Gatherers: Implications for the Evolution of Instrumental Music

Pedro Espi-Sanchis and Nicholas Bannan

INTRODUCTION: THE READING WORKSHOP-WITHIN-A-WORKSHOP

A unique contribution to the workshop held at the University of Reading in 2004 was provided by the practical session on group instrumental performance led by the South African musician and storyteller Pedro Espi-Sanchis. The purpose of this interactive session was to give participants an insider's impression of the physical, social, and structural properties of a variety of Southern African musical styles of which Espi-Sanchis is an exponent. All the contributors to this book made music together at this session, a description of which featured prominently in Michael Balter's subsequent article on the workshop in the November 2004 issue of Science. *Discussion following the session took on a different, more subjective and less theoretical slant, both as a consequence of the informative content of the presentation and because the process of making music together gave rise to an entirely different exchange of knowledge compared to the largely verbal and visual media that conventionally dominate conference seminars.*

Originally of Spanish extraction, Pedro Espi-Sanchis lived in France for eleven years and has spent more than thirty years studying and playing in South Africa, where he has a national reputation due to the popularity of his music programmes for children on television and his many recordings of African stories with music, as well as performances with other singers and instrumentalists. As an experienced educator and community musician, he has devised a research-based means of developing understanding of the culture and music-making of Southern African peoples (Espi-Sanchis 1997), presenting the musical practices of, amongst others, Namibia, Zimbabwe, Mozambique, and South Africa itself. A focus of the stories he has collected (Espi-Sanchis 1993a, 1993b) has been his informants' accounts of the origins of instrumental performance on a variety of 'found objects'.

Through stories and a participatory and creative approach to music, he communicates how to make simple and efficient instruments, and how to use them to create music by applying fundamental musical principles rather than by duplicating the music other people have composed. Most of his current work centres around African music. This term is applied very widely, much wider even than to the actual African continent where it originated, but he advocates that, at the fundamental level, one finds an amazing cohesiveness in all music emanating from 'Black Africa' (that is, African music with little or no Arabic influence). The combination of these essentially African forms of musical engagement with the employment of uncrafted instruments that arise from material available in this geographical location from times prior to the evolution of our species provides a glimpse of the kind of musical interaction that could well have played a role in the lives of our hunter-gatherer ancestors at an early stage in the evolution of human culture.

In the chapter that follows, practical observations taken from Espi-Sanchis's session at the workshop are presented in italic type and alternate with more theoretical speculations, presented in roman type, developed in collaboration with Nicholas Bannan.

HOMO SAPIENS, AFRICA AND CULTURAL EVOLUTION

If Africa was the cradle of the human species (Leakey 1981; Dawkins 1986, 1995; Johanson and Edgar 1996) then it would be logical to assume that human music-making also has African origins (Bannan 1998; Walker 2005; Merker 2000; Richman 2000, 1980; Levman 1992; Mithen 2005; Foley, this volume; Gamble, this volume; Morley, this volume). However, we have no explicit material evidence for such a claim: the earliest instruments so far unearthed and analysed are from relatively recent sites in Europe (see below). While singing would seem to have preceded instrumental performance (Darwin 1871; Bannan, this volume; Trân Quang and Bannan, this volume; Merker, this volume; Morley, this volume), since it has precedents in related species like the gibbon with which we share relatively recent common ancestry (Dawkins 2005, p. 102) and which do not employ external tools to make music, neither musical nor spoken vocalization has left any evidence apart from the fossils of hard tissue proximate to the vocal organs of our ancestors (Clegg, this volume). Indeed, singing might well be the most important human activity of which we have so little tangible evidence, since no tools or artifacts which exist in the fossil record were essential to the practice:

> not even the earliest civilizations that have left their traces in the depths of the earth are old enough to reveal the secret origins of music. (Sachs 1956, p.1)
>
> A major weakness—the major weakness—is the almost total inaccessibility of what no doubt was the majority of ancient music and/or other structured sound (signalling, and so on): tones and sounds produced by the human voice. There the only kind of 'organology' is in anatomical information from our colleagues in physical anthropology. But it gives us little help. Ancient song and speech remain beyond our horizon. (Moberg 1986, p. 238)
>
> The historical study of the voice is difficult and frustrating. As opposed to the study of instruments, there are no models to examine, and little information can be gleaned from visual depictions of singers. (Jander and Harris 2001)

But Jander and Harris's confidence that the picture is clearer for the origins of instrumental music is misplaced. Great scepticism has greeted the discovery of objects claimed as early instruments, and of proposed depictions of musical instruments in early art (Stockmann 1986). Even where the evidence that an object, such as the restored Geißenklösterle flute (Conard et al. 2004), is widely agreed to support its being a crafted instrument shaped specifically for musical purpose, and its replica is clearly capable of exquisite sounds, we can have no real idea as to how it may have been played: the performance practice and repertoire for which it was employed were rendered silent long before any means of symbolizing them for posterity became possible.

Nevertheless, new finds continue to be unearthed (Conard et al. 2009), and in assessing their significance, scientists have become increasingly interested in the place of music in human evolution. Despite the difficulties involved in harnessing physical evidence to support theoretical speculation, two sub-disciplines have come into existence during the last twenty years or so that have defined the nature of the fields they represent, and are assembling the cross-disciplinary tools they require to make progress. *Archaeomusicology* (Hickmann 2006) has grown out of the study of music in early texts and art (see Farmer 1930; Kilmer 1971; Gerson-Kiwi 1980; Mitchell 1992; Franklin 2002; Pont 2004; Scarre and Lawson 2006) and the examination of evidence of prehistoric music-making (Stockmann 1986; Schneider 1986), to embrace practical reconstructions of the acoustic properties of prehistoric sites (Watson and Keating 1999) and the musical features of prehistoric behaviours (Cross and Watson 2006, Gamble, this volume). The complementary field of *biomusicology* (Wallin 1991) has embraced human physiology and neurology in approaching musicality as an evolved phenomenon that represents a set of genetic adaptations responsible for the unique range of behaviours associated with human music-making. Work in both sub-disciplines has analysed the environmental and acoustic features which may have stimulated the development of human musical responses, as well as identifying evidence that prehistoric art may have been associated with ritual involving music and dance (Garfinkel 2003, 2001, 1998; Gamble, this volume; Mithen 2005).

Considerable attention has been paid to the presence in the fossil and artefact record of claimed musical instruments that have survived (Stockmann 1986; Kunej and Turk 2000; d'Errico et al. 2003; Conard et al. 2004), whether or not they represent objects that were intentionally constructed for musical performance (Morley 2006). This presents several catches: firstly, the fact that an object can be employed by a skilled musician today to make music does not prove that it was necessarily ever played. Percussionist-entertainers such as the group Stomp make attention-grabbing music out of dustbins and matchboxes, while one of Stradivarius's masterpieces, the *Messiah* violin, lies in the Ashmolean Museum in Oxford under the strict condition that it never be played again. Kunej's apparent capacity, complete with printed fingering-chart (Kunej and Turk, 2000), to coax a wide variety of pitches out of the Diva Babje flute is a tribute to his ability, whether or not this object is the accidental result of animal mastication. After all, the characteristic sound of Aboriginal Australia, the didjeridu, owes its existence to the capacity of humans to perform on an instrument created by termites.

So in many ways, the impact of Espi-Sanchis's practical research into the performing traditions of musicians who play on found objects in sub-Saharan Africa runs somewhat counter to the quest of the archaeologists such as Conard, d'Errico, and Turk who have sought evidence of early human symbolic thought in musical instruments and related this to theories of the development of 'modern' cultures associated with language. In turn, this assumes the 'making of marks', associated with art and artefacts as signals to posterity. Virtually none of the instrumental techniques Espi-Sanchis has studied makes use of instruments crafted from material that could have fossilized; though all of them relate to the natural environment of an Africa relatively unchanged since prehistoric times. Only objects claimed as early lithophones, amongst sound sources examined, could have survived over the ages. Instead, what we find is a variety of principally organic matter employed in the musical practices of surviving hunter-gatherer societies, such as bamboo, grass, seaweed, moth chrysalises, and gourds, that form the material from which instruments can be created with relatively little technological ingenuity.

Before looking at examples of these practices in detail, we propose that, in a simplification of the Sachs-Hornbostel organology (Hornbostel and Sachs 1914), one could commence a taxonomy of *objet trouvé* instruments by looking at hunter-gatherers' employment in everyday life of the categories: animal, vegetable, mineral. Animals surround hunter-gatherers, define the areas in which they can sustain a nomadic existence, and are prominent in their art and ritual. It is likely that animal mimicry in vocal and mimetic imitation, dance, and the spiritual states related to trance would result in an overt animal origin for musical practices. As to the employment of animals themselves in music, Dargie (1988) provides the example in Xhosa children's musical play of a live beetle impaled on a stick that is placed against the jaw to act as an oscillator whose vibrations are amplified and harmonically varied by movements of the throat and lips. This implies that, from childhood, the Xhosa employ a 'theory of music' that embraces an understanding of the properties of the harmonic series (see below; and also Bannan, this volume). When performing, Espi-Sanchis wears bracelets made from the chrysalises of moths around his ankles, a practice he encountered in many African traditions including the Swazi people, who use the Mopani moth cocoons in this way to make the effect of movement and stamping more part of the music. The body-parts of deceased animals, specifically those unfit for consumption, are employed to make instruments: examples include Kudu horns for Mhalamhala signalling horns, hide for the sounding skins of drums, and sinew for the bows of hunters employed also to make music.

So much for the great debt that musicians have throughout the world to their fellow creatures as both material for instruments and inspiration for sounds and narratives. In the category *vegetable*, wood can be employed as mallet, vibrator, and resonator: chimpanzees have, after all, discovered this (Goodall 1986). The simplest *objet trouvé* instrument, though, is the hollow gourd (hollowed by termites or just by the natural drying-out process), which can without further attention serve as a one-note ocarina-style flute. The ocarinas Espi-Sanchis employs were reported by Blacking in many of his writings (e.g. 1973), and are made out of the Maramba fruit also known as Wild Orange or Monkey Apple. They have three holes in them. One is, of course, for blowing: the other two holes provide the capacity to produce three notes; and an additional lower note is obtained when all the holes are closed by closing the blowing hole a little with your lip. Two at least of these instruments are conventionally played together, producing interlocking patterns (see CD-ROM video clip). A group of ocarinas of different sizes permits an ensemble of varying pitches to be assembled for polyphonic performance[1]. Even simpler are the ways that grass can act as a reed, or a thorn as a rudimentary form of Jew's harp (Frisbie 1971; Dargie 1988). The employment in music-making of vegetable matter from a different environment is represented by the leaf stem of the pawpaw tree, the bark of the wild cabbage or the dry seaweed which provides one of the material sources for the Lekgodilo flute.

The category *mineral* includes lithophones that employ, for instance, lengths of natural metal (they look rather like stones) that can be found lying on the surface in the Namibian desert (Mans 2004) or worked rocks that have been reported at prehistoric sites in India and North America (Boivin 2004; see also Cross 2000). Where stone

[1] A version of such polyphonic organization was performed at the Reading workshop reported in *Science* (Balter 2004), though with little plastic pipes standing in for bamboo ones or gourds.

has been worked, in making implements or carving rock art, there will have been rhythms involved in the process, and sensitivity to acoustic phenomena whereby the ear guides the hand, amongst the skills required. Indeed, one can speculate that music is in many ways an extension into ritual and rehearsed practice of the physical actions required to survive in a given environment. Acoustic response to that environment emerges in the mimicry, both in bodily/vocal performance and the use of extrasomatic tools represented by musical instruments, of its actual and potential sound-producing phenomena, whether they be live animals or inanimate matter. In conclusion, it would seem likely that the kinds of musical *objets trouvés* described extended existing musical functions (vocalization, clapping, stamping) modelled on imaginative exploitation of the resonances and capacities for sound production of the human body itself.

EARLY INSTRUMENTAL MUSIC: ANALOGIES WITH THE STUDY OF ROCK ART

Speculation on the origin and development of instrumental music in southern Africa encounters both parallels and overlaps with analysis of the history and meaning of creative work in the visual domain: rock art, especially that of the San people (Lewis-Williams 1990; Dowson 1988). Lewis-Williams's investigations have combined close study of the sites themselves, whose ages range from 26,000 BP to *c.*300 years ago, with comparative accounts of San beliefs and symbology that draw on existing documentation amassed by researchers in the late nineteenth and early twentieth centuries, as well as those of surviving members of these communities in the more recent past. He refers to these complementary methods as, respectively, the 'aesthetic approach' and the 'narrative approach', arguing that only a combination of the two begins to reveal what the rock art figures mean, and provides an insight into their role in cultural practice.

A key feature of the interpretation of much of the rock art is that it represents pictorial depiction that recaptures trance experience associated with animist ritual (Dowson 1988). An example of this (Dowson 1988, p. 123) is the connection between engravings of eland, of which there are widespread examples, and the danced imitation of eland mating behaviour that features in girls' puberty rituals. In the rituals associated with rain-making, an imaginary sacrifice is enacted by the shaman while in the spirit world of trance (Lewis-Williams 1990, p. 46). 'Rain-people' are depicted in art as having fish tails, like mermaids, and their presence protects against the dangerous effects of rain such as flooding.

It seems highly likely that the connection between rock art depictions of these kinds and San life is that they represent an element in ritual enactment. They are not, according to such an interpretation, self-sufficient as symbolic narrative, but contributory to cultural practice in which other features, such as music and dance, would also have played a part. Evidence for this comes partly from the dance-like postures of human subjects in the rock art, as well as the interpretation that those depicted may be in a trance-state, conveyed by bleeding from the nose, or with arms extended backwards, or wearing animal headdress (Lewis-Williams 1990, pp. 51–65). There are also depictions of humans playing musical instruments (p. 30) such as the musical bow (see below) as well as stronger bows clearly intended for hunting (p. 24).

A similar connection between depiction and enactment is evident in rock art studied by Boivin (2004) in Kupgal, southern India. Alongside Neolithic petroglyphs are worked boulders referred to by Boivin's guide as 'musical stones'. Striking the grooved impressions in these rocks produces a variety of 'bell-like' tones. Boivin suggests that art and lithophone music may have combined as elements in ritual enactment.

SOME OBSERVATIONS ON RHYTHM AND MOVEMENT

In looking analogously for the origins of music in the African practices of which we have more recent evidence, the first element would seem to be a sophisticated, arguably instinctive, awareness of the potential of rhythm. A working definition of rhythm in the activities that comprise the daily lives of traditional African cultures would be: the division and perception of time in patterns of stress and release. With this in mind, one might speculate that a great motivator of cultural evolution would be the search for pleasure (at all levels), both at times of relaxation and in combatting the fatigue of physical exertion. As an accompaniment to arduous tasks, music has a widespread role in fetching-and-carrying, threshing and milling, long-distance travel ('song as the earliest form of public transport') and so on. One can easily imagine early hominins walking long distances in their hunting and gathering lifestyles, falling into a common step buoyed up by the percussive symphony produced by their coordinated footsteps and the related rattling of whatever they carried—also bringing in clapping and vocal sounds contrasting, through their placement at differing points in the beat cycle, with the binary rhythmic framework of their gait.

How can one speculate about such things with any confidence? One can discern in the widespread distribution of such practices in populations that have been written about since European colonization began (Okumu 2000) parallels independent of other features of culture (such as language or social organization) that point to an ancient common origin. One observation here relates to the music of the San in particular and sub-Saharan African music more generally. The preponderance of additive time,[2] *to be found in San music, in which cycles of 5, 7, 9, and higher odd numbers of pulses are repeated, appears inconsistent with the idea of music derived from natural cyclic patterns, such as walking or breathing, which are binary in form. However, it is a central principle of African music not to repeat or stress something that is already expressed in the structure. Instead, performers have to provide a contrasting part which enriches the music by being different, and that uses some of the pulses left 'open' (unused) by the other parts happening in the music. Following that principle, if a song is in 5 time, instead of beating (walking) to the song in that time, say:*

```
* . . * . | * . . * . | * . . * . | * . ...
```

which would slavishly mirror the musical organization of the piece, the preferred beat expressed is binary, that is the walking beat.

```
R . L . R . L . R . L. R . L. R . L . ...
```

[2] Additive rhythm refers to the groupings of beats that can be derived from a relatively fast pulse that is accentuated in asymmetric patterns, as in Stravinsky's *Rite of Spring*.

This is quite simple to do once you realize that the combination of a 5-pulse musical clap cycle with a 2-cycle foot beat resolves into a clapping pattern where you do two 'on beat' claps, together with the foot beat and the next two claps on the 'offbeat', that is, between the foot beats.

To get a better feeling and understanding of this, one needs to try it out: walking normally and then clapping the asterisks as in the example above. Since some Europeans find this difficult, we can work up to it through a few simple exercises:

First walk on the spot Right, Left, ...

```
R . L . R . L . R . L . R . L . R . L . ...
1 2 1 2 1 2 1 2 1 2 1 2 1 2 1 2 1 2 1 2 ...
```

Next clap together with your foot beat ...

```
R . L . R . L . R . L . R . L . R . L . ...
*   *   *   *   *   *   *   *   *   *   ...
1 2 1 2 1 2 1 2 1 2 1 2 1 2 1 2 1 2 1 2 ...
```

Next clap on the offbeats, that is, between each foot beat.

```
R . L . R . L . R . L . R . L . R . L . ...
  *   *   *   *   *   *   *   *   *   *   ...
1 2 1 2 1 2 1 2 1 2 1 2 1 2 1 2 1 2 1 2 ...
```

This may already present some difficulty for most people brought up outside cultures in which this is perfectly normal. Remember that when you are training in rhythm, what is important is the accuracy, not the speed. So, if an exercise proves difficult, you just have to slow it down until it is manageable, and then speed it up a bit once you have the hang of it.

Now we can clap the five pattern by clapping twice on the beat (with the foot) and then twice on the offbeat.

```
R . L . R . L . R . L . R . L . R . L . ...
*   *     *   *     *   *     *   *     ...
1 2 1 2 3 1 2 1 2 3 1 2 1 2 3 1 2 1 2 3 ...
```

You can now do the seven pattern by just adding one 'on-beat' and one 'offbeat' clap which is equivalent to adding two pulses to the cycle to go from five to seven pulses.

```
R . L . R . L . R . L . R . L . R . L . ...
*   *   *     *   *   *     *   *   *   ...
1 2 1 2 1 2 3 1 2 1 2 1 2 3 1 2 1 2 1 2 ...
```

The five pattern is made of two groups of pulses: one group of two pulses; and one of three pulses, which is the group which puts the clap off the beat and then back on the beat alternatively. Most of the so-called 'time line' patterns used as rhythmic references in African percussion ensembles (mostly played on bells) consist of such on- and offbeats with an uneven (1.2.3. or 1.2.1.) group serving to put you off and then back on the beat. For example, here is the Kon-ko-lo pattern:

```
R . L . R . L . R . L . R . L . R . L . ...
*   * *   *   *   * *   *   * *   *   * ...
```

Applying an underlying beat to music is an unconscious act/process we go through in order to make sense of the musical/rhythmical stimuli. This is so very important that a rhythm pattern without a beat reference is meaningless: you cannot, in isolation, appreciate how 'good' it is, or whether it makes sense. One might consider this also to be a function of the social fabric through which music arises: one entrains rhythmically as a means of relating to others. We go through a very similar process when communicating, and one might consider that the underlying patterns of language rely on this rhythmic sense that was developed earlier for other functional reasons of doing work more easily, or providing motivation for the lengthy journeys on foot involved in hunting and migration; in both cases, rhythmic entrainment can provide a sense of belonging to the various individuals in the social group.

The focusing of rhythm along different organizational principles can be seen as a compositional technique in much African music. A piece of music will sound one way when felt in threes and completely different when felt in fours.

A practical example of this is played on the little eight-key Kalimba [CD]. The music is constructed in cycles of 12 so that it can be felt in 2s, 3s or 4s at will. Just keeping to a three-pulse organization (by 'counting', or by moving in such a way as to feel this), one can hear already several melodies. Then, if you apply a four-beat interpretation to the music, everything seems to change just as with the alteration in perceived colours and shapes of a kaleidoscope. Indeed, one could use the term 'kaleidophonic' music. This is a phenomenon that has been very well described by many ethnomusicologists, including Blacking (1973) and Kubik (1994).

Ian Cross puts this cyclic relationship between intention, production, and perception very well (see Cross, this volume ch. 10): 'Physical movement while producing the music mediates the meaning and appreciation of the music'. Movement is used both as a 'filter', and as a means to increase the melodic/rhythmic patterns produced by the music. This 'movement' can range from simply stamping one's foot in a regular manner to intricate dance patterns using many different cycles.

INSTRUMENTS AS TOOLS THAT REPRESENT A 'THEORY OF MUSIC'

Early commentary by European colonists failed entirely to detect sonic organization or beauty in African music, and the term 'primitive music' was coined in parallel with that of 'primitive art' (Price 1989) to embrace the otherness and perceived lack of sophistication of what was encountered. The following is from the correspondence of a Dutch official in the Guinea region in 1705:

> Their musical instruments are various and very numerous, but all of them yield a horrid and barbarous shocking sound: The chief of them are the mentioned Horns, made . . . of small Elephant's teeth . . . Sometimes they blow these Horns so well, that though it is not agreeable, yet it is not so horrid as to require a whole Bale of Cotton annually to stop ones Ears . . . Their second sort of instruments are their drums . . . all which produce a dismal and horrid noise. . . . They always set a little boy to strike upon a hollow piece of iron with a piece of wood; which alone makes a noise more detestable than the drums and Horns together. (Okumu 2000, p. 309)

African influences spread swiftly to the Americas with the slave trade, and music was no exception. Colonists' responses to indigenous and non-European music were equally condescending; the following is a report of syncretic music provided by black musicians for a ceremony in Portugal in 1730:

> one has never seen so peaceful a feast with so many blasts: there was a sequence of minuets, and many discant singers because for their part, the Blacks of the Rosary, making a well concerted dissonance with guitar, *gral* and violin, and playing the horn by the *cumbe* . . . for at the same time they played among them *pandeiros*, stones, *arranhol* [maybe a scratching-board, or *reco-reco*], board-shaped guitar, cocoa violin, whistle, *berimbau*, and rattles, and the mixture made such a disconcert that, if it was that way in the village, one wonders how it would be in Hell! In short, the feast was grave but God protect us from it next year. (Budasz 2004)

Ethnomusicologists such as Hugh Tracey (1961), John Blacking (1967) and Simha Arom (1991) have made considerable contributions to revealing the significance of music in African culture, as well as the complex tonal and rhythmic organization of the music itself. Working within this tradition, Dargie's (1988) comprehensive analysis of the employment of pitch systems in Xhosa vocalization and instrumental practice illustrates how sensitivity to the properties of the harmonic series underpins a wide variety of behaviours (see also Bannan, this volume ch. 12). As with the investigations of Arom and his students into tuning systems and rhythmic interaction (see below), the problem Dargie faced was that, while it was clear in performance that conventions were being followed which conferred authenticity, and that these had clearly been transmitted to the current generation of players from their predecessors, these processes were seen by his informants as mysterious and beyond description. Dargie was therefore unable to discern that the Xhosa performers he studied possessed an explicit theory of music that could explain these phenomena, though it is clear that implicit familiarity with the categorical boundaries of harmonic series relationships informed their musical productivity, as shown in their vocal, linguistic, and instrumental practice:

A similar problem of musical analysis confronted Arom and his circle. Their recent research has employed computer synthesis to test the perceptual tolerances of informants able to interact with digital versions of their music in order to reveal how tuning and rhythm may be organized (Arom, Fernando, and Marandola 2003). An approach

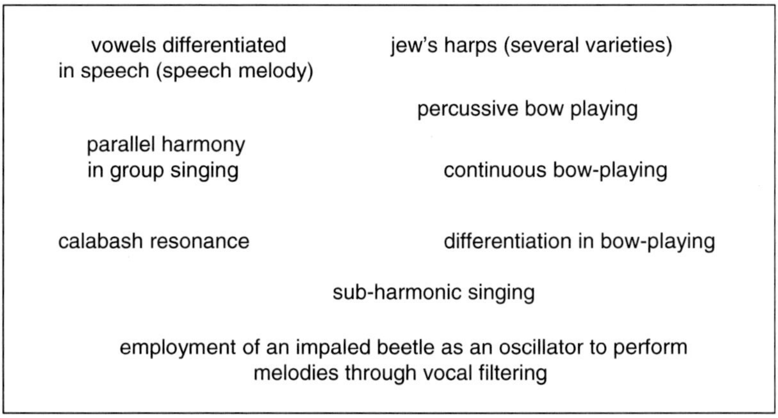

on these lines has illuminated the stylistic features of complex polyphonies, firstly in the instrumental music of the Oudémé people, and then in the close-meshed vocal style of the Bedzan pygmies. An emerging agreement on what was authentic and where the thresholds of temperament or duration were crossed emerged from this laboratory approach, making explicit through Western scientific investigation those features able to be transmitted within Bedzan and Oudémé culture through immersion and implicit knowledge.

Implicit 'theories of harmony', and of rhythmic organization, on these lines may well have arisen early in human evolution. Harnessed to a 'theory of mind', a theory of harmony acts as a bridge between musical vocalization and the properties of categorical discrimination that unambiguous language demands. Acute aural engagement with the environment, in listening to the sounds of birds and animals, mimicking them, and conveying emotion and intention through tone of voice in both speech and song (Lewis 2009) involves sensitivity to both the frequency contour of fundamental tones and the relationship between these and overtones (Bannan, this volume). Given the universality and ubiquity of music in sub-Saharan Africa, and its performers' informal sensitivity to these features, it would seem that an implicit theory of harmony may represent a more robust model for transmitting musicality of this kind to the next generation than the complex abstractions of Western theory that tend to be seen as the preserve of the specialist. But every African, it would seem, understands that there is music in language, and music in the varieties of sounds that can be produced from a wide variety of sources. They remain able to hear what many have become deaf to in 'advanced' societies.

SOME OBSERVATIONS ABOUT PITCH AND TIMBRE

In terms of how organized melody developed, one imagines that there must have been a watershed between emotionally driven fundamental frequency contours, such as those represented in the perceived pitch of animal calls, and a more categorical employment of specific intervals once the properties of the harmonic series had been 'discovered'. It would seem likely that, in adapting the body to play musical instruments, whether crafted or objets trouvés, *an existing template of meaningful pitch relationships may have guided performance: that is, musical experimentation derived from the acoustic properties of instruments may have mapped onto recognized structures that were familiar from acoustic experience in the sonic environment—in language, song, and acute aural response to natural phenomena. In this sense, one can characterize successful experimentation with an unfamiliar playing technique as 'permitting the instrument to play you', in a manner that replicates a child's early and chaotic progress in developing mastery of his or her own biological voice in the first year or so of life (Locke 1996; Halliday 1979; Trevarthen 1999; Bannan 2000).*

We can only speculate on whether melody assembled itself into harmony, or harmony 'deconstructed' into melody, though the latter would seem the more straightforward account[3]*. This is what makes it difficult with certainty to adjudicate whether ancient artefacts were musical instruments or not: it is quite possible to get elaborate sounds out of the most unpromising material, including things that were never intended to be musical instruments. Whatever the outcome of debate on the early European flutes with apparent mouthpieces and fingerholes, these are not the defining features of a fully functioning instrument. Much more likely, and one might argue inescapably in terms of musical evolution, is that the ancestor of all more recent flutes with holes that suggest intentional crafting is the 'harmonic' flute played in many (most?) early human societies: and these have no holes at all.*

In the coastal areas of Namibia and the Eastern Cape, such flutes are derived from nothing more than a length of dry seaweed, a material that has been part of the natural environment since long before a member of our species made it sing. An example is given on the CD of a melody such an instrument can play (CD 2, Lekgodilo tune).[4]

The seaweed in question is the Euclonia maxima, *which can grow up to five metres long and is hollow, which makes it float upwards from its roots in the sea bed. The plants are uprooted by the waves and end up*

[3] One suspects that it is neurology that will provide the answer, or at least some illumination of this issue.

[4] Each participant at the Reading workshop was given a 'seaweed' flute to practice on and take home (in fact international quarantine law determined that these were plastic hosepipe replicas).

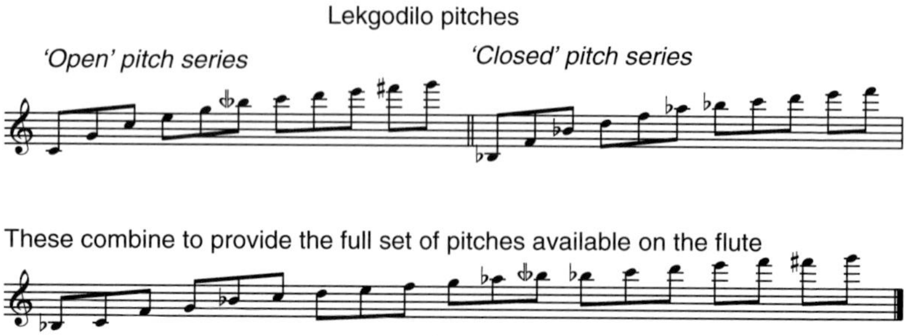

Figure 7.1. The pitch system of the Lekgodilo flute obtained from two overlapping harmonic series a whole tone apart.

on the beach where they dry into all kinds of shapes. As the plant dries, it contracts dramatically and remains hollow and smooth.

The Lekgodilo is probably the first flute to be 'found' by humans, and its equivalent, fashioned from a variety of available materials, is still traditionally played all over the world, with different mouthpieces. Lekgodilo is the name it is given in Lesotho where the angle-cut mouthpiece is used (possibly the mouthpiece with the best tone quality and control).

This instrument, which is basically just a section of tube, has no lateral finger holes, no discernible mouthpiece, and uses the harmonic series of the open and closed pipe to produce a Lydian mode, turning chromatic after the 6th degree (CD 3).

The playing technique of the Lekgodilo is related to brass instruments such as the trumpet in that it relies wholly on breath control to determine which notes of the harmonic series are sounded. Closing the hole at the opposite end of the instrument to the mouth elicits a second harmonic series based on a fundamental a whole tone lower than the open pipe harmonic series (Figure 7.1). *This harmonic relationship between two series a whole tone apart is replicated in Xhosa music in the Western Cape (Dargie 1988), where it defines the Lydian mode quality of music played on strung bows, as well as solo and choral vocal styles.*

Despite its apparent simplicity, the Lekgodilo, and its widespread cousins manufactured from whatever material is available in a variety of ecologies, can be employed to play traditional music from all over the world; and this most ancient instrument has been put to use in contemporary 'urban' musical forms such as Kwela, Marabi, and Mbaqanga, as well as recent and cosmopolitan styles of music such as R&B, bossa nova, blues and jazz. Espi-Sanchis has developed a chromatic form of the instrument which can play any music and can modulate to different keys. One needs to master the diatonic version of the instrument before going on to the chromatic version.

Various materials are used traditionally all over the world to make an instrument of this type: the stem of the pawpaw leaf, river reed, bone, tree bark, and dry seaweed, to name a few. In South Africa, every culture has a name for the instrument—Mtshingo (Zulu), Icilongo (Xhosa), Mokorenyane (Tswana)—yet the instrument itself has more or less disappeared. It is, as we have seen, closely related to southern African musical bows, and is certainly the ancestor of the penny-whistle tradition (the Irish tin whistle which is given the 'goat' sound the Lekgodilo possesses naturally by blowing over the mouthpiece as well as inside it); and several of the famous South African penny-whistlers played Lekgodilos before going on to the penny whistle and brass instruments. (Various examples of Lekgodilo music are on the CD).

SOUND AND SOCIETY: VARIATIONS ON BLACKING

In his seminal book *How Musical is Man?* (1973), John Blacking aptly played on words in coining the phrases 'humanly organized sound' as a definition of music, and 'soundly organised humanity' as a description of music's function. The developmental role of music that he illuminated in his earlier investigation into Venda children's songs represents a reminder that music plays its role in forming the members of a society (and, sadly, its absence or withdrawal also has an equally formative, inhibiting role). In considering the means by which musical instruments evolved in harness with human musicality, we should not ignore the role of toys and playthings in this process. Indeed, it may even be that the musical function of certain tools emerged prior to their use for other purposes. Which came first, the bow as a weapon for hunting, or stretched sinew as a means of producing sound? With hindsight, we tend to assume that tools developed for their most useful purpose. But in terms of cost, benefit, and risk, a bow developed first as a toy, and a technology fully understood on these lines, would confer superior skills for the construction of a bow fit to defend one when one's life depended upon it. Small versions of the bow, *umrhabe,* are indeed played by Xhosa women (Dargie 1988). Similarly, a boat-builder might well design a safer vessel once he had mastered the skills of combining curved planks into a geometrically appropriate design by acquiring these skills making a lute or mandolin. In proposing criteria for identifying utilized bone in Upper Palaeolithic deposits, d'Errico notes the possibility that marks left on bone by 'tensors' may have resulted from 'the striking or strumming of a taut string to produce a sound' (d'Errico 1993, p. 299). The technologies involved in the design of weapons, tools, toys, and musical instruments may have remained in close parallel for centuries (see Gamble, this volume). But the important issue is always to recall that such skills had to be acquired by a painstaking process of teaching and learning. Children needed templates on which to work in order to practise their abilities in design and manufacture. Adults in turn retained a fascination for the mysterious sound-producing properties of musical artefacts able to speak to other worlds.

MUSIC AND SOCIAL ORGANIZATION

A link between certain kinds of animal communication and human music-making is that both can be seen as dedicated, non-verbal means for group identification and reinforcement. Knowing your musical culture has always been a very subtle and deep way to 'belong', which could arguably predate the more verbally dependent personal tagging system that arises with language, vocabulary, and accent.

There is a philosophical principle in Africa which is best expressed in the following Nguni proverb: 'Umuntu ngumuntu ngabantu'. This means that each of us is a person thanks to and through contact with other persons (other languages have similar words, such as the related ubuntu in Zulu). In other words, belonging, learning through, and contributing to the social group is the only way to become human and to survive. The African pipe ensemble is, overtly, the musical illustration of that principle. Instead of one flute producing a full range of sounds like the Lekgodilo, each person in the group is limited to just one note.

Various ensembles are played in southern Africa; the Tshikona from Venda is one such ensemble where a group of people get together, each with a different note, to play and dance. This dance could be considered the candidate 'National Dance', together with the other related dances of Dinaka (BaPedi) and Dithlaka (BaTswana).

In order to perform in this way, each participant has one pipe producing only one note—so without the other members of the group no music can happen. All have to respond to one another as individuals, while working together as a group, in order to create music. Working together does not mean everyone plays the same rhythm pattern on the various notes, as in the unison and homophonic practice of European social music-making.

Instead of playing all together each participant must play their own pattern (see CD). *This, of course, does not mean going into the kinds of cadenzas associated with improvisation in certain other cultures. To do so would make the sound much too busy and full, and, more importantly, would be perceived as anti-social behaviour, where it would be inappropriate to dominate the timeline, leaving little or no space for the contributions of others. In this music, each person has to make an individual contribution different to and fitting in between all the others.*

A significant characteristic of this musical ensemble is that it provides a way to integrate any level of expertise. Individual contributions can range from parts that concentrate mostly on offbeats and are relatively difficult to execute to parts that would consist of one blown on one easily located beat in the cycle (see CD). *It does not matter what the temporal position of anyone's contribution is, as long as everyone is able to make an individual contribution which is different to all the others. That one on-beat blow, on that unique note, put in the right place in the cycle, can make a very big difference to the music.*

Also interesting is the manner in which, from a conglomeration of basic rhythmic parts, a wealth of perceivable melodies springs up (see CD). *Again, this represents a brand of kaleidophonic music (see above) in which the player of any given note gains a perspective of the whole dependent on the pitch and rhythmic position of their contribution, though this can change as one perceives different levels within the overall structure—especially through altering one's spatial relationship with other participants through movement and dancing.*

'One Person, One Note' ensembles of this kind can be found in the cultures of the BaTswana (Dithlaka), the BaPedi (Dinaka) and the

BaVenda (Tshikona) (see CD). *Note that this music is mostly performed with all players standing in a circle blowing their individual patterns while performing a variety of dances. The two last-mentioned traditions have a full percussion ensemble in the middle of the circle consisting of several drums producing steady interlocking patterns, with one of the drums often punctuating the various dancing steps.*

The principles of this form of social and musical organization can be summarized as follows:

- *All parts should be built on the same foot-beat cycle. This happens quite naturally, with visual cues complementing the aural effect.*
- *Each person produces a rhythmic pattern on their one-note instrument.*
- *Each part is unique and not duplicated by anybody else.*
- *Parts must be quite simple and leave spaces for other participants to fit in.*
- *All parts need to be the same loudness.*
- *Various dances need to be performed as a means of shifting the musical focus both for the players and the listeners.*

If you want to try this music you can use the following process:

- *Standing in a circle, start building the music one by one if possible to make sure the interlocking is good.*
- *Once the music is going, the players start walking in a circle. At one point the circle can double back so that the players go past each other and can then compare their pattern to each of the other ones in turn to make sure there is no duplication.*
- *Once back in the circle shape, try various dance patterns which have been rehearsed beforehand. These are done without changing the pipe-blowing pattern. (These dances are made part of the audio event through the use of ankle rattles the players wear, and serve to shift the musical focus, and produce in the ears of the players a variety of melodies.)*
- *Further performance ideas can involve finding in the group the other notes that make up the melody you are hearing and coming together as a smaller subgroup, like an eddy in a larger current of sound and movement.*
- *Also interesting is to conglomerate into a very dense group just walking on the spot and playing with closed eyes to increase melodic awareness.*

- *Another thing to do is to take turns to stop playing and stand in the middle of the circle to hear more freely the other melodies that may be happening.*[5]

CONCLUSION

At the end of the performance of this one-note ensemble formed from the participants in the Reading workshop, animated discussion ensued. One player had begun to hyperventilate, and this led to reflection on the physical nature of such performing over far longer periods than at the Reading session, and its capacity to induce trance states, or to produce some of the physiological changes that have been measured in group musical performance (Moreno 1995; Dunbar, this volume). Feedback also related to perception of the musical content. What was the melody each player was hearing? What was the length of the rhythmic cycle that people perceived as operating (this could be judged by asking people to beat the footstep pattern they recalled)?

In music of this kind, a perspective of the role of music in society and the entitlement to perform actively was evident and is quite different to what most people experience through Western-style education and enculturation. In this practice from southern Africa, everybody hears a different song, yet everybody is correct: there is no right or wrong song to hear, as everyone has a different 'hearpoint' of the music through their individual part. As an important corrective to Western impressions of who is musical and who is not, and what effect such judgements have on self-image and self-esteem, experience of this kind provides an insight into how instrumental music may have developed, and to what purpose. Following in the wake of musical principles derived from acute awareness of the properties of sound based on constant listening to the environment and to one's own natural vocal expression and that of others (Lewis 2009), performance on instruments may have played a role in cultural evolution that did not depend on the exchange of language. A claim could thus be made that, as extensions of and replacements for the human voice,

[5] All of these procedures were enacted during the Reading workshop of 2004 in which contributors to this book were participants.

musical instruments, adapted simply from materials in the surrounding environment, could well have a very long history indeed.

REFERENCES

Arom, S. (1991). *African Polyphony and Polyrhythm.* Cambridge: Cambridge University Press.

—— N. Fernando and F. Marandola (2003). 'A Cognitive Approach to Bedzan Pygmies Vocal Polyphony and Oudémé Instrumental Polyphony: Methodology and Results'. In: R. Kopiez, A. Lehmann, I. Wolther et al. (eds.), *Exp[ear]ience—Music in Science: Science in Music.* Proceedings of the Fifth Triennial Conference of the European Society for the Cognitive Sciences of Music, pp. 457–66. [CD-ROM]. Hanover: Institute for Research in Music Education Monograph No. 6.

Balter, M. (2004). 'Evolution of Behaviour: Seeking the Key to Music', *Science*, 306, 5699: 1120–2.

Bannan, N. (2000). 'Instinctive Singing: Lifelong Development of "the Child Within"', *British Journal of Music Education*, 17, 3: 295–301.

—— (1998). 'Out of Africa: The Evolution of the Human Capacity for Music'. In: C. v. Niekerk (ed.), *Proceedings of the 23rd International Society for Music Education World Conference*, pp. 26–33. Pretoria: University of South Africa Press.

Blacking, J. (1973). *How Musical Is Man?* Seattle, WA: University of Washington Press.

—— (1967) *Venda Children's Songs: A Study in Ethnomusicological Analysis.* Johannesburg: Witwatersrand University Press.

Boivin, N. (2004). 'Rock Art and Rock Music: Petroglyphs of the South Indian Neolithic', *Antiquity*, 78, 299: 38–54.

Budasz, R. (2004). *A Musicá no tempo de Gregório de Mattos: Música Ibérica e Afro-Brasileira na Bahia dos Séculos XVII e XVIII*, p. 12. Paraná, Brazil: DeArtEs, Universidade Federal do Paraná.

Conard, N., M. Malina, and S. Münzel (2009). 'New Flutes Document the Earliest Musical Tradition in Southwestern Germany', *Nature*, 460: 737–40.

———— and F. Seeberger (2004). 'Eine Mammutelfenbeinflöte aus dem Aurignacien des Geissenklösterle: neue Belege für eine musikalische Tradition im Frühen Jungpaläolithikumauf der Schwäbischen Alb', *Archäologisches Korrespondenzblatt*, 34, 4: 447–62.

Cross, I. (2000). 'Lithoacoustics—Music in Stone, Preliminary Report'. <http://www.mus.cam.ac.uk/~ic108/lithoacoustics/>.

—— and A. Watson (2006). 'Acoustics and the Human Experience of Socially Organised Sound'. In: C. Scarre and G. Lawson (eds.), *Acoustics, Space and Intentionality: Identifying Intentionality in the Ancient Use of Acoustic Spaces and Structures*, pp. 107–16. Cambridge: McDonald Institute for Archaeological Research.

Dargie, D. (1988). *Xhosa Music: Its Techniques and Instruments, with a Collection of Songs*. Cape Town: David Philip.

Darwin, C. (1871). *The Descent of Man, and Selection in Relation to Sex*. London: John Murray.

Dawkins, R. (2005). *The Ancestor's Tale: A Pilgrimage to the Dawn of Life*. London: Phoenix.

—— (1995). *River out of Eden*. London: Weidenfeld & Nicolson.

—— (1986). *The Blind Watchmaker*. London: Longman.

Dowson, T. (1988). 'Revelations of Religious Reality: the Individual in San Rock Art', *World Archaelogy*, 20, 1: 116–28.

d'Errico, F. (1993). 'Criteria for Identifying Utilised Bone: The Case of the Cantabrian "Tensors"', *Current Anthropology*, 34, 3: 298–311.

—— C. Henshilwood, G. Lawson et al. (2003). 'Archaeological Evidence for the Emergence of Language, Symbolism, and Music—An Alternative Multidisciplinary Perspective', *Journal of World Prehistory*, 17, 1: 855–925.

Espi-Sanchís, P. (1997). 'Enfrentarse a los Cambios y a la Diversidad de la Música Africana', *Eufonía: Didáctica de la música*, 6: 31–6.

—— (1993a). *Cow Bells and Tortoise Shells*. Cape Town: Talking Drum Productions.

—— (1993b). *Another Lion on the Path: Stories from Zimbabwe*. Cape Town: Talking Drum Productions.

Farmer, H. (1930). 'Greek Theorists of Music in Arabic Translation', *Isis*, 13, 2: 325–33.

Franklin, J (2002). 'Musical Syncretism in the Greek Orientalizing Period'. In: E. Hickmann and R. Eichmann (eds.), *Archäologie früher Klangerzeugung und Tonorinungen. Studien zur Musikarchäologie III, Orient-Archäologie*, 10: 441–51.

Frisbie, C. (1971). 'Anthropological and Ethnomusicological Implications of a Comparative Analysis of Bushmen and African Pygmy Music', *Ethnology*, 10, 3: 265–90.

Garfinkel, Y. (2003). *Dancing at the Dawn of Agriculture*. Austin, TX: University of Texas Press.

—— (2001). 'Dancing or Fighting? A Recently Discovered Predynastic Scene from Abydos, Egypt', *Cambridge Archaeological Journal*, 11, 2: 241–54.

—— (1998). 'Dancing and the Beginning of Art Scenes in the Early Village Communities of the Near East and Southeast Europe', *Cambridge Archaeological Journal*, 8, 2: 207–37.

Gerson-Kiwi, E. (1980). 'Cheironomy'. In: S. Sadie (ed.), *The Grove Dictionary of Music and Musicians,* iv. London: Macmillan.

Goodall, J. (1986). *The Chimpanzees of the Gombe: Patterns of Behaviour.* Cambridge, MA: Belknap Press.

Halliday, M. (1979). 'One child's Protolanguage'. In: M. Bullowa (ed.), *Before Speech,* pp. 171–90. Cambridge: Cambridge University Press.

Hickmann, E. (2006). 'Archaeomusicology'. In: L. Macy (ed.), *Grove Music Online.* <http://www.grovemusic.com>.

Hornbostel, E. von and C. Sachs (1914). 'Systematik der Musikinstrumente: Ein Versuch', *Zeitschrift für Ethnologie*, 46, 4–5: 553–90.

Jander, O. and E. T. Harris (2001). 'Singing'. In: L. Macy (ed.), *New Grove Online.* <http://www.oxfordmusiconline.com/>.

Johanson, D. C. and B. E. Edgar (1996). *From Lucy to Language.* New York: Simon & Schuster.

Kilmer, A. (1971). 'The Discovery of an Ancient Mesopotamian Theory of Music', *Proceedings of the American Philosophical Society*, 115, 2: 131–49.

Kubik, G. (1994). *Theory of African Music*, i. Wilhelmshaven: Florian Noetzel.

Kunej, D. and I. Turk (2000). 'New Perspectives on the Beginnings of Music: Archaeological and Musical Analysis of a Middle Paleolithic Bone "Flute"'. In: N. Wallin, B. Merker, and S. Brown (eds.), *The Origins of Music*, pp. 235–68. Cambridge, MA: MIT Press.

Leakey, R. (1981). *The Making of Mankind.* London: Michael Joseph.

Levman, B. (1992). 'The Genesis of Music and Language', *Ethnomusicology*, 36, 2: 147–70.

Lewis, J. (2009). 'As Well as Words: Congo Pygmy Hunting, Mimicry, and Play'. In: R. Botha and C. Knight (eds.), *The Cradle of Language,* pp. 232–52. Oxford: Oxford University Press.

Lewis-Williams, J. (1990). *Discovering Southern African Rock Art.* Cape Town: David Philip.

Locke, J. L. (1996). 'Why Do Infants Begin to Talk? Language as Unintended Consequence', *Journal of Child Language*, 23, 2: 251–68.

Mans, M. (2004). *Discover Musical Cultures in the Kunene: A Guide to the Living Music and Dance of Namibia.* Windhoek, Namibia: self-published.

Merker, B. (2000). 'Synchronous chorusing and human origins'. In: N. Wallin, B. Merker, and S. Brown (eds.), *The Origins of Music,* pp. 315–27. Cambridge, MA: MIT Press.

Mitchell, T. (1992). 'The Music of the Old Testament Reconsidered', *Palestine Exploration Quarterly*, 124: 124–43.

Mithen, S. (2005). *The Singing Neanderthals.* London: Weidenfeld & Nicolson.

Moberg, C. (1986). 'On Music Archaeology'. In: C. Lund (ed.), *Second Conference of the ICTM Study Group on Music Archaeology*, pp. 237–9. Stockholm: Royal Swedish Academy of Music.

Moreno, J. (1995). 'Ethnomusic Therapy: An Interdisciplinary Approach to Music and Healing', *The Arts in Psychotherapy*, 22, 4: 329–38.

Morley, I. (2006). 'Mousterian Musicianship? The Case of the Divje Babe I Bone', *Oxford Journal of Archaeology*, 25, 4: 317–33.

Okumu, C. (2000). 'Africa Awake: Celebrating a Coming of Age of African Music in the Millennium'. In: M. Taylor and B. Gregory (eds.), *Music of the Spheres: Proceedings of the 24th World Conference of the International Society for Music Education*, pp. 308–18. Regina, Saskatchewan: University of Regina Faculty of Education/ISME.

Pont, G. (2004). 'Philosophy and Science of Music in Ancient Greece: The Predecessors of Pythagoras and their Contribution', *Nexus Network Journal*, 6, 1: 1–16.

Price, S. (1989). *Primitive Art in Civilized Places*. Chicago: University of Chicago Press.

Richman, B. (2000). 'How Music Fixed "Nonsense" into Significant Formulas: On Rhythm, Repetition, and Meaning'. In: N. Wallin, B. Merker, and S. Brown (eds.), *The Origins of Music*, pp. 301–14. Cambridge, MA: MIT Press.

—— (1980). 'Did Human Speech Originate in Coordinated Vocal Music?' *Semiotica*, 32, 3–4: 233–44.

Sachs, C. (1956). *A Short History of World Music*. London: Dobson.

Scarre, C. and G. Lawson (2006). *Archaeoacoustics*. Cambridge: McDonald Institute for Archaeological Research.

Schneider, A. (1986). '*ΑΡΧΑΙΟΛΟΓΕΩ*. Some Comments on Methods and Sources in Music Archaeology'. In: C. S. Lund (ed.), *Second Conference of the ICTM Study Group on Music Archaeology*, pp. 195–224. Stockholm: Royal Swedish Academy of Music.

Stockmann, D. (1986). 'On the Early History of Drums and Drumming in Europe and the Mediterranean'. In: C. S. Lund (ed.), *Second Conference of the ICTM Study Group on Music Archaeology*, pp. 11–28. Stockholm: Royal Swedish Academy of Music.

Tracey, H. (1961). *The Evolution of African Music and its Function in the Present Day*. Johannesburg: Institute for the Study of Man in Africa.

Trevarthen, C. (1999). 'Musicality and the Intrinsic Motor Pulse: Evidence from Human Psychobiology and Infant Communication', *Musica Scientiae*, special issue: Rhythm, Musical Narrative and Origins of Human Communication, pp. 155–211.

Walker, R. (2005). 'We All Came Out of Africa Singing and Dancing and We Have Been Doing It ever since', *Research Studies in Music Education*, 24: 4–16.

Wallin, N. (1991). *Biomusicology: Neurophysiological, Neuropsychological and Evolutionary Perspectives on the Origin and Purposes of Music*. Stuyvesant, NY: Pendragon Press.

Watson, A. and D. Keating (1999). 'Architecture and Sound: An Acoustic Analysis of Megalithic Monuments in Prehistoric Britain', *Antiquity*, 73: 325–36.

Part IV

Perspectives from social and cognitive psychology

8

On the Evolutionary Function of Song and Dance

Robin Dunbar

INTRODUCTION

Among the many aspects of human behaviour and cognition that can be counted as universals, music and dance are perhaps the most striking. Every culture boasts something that is instantly recognizable as music; every known culture dances.[1] While there has been considerable interest in the psycho-biological underpinnings of musical ability and appreciation, and some interest in the timing of its origins, the one aspect that has proved unusually intractable has been its function. Why should humans have evolved the capacity to be moved by music? Pinker (1997) has famously dismissed music as a trivial by-product of having a large brain, despite the fact that such a claim flies in the face of basic evolutionary principles. No phenomenon on which individuals are prepared to lavish so much time (and, in our case, money) could possibly be an evolutionarily trivial thing. As a matter of heuristic principle, evolutionary biologists take the view that the evolutionary processes are not so profligate as to tolerate activities that incur any kind of cost unless they provide commensurate benefits in terms of individuals' survival or reproductive success. In purely evolutionary terms, Pinker's (1997) conclusion is just odd; it falls foul of what we might refer to as the Spandrel Fallacy: 'I haven't

[1] Or, it should be noted, has a concept of dance even if this represents, on religious or political grounds, a repression of the phenomenon.

really had time to determine empirically whether or not something has a function, so I'll conclude that it can't possibly have one'.

Those who, like Pinker, regard music as without functional benefit make this claim on the grounds that it does not seem to aid survival in any obvious way. It does not, for example, improve foraging efficiency, and it can hardly be said to benefit the individual's ability to avoid being taken unawares by a predator. Indeed, if anything, it might be said to have the opposite effect: wasting time in musical activity detracts from the time that could be devoted to foraging, and the noise and/or movement so generated can only attract predators rather than help the performers to avoid them. Curiously, perhaps, those who advocate this view have never asked whether music and, in particular, musical ability might benefit individuals in the eternal campaign for reproductive success. Survival is surely one important element of fitness, but evolutionary biologists from Darwin himself onwards have been acutely aware that survival is only one component in the complex business of leaving descendants. Mating (and hence the business of mate choice and courtship) and the processes of parental care play a crucial role, especially in long-lived species in which, like most birds and mammals, offspring require careful nurturing in order to ensure that they successfully reach adulthood and reproduce in their turn. Evolution, it is often said, is interested not in children, but in grandchildren. In addition to sexual selection, there are other fairly conventional functional considerations at the group level (sometimes referred to as *social selection*) that have never been considered in this context.

In this chapter, I will consider two possible alternative functional explanations for the evolution of music within the hominin lineage. One is its role in mating and mate choice, the other is its role in social bonding.

A ROLE FOR SEXUAL SELECTION?

At a very early stage in developing his ideas, Darwin recognized that pure natural selection (conceived mainly in terms of survival in a world that is not always conducive to survival) could not account for all the phenomena with which biologists were familiar. Many phenomena in the natural world were anything but conducive to the survival of the individuals concerned. Examples included the excessively elaborate tail feathers of many birds, and the exaggerated and sometimes complex horns of many mammals and insects. Most of these are extremely costly to produce (so demanding, in fact, that a great deal of time has to be devoted unnecessarily to extra foraging) and are generally disadvantageous for survival (either by handicapping an animal's ability to escape from predators or by being functionally useless in defence).

Recognizing that many of these phenomena played a crucial role in courtship, Darwin proposed his theory of sexual selection as an essential companion to natural selection in his grand theory of evolution. In doing so, he implicitly recognized that evolutionary fitness (an organism's capacity to leave descendants) is a product of both its ability to survive *and* its ability to find mates and produce viable offspring. Mere survival is not enough. Though the significance of sexual selection remained largely unrecognized within biology for the better part of a century after Darwin first proposed it, this arm of Darwin's grand theory has grown to increasing prominence within the last three decades. Indeed, some now argue that sexual selection has been a more important factor promoting speciation than conventional natural (or, perhaps more correctly, environmental) selection.

Miller (2000) has drawn on Darwin's theory of sexual selection to provide an explanation for many aspects of human behaviour, including language and art. Musical ability provides a natural example of his claim that many aspects of human behaviour relate explicitly to the business of mate choice. His focus is principally on the way in which mate choice has provided powerful selection pressures favouring displays that one sex uses to advertise its suitability as a prospective mate for the other sex. The most extreme forms of sexual selection (resulting in the evolution of the most elaborate and functionally purposeless adornments) are found in those species where males contribute little by way of parental care. In these cases, the only thing that differentiates one male from another as a mating partner is the quality of his genes.[2] One of the principal ways in which males advertise the quality of their genetic constitution is to carry

[2] In this context, *quality* refers to the extent to which the male's genes complement those provided by the female and the conjoint effect that these then have on the survival—and, eventually, successful reproduction—of the resulting offspring. The term has no judgmental or moral implications.

ornaments or behave in ways that are detrimental to survival. Usually referred to as Zahavi's 'handicap principle' (Zahavi and Zahavi 1997), this phenomenon emphasizes the role of honest signalling as a basis on which females can make more informed choices. In effect, by putting their own lives at risk, males demonstrate the fitness of their genetic constitutions in a way that cannot easily be faked. The message is: just look at how good I really am—I can fly well enough to escape predators despite handicapping myself with this enormous and ungainly tail. Since predation is a genuinely serious threat that cannot easily be avoided, the handicap principle provides a bona fide test of males' advertising honesty.

The essence of the handicap principle lies in honest signalling: signals of quality can be taken at face value because they are difficult to cheat. The concept of honest signalling applies more broadly because it can easily be extended to production capacity as well as survival abilities. In other words, any process that tests a biological ability can in principle be selected for and used in mate choice. Indeed, it is possible for traits that are completely functionless to be selected for in this way. In humans, the same principle applies with respect to wealth: parental care is such a crucial component of successful reproduction (thanks to our unusually large and expensive brains) that males may be prompted to display their inherent wealth as well as their good looks. Once again, honest signalling is the key to success for both parties: only the genuinely wealthy can afford to spend money profligately on expensive fur coats or lavish entertainments.

The companion mode to Zahavi's handicap principle is known as Fisher's 'sexy sons hypothesis' (or, alternatively, 'runaway selection'). Fisher (1930) showed mathematically that it was in principle possible for any trait to be driven very rapidly to fixation (i.e. spread through a population and be adopted by all its members) providing it resulted in females acquiring a preference for mating with males that possessed it. Fisher pointed out that this process could select traits that were entirely arbitrary: it merely needed something that appealed to females' sensibilities. If females began to converge on some such trait, then intense selection pressure would kick in for females to produce sons who exhibited that trait, since these would fare best in the mating market of the next generation (hence the technical name that the phenomenon has since acquired). Fisher's process can be expected to work especially effectively if it focuses on some aspect of females' natural sensory mechanisms. What has become known as receiver psychology theory (Guilford and Dawkins 1992) argues that traits are especially likely to become exaggerated under the force of sexual selection if they target the natural design features of the perceptual systems of the intended recipients (in this case, of course, the females).

We could argue a case for music and musical ability under either of these mechanisms. On Miller's view, musical performance would be an honest signal of gene quality if capacity to perform was a genuine test of how well an individual's musical capacities could be exercised (and his musical capacities were, in turn, a good test of his more general levels of mental and physical coordination). Anyone can blow a horn (after a fashion), but can everyone produce a pleasing tune in the process? Likewise, anyone can sing, but can everyone sing beautifully enough to make the listener's heart soar? Certainly, the capacity to compose and/or perform is a test of some physical (i.e. motor), and even perhaps cognitive and emotional, competence. Of course, whether this is an adequate test of gene quality in the broader sense required by sexual selection theory is a moot point that would have to be decided by relevant empirical evidence. One can easily see the peacock's tail in terms of a real and meaningful test of one useful everyday capacity (the ability to escape predators), but musical ability is not self-obviously in the same league. Miller (2000), however, would justify his suggestion largely in terms of its significance as an index of mental competence, on the grounds that it is our mental abilities that have, above all, been the key to our evolutionary success as a lineage.

Miller (1999, 2000) has argued that the sexual asymmetry in both music production and the age distribution of production (males produce more musical works than females, and male production peaks during the third and fourth decades when mate searching activities are at their peak whereas female output peaks much later in life) provides prima facie evidence in support of the sexual selection hypothesis. Miller (1999) demonstrated these effects for jazz musicians, and we have replicated them for both classical nineteenth-century composers and contemporary pop musicians (Kaskatis, 2006). However, these analyses are by no means conclusive: a number of alternative explanations can be advanced to explain both outcomes that have little to do with sexual selection or the mating market. For example, sex differences in output may simply represent time constraints: women settle to the business of composition much later than men because they are (or, at least in the past, have been) distracted during their reproductive years by childcare. Similarly, the slow decline of male output during midlife decades could be due either to natural aging processes (laziness, loss of mental agility), increasing involvement in the business of earning sufficient money to cover family needs or, dare one say it, even increasing involvement in the business of childcare. It has not been easy to find suitable ways of testing between these alternatives.

Some kind of Fisherian receiver psychology process might seem an attractive alternative in this context. In this case, one could argue that musical ability targets some kind of natural resonance in human (and especially female) auditory mechanisms, perhaps some mechanism(s) derivative of the capacity for speech. This would relieve us of the burden of having to demonstrate that musical abilities are a reflection of genetic quality in a more general sense, since Fisherian processes can pick up and exaggerate any trait, whether or not it is correlated with fitness-enhancing abilities. So far, this suggestion has not been explored in any detail with respect to musical ability.

On balance, it is probably fair to say that, while there is some circumstantial evidence in support of the sexual selection hypothesis, it has not yet been possible to advance a really convincing case. This could mean that sexual selection *is* part of the evolutionary story for music, but that its role has been a secondary one rather than a primary one. In other words, sexual selection may have exploited musical capacity once it was in place, having been first selected for some other reason. The most likely alternative reason is something to do with bonding social groups.

A ROLE FOR SOCIAL SELECTION?

In recent years, it has become apparent that our conventional view of selection, with its focus on individual fitness, is missing an important element in the context of socially more advanced species, particularly those in which some kind of cooperation is used to solve the problems of survival and reproduction more successfully. This recognition has led to the development of what has come to be known as multilevel selection theory. In multilevel selection theory, selection may operate on individuals' capacities to form higher order social groupings, and individual fitness (or, more properly, the fitness of the genes that an individual bears) is the product of the accumulation of fitness through a number of different channels acting on the individual (its performance at a strictly individual level and the benefits that filter down to influence these or other aspects of individual reproduction through more efficient cooperation at group level).

Unfortunately, the waters have been somewhat muddied in recent years by labelling this process 'group selection'. Although selection is acting through the efficiency with which groups act, it must be clearly understood that this is not group selection as traditionally understood by evolutionary biologists: in this case, selection ultimately accumulates at the level of the individual's genes, and not at the level of the group as a biological entity in its own right (as is required by formal group selection). Hence, the term *social selection* seems infinitely preferable.[3]

Multilevel selection becomes a promising candidate for explaining the evolution of music in the context of the fact that primate sociality seems to be of a rather unusual kind. Primate societies are essentially cooperative solutions to the problem that individuals have of maximizing their fitness (Dunbar 1988). As such, they are implicit social contracts that require individuals to sacrifice some personal benefit in the immediate term in the interests of greater advantage accruing via the group in the longer term. However, social contracts of this kind are always at risk of being undermined by 'freeriders' (those who take the benefits of the compact, but decline to pay some or all of the costs) (Dunbar 1999). Some mechanism is therefore needed to enforce the community line, otherwise it is inevitable that individuals will gradually withdraw from the social contract as they find themselves exploited. If this happens, the benefits that would have accrued through cooperation will be lost. If these losses are significant enough (compared to what an individual could achieve by his or her own efforts), there will be selection pressure in favour of mechanisms that allow cooperation to be enforced. In fact, we seem to have a number of mechanisms that do this job, ranging from direct coercion (policing), to sensitivity to those who cheat on the social contract, to the use of language to cajole and encourage (Dunbar 1999; see also Barrett et al. 2002). I would suggest that our susceptibility to music is another such mechanism.

[3] My recollection is that this term was first used in exactly this context by John Crook in the early 1970s, but I have been unable to locate anything in his writings from this period.

My proposal is that music arose originally because it allows individuals to become more group-oriented. Music seems to achieve this through a capacity to produce endorphins which have a positive effect on our attitudes towards others (Dunbar et al. under review). As a by-product of their role in pain control, endorphins have the property of making us feel warm and generally well disposed towards anyone with whom we share the experience that stimulates their production. The suggestion is that the surge of endorphins triggered by engaging in communal musical activities makes us feel positively disposed towards those with whom we engage in this activity. Everyday folk wisdom suggests that this is so, and we have some evidence that music (and/or dance) is effective at stimulating the production of endorphins (though these results are at the moment only preliminary and have yet to be confirmed in detail).

A regular round of music and dance can thus be seen as an essential prerequisite for maintaining social cohesion in the face of all the many everyday irritations that threaten to destabilize and overwhelm our carefully crafted social arrangements. As such, it is likely to have a relatively ancient evolutionary origin (at least by comparison with language). The reasons for thinking this might be so stem from the ways in which primates bond their social groups.

For most Old World primates (monkeys and apes), grooming is the mechanism that is used to cement social relationships in sufficient depth to allow them to carry the weight of social coalitions. Grooming, too, is an extremely efficient way of stimulating the release of endorphins. Monkeys visibly relax while being groomed, and show heightened β-endorphin levels (Keverne et al. 1989). Building on Dunbar's (1993) suggestion that, in the human lineage, language ultimately replaced grooming as the principal bonding mechanism, Aiello and Dunbar (1993) used the series of relationships between cranial volume, neocortex volume, social group size, and grooming time in anthropoid primates to predict the required grooming time for different fossil hominin populations. By interpolating hominin cranial volumes into these equations, they were able to show that language probably arose quite late in hominin evolution (not earlier than the appearance of archaic humans around 500,000 years ago). Aiello and Dunbar drew an important distinction between speech and language, suggesting that the evolution of language may have occurred in a minimum of two distinct phases, the first of which may have been more akin to chorusing (wordless vocal exchanges—in effect, a form of 'grooming-at-a-distance').

There is some anatomical evidence to suggest that this chorusing phase first appeared with late *Homo erectus* (and certainly not later than the appearance of *Homo heidelbergensis*). The two key pieces of anatomical evidence are the dramatic expansion of the vertebral column in the thoracic region (thought to be related to breath control: Maclarnon and Hewitt 1999) and a similar increase in the size of the hypoglossal nerve (thought to be related to control of the tongue: Kay et al. 1998[4]). Although the fossil record is patchy, between them these two sources of evidence suggest that articulate speech (as reflected in increased control over breath and the vocal space in the mouth) is likely to have coincided with the appearance of archaic humans around 500,000 years ago.

This would seem to contradict the (admittedly as yet rather limited) genetic evidence for the evolution of grammar (the so-called 'FOXP2' gene), which suggests that the crucial mutations for full grammatical language did not occur until around 200,000 years ago (coinciding with the origins of modern humans, *Homo sapiens*). However, the capacity to articulate human-like sounds is not the same thing as having language, since the latter requires additional cognitive mechanisms associated with syntax.

For the moment, the most parsimonious interpretation of these various data is that some kind of chorusing (wordless singing) evolved at a relatively early stage as a form of grooming-at-a-distance in an attempt to supplement the more conventional bonding process of social grooming. Since singing requires exactly the same level of voice control as speech (control over both articulation and exhalation), singing would

[4] Although this claim of Kay et al. has been disputed by DaGusta et al. (1999), the latter's argument is based on the assumption that the smallest hypoglossal canal value for modern humans is sufficient to define the capacity for language (on the grounds that the individual concerned could speak). Hence, they argued that since some australopithecines had relatively larger hypoglossal canals than the smallest modern humans, we should conclude that all australopithecines had speech. While there is no intrinsic reason why language (or speech) might not have originated with the australopiths, DaGusta's argument is at best neurologically rather naïve. Using anything except average values risks confounding individual variation with species-typical capacities: the fact that one exceptional individual has more tongue control than others of its kind does not imply that the entire species has the full vocal capacities of modern humans. If we confine our attention to population or species average values, the analyses of DaGusta et al. (1999) simply confirm those of Kay et al. (1998).

have provided a natural precursor for the subsequent evolution of language by fine-tuning the articulatory machinery. The appearance of full-blown language would then only require the cognitive components of language (initially meaning, and subsequently grammar) to be linked up with articulation. On this view, musical ability probably long predates language (and hence speech in the formal sense) in the hominin lineage, even though both arose as attempts to solve continuing bonding crises as group sizes were forced inexorably upwards by whatever ecological processes were responsible.[5]

The significance of music in this context is that it bridges a major gap in the bonding process between what conventional primate social grooming can manage (groups of 60–80 at a push) and the point at which language seems to have kicked in (group sizes of around 120 or so) (Aiello and Dunbar 1993; Dunbar 1996). It does not seem to me especially difficult to derive some kind of intensely musical chorusing from existing primate vocalizations. Many of these have marked musical overtones, and this seems to be especially true of the vocal exchanges of those species which live in unusually large social groups (gelada baboons and bonobos being two classic examples).

A SEMINAL ROLE FOR LAUGHTER?

I will conclude with a final observation that might help develop this line of thinking a little further. Modern humans have one other form of behaviour that is all but unique: laughter. Laughter is something we share with chimpanzees (see Waller and Dunbar 2005), and probably only with chimpanzees in the form of real laughter (although, as part of the complex of play and play-invitation signals, it is something we share widely with other anthropoid primates, and possibly other species). However, we have developed and exploited this behaviour to novel heights during the course of our evolutionary history: we use

[5] We have little idea what these might have been, although three possibilities have been suggested: higher predation risk in the more open habitats occupied by later hominins; increased risk of conflict with neighbouring groups of hominins (a consequence of higher densities as the species became more successful); and increased demand for geographically extended trading networks (as hominins became more nomadic) (Dunbar 1996).

it in wildly exaggerated forms, in many more circumstances, and a great deal more often (Provine 1997).

Despite this, very little attention has been paid to laughter as a phenomenon for scientific investigation, and most authorities—if they think about it at all—would assume that Pinker's view of music applies, *mutatis mutandis*, to laughter: it is a phenomenon of little functional relevance, just another trivial example of Gould's class of evolutionary spandrels (Gould and Lewontin, 1979). But the fact that laughter is such a ubiquitous form of behaviour, and one that we devote a great deal of time seeking to stimulate, should alert us to the fact that neither its evolutionary origins nor its current function are likely to be trivial.

In fact, laughter begins to make sense when we see it in its social context. It is a facilitator of social interactions (Dunbar 2004), something that increases the intensity of an interaction and, ultimately, the relationship built up from these. Conversation typically lasts longer following laughter than it does if no laughs occur (Dunbar 2004). Whatever the cognitive mechanism involved here, the important thing is probably the fact that laughter seems to trigger the release of endorphins: we have now demonstrated this in a series of experimental studies on pain threshold elevation after laughter (Dunbar et al., forthcoming).

The presence of endorphins once again suggests a possible role in social bonding. Indeed, even a cursory consideration of laughter's evolutionary roots would seem to suggest as much: in chimpanzees, the homologue of laughter (associated with the ROM or 'round open mouth' facial expression) is principally given in play contexts, often as an invitation to play (Waller and Dunbar 2005). As with laughter in humans, interactions which involve 'laughter' or ROM face in the chimpanzee last longer than those that do not (Waller and Dunbar 2005).

What role might laughter have played in human social evolution? My suggestion would be that it was a precursor to musical chorusing, the first important stage in the gradual expansion of the hominin vocal repertoire towards true language. I would argue this for two reasons: first, its relative simplicity in structural and performance terms compared with both music and speech, and, second, its natural properties of segmentation and synchrony. The latter I see as being of fundamental importance in laying the groundwork for music. Laughter is uniquely well suited to this since it seems to depend for its effectiveness on the degree of synchrony between interactants in the cycle of exhalations that gives human laughter its unique and characteristic features. This level of synchrony does not occur in any other form of human activity outside of music (and, in a slightly different form, perhaps, in conversation where the alternation of speaking roles is very closely synchronized).

If laughter is the precursor to music, then we must assume that it arose even earlier than music did. If singing first appeared with archaic humans around 500,000 years ago, then laughter presumably has its origins in the *Homo erectus* phase of our history. We would not want to place it any earlier than this because, assuming that Aiello and Dunbar's (1993) calculations are even roughly correct, it is unlikely that australopithecine group sizes ever got above those observed among living catarrhine primates (i.e. groups of fifty to seventy). Even those of *Homo habilis* and early *Homo erectus* are unlikely to have been significantly above eighty or so, and probably just about within the limits of what can be bonded by social grooming (i.e. in terms of the amount of time that would have to be devoted to grooming). This is elaborated in more detail in Dunbar (forthcoming).

However, laughter as a form of chorusing seems to make a neat bridge between conventional primate grooming in the australopithecine phase and the appearance of music as a bonding mechanism with the larger groups that seem likely to have characterized archaic humans (and maybe some of the later *Homo erectus* populations). It thus seems to fall into a natural sequence running from grooming to laughter to music and finally to language, all underpinned by endorphin production as a bonding mechanism.

IN CONCLUSION

Disentangling the evolutionary history of music will not be an easy task. One reason it may be so difficult is that what we see now is the outcome of several different processes acting sequentially or in tandem—sexual selection exploiting opportunities provided by more conventional processes like social selection and group bonding. This layering of selection processes is far from uncommon in the biological world. Indeed, selection (and sexual selection in particular) seems to be extremely effective at exploiting windows of opportunity when these are created by more conventional natural (i.e. environmental) selection. Nonetheless, careful consideration of the role of music in human social life suggests that its original function is most likely to have been associated with social bonding once group sizes became too large to be adequately bonded using grooming alone.

In this context, music seems to provide a natural bridge to language, both in terms of allowing group size to creep up towards the levels where language seems to become inevitable (or at least essential) as a mechanism for group bonding and cohesion, and in terms of its psychophysical properties. Music itself may, in its turn, be seen as an outgrowth of laughter, a form of vocal social behaviour that is unique to humans and chimpanzees.

REFERENCES

Aiello, L. C. and R. Dunbar (1993). 'Neocortex Size, Group Size and the Evolution of Language', *Current Anthropology*, 34: 184–93.

Barrett, L., R. Dunbar and J. E. Lycett (2002). *Human Evolutionary Psychology*. Basingstoke and Princeton, NJ: Palgrave-Macmillan and Princeton University Press.

DaGusta, D., W. H. Gilbert, and S. P. Turner (1999). 'Hypoglossal Canal Size and Hominid Speech', *Proceedings of the National Academy of Sciences, USA*, 96: 1800–4.

Dunbar, R. (1988). *Primate Social Systems*. London: Chapman & Hall.

——(1993). 'Coevolution of Neocortex Size, Group Size and Language in Humans', *Behavioral and Brain Sci*ences, 16: 681–735.

——(1996). *Grooming, Gossip and the Evolution of Language*. London and Cambridge, MA: Faber & Faber and Harvard University Press.

——(1999). 'Culture, Honesty and the Freerider Problem'. In: R. Dunbar, C. Knight and C. Power (eds.), *The Evolution of Culture*, pp. 194–213. Edinburgh: Edinburgh University Press.

——(2004). 'Language, Music and Laughter in Evolutionary Perspective'. In: D. K. Oller and U. Griebel (eds.), *Evolution of Communication Systems: A Comparative Approach*, pp. 257–74. Cambridge, MA: MIT Press.

——(forthcoming). 'Bridging the Bonding Gap: the Transition from Primates to Humans', *Philosophical Transactions of the Royal Society of London B*.

——R. Baron, A. Frangou et al. (forthcoming). 'Social Laughter is Correlated with an Elevated Pain Threshold', *Proceedings of the Royal Society of London B*.

——K. Kaskatis, I. MacDonald et al. (under review). 'Performance of music elevates pain threshold and affect'.

Fisher, R. A. (1930). *The Genetical Theory of Natural Selection*. Oxford: Clarendon Press.

Gould, S. J. and R. Lewontin (1979). 'The Spandrels of San Marco and the Panglossian Paradigm: A Critique of the Adaptationist Programme', *Proceedings of the Royal Society of London B*, 205: 582–98.

Guilford, T. and M. S. Dawkins (1992). 'Receiver Psychology and the Evolution of Animal Signals', *Animal Behaviour*, 42: 1–14.

Kaskatis, K. (2006). *Searching for the Evolutionary Origins of Music*. PhD thesis, University of Liverpool.

Kay, R. F., M. Cartmill, and M. Balow (1998). 'The Hypoglossal Canal and the Origin of Human Vocal Behaviour', *Proceedings of the National Academy of Sciences, USA*, 95: 5417–19.

Keverne, E. B., N. Martensz and B. Tuite (1989). 'Beta-Endorphin Concentrations in Cerebrospinal Fluid of Monkeys Are Influenced by Grooming Relationships', *Psychoneuroendocrinology*, 14: 155–61.

Maclarnon, A. and G. Hewitt (1999). 'The Evolution of Human Speech: The Role of Enhanced Breathing Control', *American Journal of Physical Anthropology*, 109: 341–63.

Miller, G. (1999). 'Sexual Selection for Cultural Displays'. In: R. Dunbar, C. Knight and C. Power (eds.), *The Evolution of Culture*, pp. 71–91. Edinburgh: Edinburgh University Press.

——(2000). *The Mating Mind*. London: Heinemann.

Pinker, S. (1997). *How the Mind Works*. Harmondsworth: Penguin.

Provine, R. (1997). *Laughter*. London: Faber & Faber.

Waller, B. and R. Dunbar (2005). 'Differential Behavioural Effects of "Smiling" and "Laughing" in Chimpanzees (*Pan troglodytes*)', *Ethology*, 111, 2: 129–42.

Zahavi, A. and A. Zahavi (1997). *The Handicap Principle: A Missing Part of Darwin's Puzzle*. Oxford: Oxford University Press.

9

The Vocal Learning Constellation: Imitation, Ritual Culture, Encephalization

Björn Merker

INTRODUCTION

Throughout the rich diversity of animal forms and functions there is neither analogue nor homologue of human language (Maynard Smith and Szathmáry 1995). The singular novelty of this mode of cognition and communication, which is second nature to us, poses a major obstacle to approaching it via the paths of ordinary biology. Taken as a whole, human language is the ultimate derived trait, and, with an *n* of one, the comparative method falls short. Fortunately, however, language is not a unitary trait. It is decomposable into components or constituents (Fitch 2000), and for some of these, non-human analogues or homologues may provide suitable objects of comparison. An ideal candidate for such treatment would be a function or capacity without which language as we know it cannot exist, a capacity, moreover, with a specialized and distinctive rather than a broad and general function, and that has clear-cut instantiations in the animal kingdom, preferably in multiple, independently evolved forms (for ease of comparative analysis). Such a candidate capacity is *vocal learning*, as has been noted by students of bird song for more than thirty years (Marler 1970). As an 'evolutionary bottleneck' on the way to language, it provides powerful constraints on the reconstruction of language origins (Nottebohm 1976).

For every word we know how to pronounce and for every song we know how to sing we rely upon a competence that is lacking in other apes, and whose corresponding neural mechanism is, accordingly, a uniquely derived trait of the genus *Homo*. The competence in question is the capacity to *reproduce by means of the voice that which has been heard by ear*. Though a basic prerequisite for both our singing and our speaking, it is essential for little else in our behaviour. The capacity is technically known as *vocal learning*, a term I will consistently use in the sense of *vocal production learning* as defined by Janik and Slater (2000). It is rare among mammals, and absent not only from other apes, but also from primates generally, with the sole exception of humans (Janik and Slater 1997, 2000; see also Egnor and Hauser 2004). No other capacity ties together the topics of music (singing), language, and human evolution as directly and as categorically as does vocal learning. It represents, in a sense, the factor missing from the human equation, and in this chapter I will explore the possibility that it played a crucial role in the genesis of central human traits such as imitation, ritual culture, and brain expansion, besides being an essential prerequisite for our singing and speaking.

The suggestion that a mechanism which among animals most commonly serves song learning might be relevant to our understanding of the nature and origin of a human trait as distinctive as language is far from new. It was originally made by Darwin (1871) and more recently has been urged by students of birdsong (Marler 1970, 1991; Nottebohm 1975, 1976; Saito and Maekawa 1993; Doupe and Kuhl 1999; Okanoya 2002; Margoliash 2003; Wilbrecht and Nottebohm 2003; Jarvis 2004) and a few students of language (Jespersen 1922; Kuhl 1989; Staal 1989). However, with rare exceptions such as Fitch (2000), the topic has not been part of the discussion of language origins more generally. It is not mentioned, for example, in the influential treatise on natural language and natural selection by Pinker and Bloom (1990).

One reason for neglect of vocal learning in theories of language origins, as well as in the reconstruction of our evolutionary past, is that taking it seriously as a human trait requires one to abandon a common, and often tacit, assumption underpinning much work on human origins. It is the assumption that 'comparative analysis of the behaviour of modern primates, in conjunction with an accurate phylogenetic tree of relatedness, has the power to chart the early history of human cognitive evolution' (Byrne 2000, p. 543). I let this quote stand for a pervasive tendency only because of its conciseness. By gratuitously limiting the search space for evolutionary insight into human traits to other primates, such an approach accords a privileged status to homology as compared to analogy in evolutionary analysis, as if we could count on finding every human characteristic of interest at least incipiently represented among other primates.

Not every trait is, however, handed down like an heirloom in phylogeny (Lockard 1980), and, ultimately, any adaptation must at some point be founded *de novo* in a speciation event. For the analysis of such uniquely derived traits (autapomorphies), other instances of the evolution of the trait, however distant in phylogeny, and however independently originated, are invaluable sources of potential insight into the selection pressures that may underlie the trait. The comparative method accordingly gives us analogy as the companion to homology for analysing the evolution of life forms, and shows no preference for one above the other. In a sense, it is in the analysis of analogy that the comparative method truly comes into its own, and primates are no exception in this regard (Crook 1970; see also Harcourt 1998).

Non-human primates, like most mammals—and indeed like all vocal animals except humans—basically communicate by means of a repertoire of innate calls (Jürgens 1992). There are various ways in which such calls undergo learned modification without, for that reason, being instances of vocal learning (see Janik and Slater 1997, 2000 for details). Among primates, there is scattered evidence for modification of the details of innate call patterns through context (e.g. social context) and learning, as well as evidence for the learning-based preferential use of certain call variants for particular purposes (Crockford et al. 2004; Mitani and Brandt 1994; Mitani and Gros-Louis 1998; Seyfarth and Cheney 1980; Snowdon and Elowson 1999; Sugihara 1998). None of this resembles, of course, our massive reliance on learned acquisition, from heard models, of the vocal patterns of song and speech. Human mastery of that process is not confined to the modification of innate calls. We excel, rather, at the learned acquisition of a repertoire of biologically arbitrary (see next section) and complex vocal patterns that make up the song and speech traditions of our natal culture, down to its fine dialectal nuances.

It is not until we abandon a homology-inspired focus on non-human primates, and follow the comparative method wherever it may lead us, that we encounter true analogues of human vocal learning among the vocal traditions of animals such as birds and cetaceans. In so doing, we meet with biological settings featuring selection pressures and adaptations in some respects quite different from those that shaped our closest primate relatives but, as we shall see, potentially capable of illuminating those very traits in which we depart most decisively from primate and ape patterns. Indeed, I will propose that something unique happened when vocal production learning, for the first and only time, evolved in a primate; and that a good portion of our divergence from our closest ape relatives is intimately entwined with our possession of vocal learning and its neural underpinnings. Thus, there are reasons to believe that vocal learning played a key role in making us big-brained carriers of a type of historical culture hardly found among the great apes, but instantiated in the cultural traditions of some of the more accomplished vocal learners in the animal kingdom. As we shall see, they also tend to be big-brained. These circumstances, in turn, could have provided the setting in which our ancestors found their way to human language—as I will suggest, unwittingly.

IMITATION AND RITUAL CULTURE IN NATURE

> To make a complete study of the Mocker's powers of imitation, one would need to know: First, the songs and calls of birds now extinct in this section of the world, which may have been imitated by the ancestors of the present Mocker; second, songs and calls of all contemporary species in this part of the United States; third, the songs and the calls of all Mockingbirds now singing. (Mayfield 1934, p. 17).

Though another student of mockingbird song, J. R. Baylis, has remarked that Mayfield overstates the case a bit, he agrees that the difficulties involved in accounting for the song repertoire of virtuoso vocal learners are formidable (Baylis 1982). Here I wish to note the striking parallelism apparent in the sources contributing to the mockingbird's song repertoire and those contributing to the forms of spoken human language: one need only substitute 'other languages' (including extinct ones) for 'other species' in Mayfield's complaint to realize the closeness of the correspondence. The resemblance, I suggest, is not fortuitous but generic, and finds its explanation in a particular form of cultural transmission which has not so far been recognized as a distinct mode of culture, though it has a name of its own: *ritual culture*. Ritual culture is distinct from what I call *instrumental culture*, and the distinction is a fundamental one (Merker 2005, 2009).

The two publications just cited jointly provide a comprehensive exposition of the contrast between instrumental culture and ritual culture that will not be repeated here. Suffice it to say that instrumental cultural traditions are set apart from ritual ones along all defining dimensions, including form, function, and natural history. The difference turns on whether invariant aspects of the performance or patterning of the cultural behaviour in question are dependent upon—in the sense of being causally constrained or shaped by, derivable from, or obligatory for—the function or purpose of the behaviour or not. This is the case for instrumental behaviour, which, when transmitted culturally on a sub-population-specific basis by non-genetic means, defines a tradition of instrumental culture. Its invariant outcome or practical end result being primary, context-attuned flexibility of performance is its hallmark. Efficiency may favour some modes of execution over others, but literal duplication of the actual details of execution as performed by an observational model can impede instrumental success and is neither obligatory for, nor a defining feature of, instrumental culture. Examples of instrumental culture abound: they include all the human 'useful arts' and most cultural traditions of non-human primates (Byrne and Byrne 1993; Imanishi 1957; Itani 1958; Whiten et al. 1999), while striking non-primate examples can be found in the anthology edited by Fragaszy and Perry (2003).

By sharp contrast, in ritual culture the exact form and particulars of execution are primary, while the achievement of an invariant outcome counts for nothing if the *correct form* of the ritual is violated. The obligatory particulars of this form are *arbitrary* with regard to the ritual's purpose (whatever it may be) in the sense that a hypothetical pattern substitution (such as a sequence reversal or inversion) does not on those grounds alone allow one to predict that such a sequence would in principle be any less fit to serve the ritual purpose than the canonical version (Merker 2005). Indeed, the alternate version may be obligatory for another sub-population of the species. Nevertheless, performing such an alternate version would constitute a violation of the ritual. In this sense, then, *obligatory adherence to arbitrary form* is the hallmark of ritual culture, examples of which abound in human ritual (see Merker 2009) and learned animal song (see Catchpole and Slater 1995; Payne 2000).

The only motivation needed to sustain a tradition of instrumental culture is a desire to partake of the outcome its adoption brings within behavioural reach (feasting on nettles, say). Driven by utility, it 'pays its own way', and the learning capacity required to sustain it accordingly has natural selection as its evolutionary setting. Observational learning usually suffices for purposes of acquisition (a number of human examples can be found in Rogoff et al. 2003), and the achievement of the goal or purpose of the transmitted behaviour provides an objective check on the adequacy of execution. Crucially, this prevents non-functional variants from invading the tradition, ensuring its stability over time. No such external constraint exists to stabilize the pattern specifics transmitted by a tradition of ritual culture. Its formal details being arbitrary, it is under constant threat of deterioration through errors of copying and idiosyncratic variation. *Fidelity of copying by imitation (in the dictionary sense of imitate) is therefore essential for maintaining a ritual tradition.* This in turn implies that the carriers of a ritual tradition are possessed of a *conformal motive* which ensures that the practice and rehearsal that is needed to achieve fidelity converges in the end on actual duplication of the transmitted model (see Merker 2005 for a fuller account).

The ritual requirement of fidelity of copying helps explain the concentration of animal instances of ritual tradition to the medium of vocal learning, that is, to the duplication by voice of a sequence of sounds received by ear. As pointed out by Thorpe, the auditory-vocal channel has an advantage over other expressive modalities in allowing more complete monitoring (by ear) of self-produced output for purposes of comparison with model patterns than do its alternatives (Thorpe 1961, p. 69). Indeed, all oscine songbirds use auditory feedback of their own voice to perfect their singing (reviewed by Konishi 2004). Highly specialized telencephalic mechanisms are dedicated to the duplicative feats that sustain a cultural tradition of vocal learning. Following the pioneering studies of Thorpe (1961), half a century of intensive study of vocal learning for song in birds has laid bare a wealth of detail regarding its phylogeny, ontogeny, function, and neural mechanisms (a comprehensive overview is available in the anthology edited by Ziegler and Marler 2004).

Though there is no standard trajectory for the developing bird's duplication of adult song (see Beecher and Brenowitz 2005; Devoogd

2004; Freeberg 2000; King and West 2002; Williams 2004), it typically includes an initial period of auditory learning resulting in memory storage of the learned pattern of song produced by a conspecific model, followed by a protracted period of production learning. An initial practice phase called subsong features highly variable jumbles of low-intensity sounds. It has been analogized to the babbling phase in the speech development of the human infant (Doupe and Kuhl 1999; Goldstein et al. 2003; Marler 1970), and presumably serves primarily to practise mastery of vocal control itself. This is followed by so-called plastic song, in which the animal produces increasingly competent and elaborate song, but without adhering closely to the pattern of the model. Variability at these stages appears to be integral to the process of pattern learning itself (Ölveczky et al. 2005). The model template stored in memory is nevertheless in the background throughout, because plastic song eventually issues in 'song crystallization' and full adult competence, at which point the animal produces a high-fidelity replication of the model pattern in its own vocal output (cf. Gardner et al. 2005).

At song crystallization, a young bird achieves a faithful reproduction of the structural details of song it heard and memorized many weeks or months earlier. The process of vocal learning is thus an instance of true and spectacularly deferred imitation in the ordinary, dictionary, sense of the word. The structural content of the resulting song tradition is sub-population specific, and accordingly has a form whose pattern details are not dictated by the purpose served by the behaviour. Yet for a given individual in a given population, those particulars are obligatory. The behaviour has, in other words, all the hallmarks of cultural ritual identified in the foregoing. Vocal learning for song accordingly presents us with paradigmatic examples of ritual culture in animals, that is, *forms of imitative learning that faithfully duplicate a model to result in the lasting acquisition of a formally patterned behaviour as an integral part of a cultural tradition with ritual characteristics.*

Not all vocal learners end the acquisition process upon mastery of a canonical pattern supplied by a conspecific model (see Slater 1983; Beecher and Brenowitz 2005; and discussion in Brenowitz 2004, pp. 576–579). Species with 'open-ended' vocal ontogeny incorporate new material into their repertoire as adults. The species that do so include, but are not limited to, the true mimics among birds, who add new songs/sounds learned from arbitrary models (other species and environmental sounds) to a repertoire largely learned from local conspecifics who do and always have done likewise (mockingbird: Baylis 1982; superb lyrebird: Powys 1995; Robinson and Curtis 1996; butcher birds of Australia: Johnson 2003). Thus, the sounds of as many as seventy-six different model species from two continents may be identifiable in the song repertoire of a single marsh warbler individual (Baylis 1982; Dowsett-Lemaire 1979). Species that engage in invention furnish an additional path to open-ended repertoire growth. Invention is known to be part of the progressive turnover of the complex, group-specific song repertoires of humpback whales (Payne 2000), and may feature in a brown thrasher's accumulation of a repertoire estimated to contain in excess of 1,800 melodies (Kroodsma and Parker 1977).

Open-ended vocal learning and invention imply emancipation of vocal production from innate constraints that limit the outcome of song learning to variations on a structural schematism or schematisms common to the species. In such 'emancipated' learners, the specific structural content of their vocal production cannot be predicted, even to a rough approximation, from knowledge of the patterns prevalent in another, randomly chosen, sub-population or individual of the species. Additional paths to non-predictability are modes of production in which the sequencing of a set of song elements is varied on a non-deterministic basis (presumably the sedge warbler: Catchpole 1976; Bengalese finch: Okanoya 2004), and modes of acquisition in which model patterns are disassembled into phrases and fragments and reassembled into new unique song types on an individual basis, as reported for the winter wren (Kroodsma and Momose 1991). Since these paths to pattern novelty are available even on the basis of a closed vocal ontogeny (indeed, the sedge warbler requires no exposure to song at all to develop a large acquired repertoire: Leitner et al. 2002), the occurrence of genuine pattern novelty in vocal learning traditions cuts across the distinction between open-ended and closed ontogenies. For present purposes I will therefore let the notion of 'vocal emancipation' stand for the full range of devices by which vocal production is released from innate constraints to achieve genuine pattern novelty. Two or more of these various paths to 'protean pattern richness' may of course be combined in a given species. In addition to the song virtuosos already mentioned, parrots furnish striking examples of emancipated vocal learning capacity (Cruickshank et al. 1993; Farabaugh et al. 1994; Gramza 1970; Kalischer 1905; Pepperberg 1994).

Needless to say, the kind of extravagance embodied in the above examples does not characterize every animal singer reliant on vocal learning, and vocal learning may take different forms depending upon whether it functions to attract potential mates or to fend off rivals (for this, see Beecher and Brenowitz 2005). Moreover, vocal learning may be put to uses other than song (cf. the 'signature whistles' of dolphins (Caldwell et al. 1990) and the recognition calls of some bats (Boughman 1998; Esser 1994; Knörnschild et al. 2006)). We ourselves, however, with our culturally varied song repertoires and vocabularies numbering in the tens of thousands of words acquired by vocal learning from our natal speech community (down to its fine dialectal details), as well as our universal capacity for the imitation of other species, obviously belong among the more accomplished of emancipated vocal learners. Hence my emphasis here on virtuoso capacities and vocal emancipation, since this appears to be the domain of comparative phenomena that by evolutionary analogy might shed light on the nature and origin of our own advanced capacities in this regard.

How, then, are we to understand the extravagance and lavishness that meet us in the song productions of emancipated singers? Among birds, there tends to be an association between a high duty cycle for song (i.e. a large amount of continuous singing per day), large song repertoires, and high production variance in song (Baylis 1982, p. 64, and references therein). This association between persistent singing and virtuosity is presumably necessary, since protracted production of a monotonous pattern loses the attention of its audience by the ubiquitous mechanism of habituation (Hartshorne 1956; Kroodsma 1978). But why, then, persist in continuous song, which is energetically costly and takes place at the expense of alternate activities—such as foraging—to which the singer might devote the time and effort invested in singing? Why prolong the song display beyond the boredom threshold, thus incurring the additional cost of acquiring the means to produce elaborate song? The utility which we saw drive the evolution of a capacity for instrumental culture by natural selection is not apparent in the case of these extravagant displays, whose production consumes rather than creates resources. Whence the waste and frivolity of virtuoso performance? The question goes to the heart of not only emancipated song, but of ritual culture more generally.

SQUANDERING AS ASSET: MANIPULATIVE SIGNALLING AND COMPREHENSIVE FITNESS INDICATORS

When time is the medium of trade-offs, the present exerts pressure on decision-making, because the opacity of the future makes holding out for an improved outcome an uncertain prospect (Parker 2006). Under such circumstances, conflicts of interest among interacting agents will tend to ensure that signalling between them is employed manipulatively, by exploiting pre-existing perceptual biases of receivers (Enquist et al. 2002). In the face of countermeasures against manipulation, this will drive signalling towards conspicuousness, variability and exaggeration in purely formal terms, i.e. without a corresponding increase in the extent to which the signal is informative to receivers, say, with regard to sender qualities (Wachtmeister and Enquist 2000; Wachtmeister 2001).

Hypothetically, an exploited party to such signal interactions might escape the effects of exaggerated signalling by evolving an inflexible, fixed response to any adequate stimulation, a response set to the animal's own optimum. The fact that nature abounds in instances where this has not occurred suggests that it is in the 'exploited' party's interest to remain open to manipulation, and the reason for this is not far to seek: all signalling takes place at some cost; and elaborate, exaggerated signalling all the more so. Exaggerated signalling is therefore never without informative value for its audience: it supplies proof positive that the signaller is capable of sustaining the burden of exaggerated signalling. The capacity to sustain such a burden is thus necessarily an aspect of signaller quality, falling under a broad construal of the so-called handicap principle (Zahavi 1975; Zahavi and Zahavi 1997).

The logic of the handicap principle is quite general. Though there are circumstances under which its operation may inform a female regarding heritable male characteristics (Iwasa et al. 1991), its significance extends farther afield (cf. Møller and Jennions 2001). The agents of signalling are phenotypes, and a great many epigenetic and ontogenetic circumstances have an impact on an individual's ability to sustain elaborate signalling. That ability can therefore be informative regarding such factors, and these in turn may be of interest to any receiver that may be affected by behaviour involving those phenotypic traits, quite apart from whether they are heritable or not. Thus, a male in an agonistic interaction with a conspecific needs to assess the actual fighting ability of the rival, and not his genetic potential. Similarly a female in a species with obligate bi-parental care must assess the extent of a prospective mate's potential investment in the care of offspring, a quantity which, for a variety of reasons, can and does vary independently of the genes he contributes to those offspring. In these cases and others, a display of excess capacity in the form of elaborate signalling can indicate capacity in the relevant behavioural dimension, provided such signalling actually relates, directly or indirectly, to abilities and resources employed in the behavioural dimension under assessment.

An example involving a direct relationship between signalling and a desired or feared quality in the signaller would be physical fitness itself. Loud singing for many hours on a daily basis proves that the singer has the energy reserves, stamina, foraging ability, and predator vigilance to sustain such behaviour. The import of this for a receiver is that those same resources are available for other uses should the animal's circumstances or needs require it. The evidentiary value of a long, loud song bout assumes particular immediacy in small homeothermic animals such as birds. Their high metabolic rates, coupled with constraints on energy reserves such as fat stores, put them in constant danger of starvation unless they feed every few hours (Lindstedt and Boyce 1985).

Indirect linkages between signalling and singer qualities may be quite remote, as illustrated by the infelicitously named 'developmental stress hypothesis' proposed to account for the elaborateness of song in birds with vocal learning (Hasselquist et al. 1996; Nowicki et al. 2002a; Soma et al. 2005; Ball et al. 2006). Many bird species are highly altricial, undergo prolonged post-hatching development, and complete significant portions of their brain development after hatching (Emery 2006; Iwaniuk and Nelson 2003; Ricklefs 2004). This post-hatching neural development specifically includes the telencephalic system of song control nuclei that is essential for, and co-varies with, song learning, both as regards its presence or absence and its anatomical size (Catchpole and Slater 1995; Gahr 2000; Garamszegi and Eens 2004; Nowicki and Searcy 2004, pp. 711–12; Szekely et al. 1996).

The crux of the developmental stress hypothesis is the fact that the acquisition of complex song is an exacting as well as prolonged postnatal process involving perceptual-attentional acuity, memory resources in turn dependent on brain growth, hormonal-motivational factors, and extended rehearsal/practice. Each of these unfolds in dependence on, and in interaction with, the details of post-hatching neural development (Buchanan et al. 2004; Nowicki and Searcy 2004, pp. 711–12; MacDonald et al. 2006), and each obligatory stage in the sequence may be disrupted by a variety of external stressors and disturbances with which it competes for developmental resources. The stressors include nutritional stress/status (Buchanan et al. 2003; MacDonald et al. 2006; Spencer et al. 2003), exposure to parasites (Spencer et al. 2005), pollutants (Gorissen et al. 2005), and disruptions at the nest (nest security). The nutritional prerequisites for song-learning include an early phase of parental provisioning (Buchanan and Catchpole 2000), followed by the young bird's own foraging.

Shortfalls in such developmental prerequisites lead to shortfalls in adult song performance, and proficient song accordingly summarizes successful passage through a demanding and varied developmental obstacle course. Such passage, moreover, can hardly proceed unaffected by a bird's (and its parents') genetic makeup. *In this multifaceted manner, then, the level of song proficiency in effect sums up, in a single performance, the entire developmental history of the singer, and as such provides an all-round certificate of competence.* A proficient song bout tells its audience, in a way hard to counterfeit, that the singer comes from, as it were, 'a good background': it broadcasts the fact that as a nestling this singer was cared for by well-functioning parents, that then and later it was equipped with competent faculties and patent abilities, and that in general it flourished in a way that allowed it to master the demanding details of the extended learning task posed by song acquisition. Needless to say, such an indicator is bound to be of particular significance in K-selected species, whose biological capital is disproportionately invested in the phenotypic quality of individuals (their experience included; see Merker 2004) rather than in rapid expandability of numbers.

In sum, proficient learned song, and exceptional performance above all, provides something amounting to a 'comprehensive fitness indicator', and it is worth noting that, to function as such, the intended audience must have means to assess levels of proficiency according to an unambiguous and generally accessible criterion of judgement. Fidelity of copying would seem to provide an exacting criterion of adequacy that is also difficult to subvert by cheating, whether what is being copied is the style and repertoire of the local song tradition of conspecifics, or other models. As already mentioned, the latter may include not only the calls and songs of other species of birds, but—as in the case of lyrebirds—wingbeats and alighting rustles of various birds and assorted other ambient sounds, all imitated by voice, even in the case of wingbeats. The intended audience—whether male or female—has ample exposure to the same models, recognizes them, and discriminates between different performance versions (Searcy et al. 1997; Nowicki et al. 2002b).

The need for a shared criterion helps explain the fact that, after a young bird has produced competent song featuring idiosyncratic variations during the 'plastic' stage of acquisition, this variance is eliminated on the transition to mature song, to yield a high-fidelity rendition of the model at song 'crystallization' (Gardner et al. 2005; Tchernichovski et al. 2004). Fidelity as performance ideal is also indicated by the fact that high alertness levels or testosterone titres make song production less, rather than more, variable (Williams 2004; Theunissen et al. 2004). The stereotypy of mature performance is in other words *not* a stereotypy of limited capacity for alternate patterning. Rather, it appears to be 'self-imposed' in the sense of representing adherence to the ideal of fidelity (Gardner et al. 2005), and so requires a 'conformal motive' as its motivational underpinning (Merker 2005). The duplicative logic of such a motive sustains the stability of transmitted patterns without which a ritual tradition perpetuated through intergenerational learning degenerates into idiosyncratic caprice. Stabilized through the criterion of fidelity of copying resting on a conformal motive, the relative temporal inertia of the resulting cultural tradition provides individual learners with a vehicle by means of which to advertise the quality of their developmental history through the quality of their singing, and their audience with the standard by which to assess the performance.

Thus, a proficient singer is indeed worth taking seriously by conspecifics—be they potential mates or rivals—and the circumstances just reviewed, however summarily, go some way towards accounting for the resources expended on the apparently frivolous behaviour of elaborate learned song among birds and a few other animals, including, perhaps, ourselves. The behaviour serves to impress, and it does so for solid reasons: the competence expressed in song has been assembled through a biologically costly developmental trajectory tapping a wide range of abilities and resources. That trajectory is held together by an elaborate neural system specifically dedicated to vocal learning, a system which in birds extends from multiple telencephalic nuclei via thalamus and midbrain to the medulla. It is even possible that the significance of this system goes beyond that of song learning proper, because, as we shall see, it is so constituted as to provide an ideal lever for brain expansion or encephalization.

VOCAL LEARNING AND ENCEPHALIZATION

The size of animal brains varies non-linearly with body size in various taxa, and thus constitutes a so-called allometric trait. A comprehensive survey of the quantitative facts of vertebrate brain size and some of the factors behind variance in relative brain size has been provided by van Dongen (1998). From it we learn that, to a first approximation, vertebrates as a whole exhibit two major 'grades' of brain-body scaling, which we might call fish-like and mammal-like after the taxons with the widest range of brain-body variation within the two grades. The former includes all fishes, amphibians, and reptiles, and the latter all mammals and birds. Only sharks and rays protrude to any appreciable extent from the fish-like into the mammal-like range of relative brain sizes.

The larger body sizes of some mammals are, of course, missing among birds, but at comparable body size the relative brain sizes of birds and mammals largely overlap, with but a slight advantage for mammals. This overlap holds true as well for the proportion of overall brain volume occupied by the telencephalon, which ranges from 45 to 82 per cent in birds and from 50 to 85 per cent in mammals (Burish et al. 2004; Clark et al. 2001). Thus, at the low end of the avian range, the proportion of telencephalon for a pigeon roughly matches that for a rodent like the rat (Johnson and Whitney 2005; Swanson 1995). The zebra finch, at mid range with 64 per cent telencephalon, compares with New World monkeys, and the telencephalic percentage for an African grey parrot, at the high end of the avian range, lies close to that of of modern humans. The three birds represent a non-singing bird, a songbird with vocal learning, and an emancipated vocal learner, respectively. Might more than sampling bias be involved in the striking range of encephalization they cover?

The two avian orders that contain most of the vocal learners among birds, namely perching birds (*Passeriformes* consisting of oscine and suboscine songbirds) and parrots (*Psittaciformes*), dominate the high end of relative brain size across the full range of body weights represented among them (van Dongen 1998; see also Iwaniuk et al. 2005 for some allometric comparisons). The same is true of altricial birds, while precocial birds cluster towards low relative brain sizes, for all body weights represented in either grouping. The coincidence derives from the fact that parrots and perching birds represent the altricial extreme of the altricial-precocial spectrum in birds (Starck and Ricklefs 1998), and parrots count among the slowest-growing of altricial species (Koutsos et al. 2001).

The fact that the high end of avian encephalization is made up of orders that contain most of the bird species exhibiting vocal learning raises the possibility that vocal learning itself may be a factor in encephalization. The division of passerines into songbirds without vocal learning (the suboscines or tyranni) and songbirds with vocal learning (the oscines or passeri) provides a natural experiment that cries out to be analysed by comparing the respective encephalization quotients between the two sub-orders. Yet even short of that crucial comparison (which, curiously, has not been made) there are reasons to believe that oscines generally are more encephalized than suboscines. The system of song control nuclei is a dedicated set of telencephalic neural structures present in birds with vocal learning but not in those without, even when they sing, as do suboscines (Gahr 2000). Moreover, the size of its principal nuclei correlates with the size of the song repertoire across species (Devoogd et al. 1993). The neural apparatus of the song system is, in other words, something vocal learners have over and above the neural equipment of other singing birds, and this in itself would lead one to expect somewhat larger brain size in species with vocal learning than in otherwise matched songbirds without vocal learning. Yet the difference may go beyond simple addition of a set of telencephalic 'song nuclei' to an otherwise equivalent brain, because the nature of vocal learning predisposes it to act as a particularly effective lever for encephalization itself.

In a seminal analysis of allometric relationships in the evolution and development of mammalian brains, Finlay and Darlington (1995) showed that developmental constraints ensure that an evolutionary expansion of a functional capacity served by a given brain subsystem or compartment tends to be accomplished by expanding the brain as a whole, that is, by encephalization. Selective expansion of single components of brain macrostructure mandates alterations in the timing of yoked schedules of neurogenesis, and the balance of competitive processes involved in the subsequent formation of neural connections. Apparently, the costs of such alterations often exceed those of simple allometric expansion of all brain compartments in concert, a process largely involving global scaling rather than selective revision of delicately tuned developmental programmes.[1] Under such circumstances a selection pressure for a capacity expansion of a given functional circuit or component of brain macrostructure tends to be realized by expanding that entity as part of an allometric expansion of the brain as a whole, that is, by encephalization.

If a need to expand a single functional capacity can drive encephalization, then a requirement for conjoint expansion of two separate capacities represented in different parts of the telencephalon would seem to virtually guarantee it. But that is exactly the requirement imposed by a selection pressure for vocal learning. It must put in place an expanded capacity for auditory pattern analysis and auditory memory in posterior sensory and association areas, and it must also put in place the circuitry for higher-order learned and auditory feedback-regulated control of the vocal and respiratory apparatus in frontal premotor areas (Jarvis et al. 2000). Neither of these functional capacities is likely to be supplied by adaptations other than vocal learning: that is, they are specific to vocal learning, and both must be present and appropriately linked to enable the animal faithfully to reproduce by voice that which it once heard by ear. A direct projection from the primary telencephalic motor region to the motor nuclei of vocal and respiratory control in the medulla is an additional specific requirement for vocal learning (Jarvis 2004; Okanoya and Merker 2007)

Vocal learning, then, would seem ideally suited to act as a powerful lever for encephalization, and evidence that encephalization correlates with song complexity in songbirds has been reported (Airey et al. 1996). Needless to say, there are factors other than vocal learning capable of promoting encephalization in birds, including songbirds, but it should be noted that the analysis performed by

[1] If these developmental schedules and balances are regulated by genes with even moderate pleiotropy, the constraints they impose on development are both fundamental and formidable. See Waxman and Peck (1998).

Burish and colleagues is not a valid indication in this regard (Burish et al. 2004). They based membership in the top category of their statistical variable 'social structure' on drastically different grounds than those employed for the rest of that variable, and thus this key construct of their analysis does not represent a coherent parameter. A study basing this parameter strictly on differences in group structure found no overall effect of social structure on encephalization in birds (Emery 2004). As already mentioned, studies exploiting the contrast between suboscine and oscine songbirds with respect to vocal learning should be performed to define in quantitative terms the contribution of vocal learning to encephalization in birds.

VOCAL LEARNING IN HUMAN EVOLUTION

In light of what has gone before, the fact that humans, alone among the primates, are vocal learners, and emancipated ones at that, carries far-reaching implications for the reconstruction of our evolutionary history generally, and for conceptions of the biological background to human language more specifically. Though possession of the capacity to reproduce by voice that which has been heard by ear (vocal learning) is an absolute requirement for human spoken language (Nottebohm 1976), nothing compels us to assume that it evolved for that purpose. As we have seen, the capacity has evolved independently and repeatedly for other purposes in the two large-brained classes of vertebrates: three times in birds with thousands of species exhibiting the trait; and at least as many times in mammals, though mammalian species exhibiting the capacity are few (Janik and Slater 1997).

In both birds and mammals, the most common function of vocal learning is learned song. Humans are a singing species (Humboldt 1836); so, in the absence of reasons to think otherwise, song presents itself as the function for which the capacity evolved in our ancestry. Given that we are vocal learners, and that song may be the original function of our vocal learning, a number of issues considered in relation to the origin of human language appear in a new light. More than thirty years ago Nottebohm pointed out that the vocal learning ability of an Amazonian parrot allows it to pronounce phrases of human language with a fidelity of diction that fools the human ear (Nottebohm 1976). He went on to note that it does so using a vocal anatomy without the slightest resemblance to the human, and that it is therefore the neural control of the vocal apparatus, and not its gross anatomy, that principally determines its production characteristics (see further Titze 2003). This eliminates the presumed significance of the shape of the human pharynx and the descent of the larynx for the human transition to language (see further Fitch 2002). It is not because the vocal tract of a chimpanzee has the wrong shape that it cannot do what parrots do—learn to pronounce human words (Kellogg and Kellogg 1933; Hayes and Hayes 1951)—but because chimpanzees lack the capacity for vocal learning that allows parrots to perform their imitative feats (Janik and Slater 1997).

Moreover, on a number of measures such as tidal volume, range of sub-glottal pressure and muscular control, the biomechanics of song are more demanding than those of conversational speech (Sundberg (1987) provides quantitative data on a set of these parameters). As noted by Fitch (2006) this means that fossil indicators related to vocal biomechanics that traditionally have been taken to reflect the emergence of human spoken language may reflect the emergence of human song instead. Cases in point are the size of the hypoglossal canal carrying the principal motor innervation of the tongue (Kay et al. 1998; DeGusta et al. 1999; Jungers et al. 2003), and the width of the lumen of the thoracic spinal cord related to respiratory motor control (MacLarnon and Hewitt 1999). In comparative perspective, song evolves far more readily than speech, so an unusual size or shape of such fossilizing markers does not begin to bear on the issue of speech until the explanation 'song' has been eliminated (Merker and Okanoya 2007).

Similar considerations apply to potential fossilizing cerebral markers for human speech: as already mentioned, the neural implementation of vocal learning necessarily includes a posterior auditory perception-and-memory system working in functional concert with a frontal system for learned patterning of vocal and respiratory output. This organizational theme is apparent in each of the three avian orders that independently evolved vocal learning (Jarvis et al. 2000). As a first approximation to relating the avian pattern to the functional organization of the human brain, the posterior system plausibly includes human auditory association areas and Wernicke's area, while the frontal system would include Broca's area and additional frontal territories (see Jarvis 2004; Merker 2000, p. 321). Indeed, when subjected to neuroimaging, human singing activates cortical territories that show appreciable overlap with those involved in speech, Broca's area included (see, for example: Brown et al. 2004; Callan et al. 2006; Özdemir et al. 2006). This overlap is not particularly surprising from the present perspective, since spoken language no less than song is dependent upon vocal learning.

Taken together, these circumstances propel human song and the vocal learning on which it depends to a superordinate strategic position in the study of language origins. Answers regarding the latter depend on answers regarding the former, and we are accordingly faced with a need to determine the time and manner of appearance of vocal learning in our past, a task concerning which Nottebohm once warned that 'you might find it much harder to explain this first step, vocal learning, than the latter acquisition of language' (Nottebohm 1976, p. 645). All we know with any certainty in that regard is that we evolved our capacity for vocal learning at or subsequent to our evolutionary parting of ways with the common ancestor we share with chimpanzees, but we do not know when and how.

The potential role of vocal learning as a lever for encephalization outlined in the previous section suggests that our ancestral history of brain expansion might offer clues to adaptive events in which vocal learning played a part. The first of these expansions marks even the earliest remains about which we can be reasonably certain that they represent actual ancestors of ours. They belong to *Homo erectus* making a rather sudden appearance in the fossil record just before one and a half million years ago (Gathogo and Brown 2006; Walker and Shipman 1996; see also Hawks 2006). The second is the brain expansion that accompanied the emergence of our species *Homo sapiens* from *Homo erectus* over the past half million years or so. A period of relative stasis (i.e. adaptive success) separates the two (McHenry 1994; Ruff et al. 1997), making it tempting to associate the first with learned song and the second with language. Additional circumstances surrounding the emergence of *Homo erectus* are interpretable as constituting a coherent adaptive syndrome in which vocal learning for song could have played an integral part, as follows.

Along with brain expansion the emergence of *Homo erectus* is marked by increased body size, decreased sexual dimorphism in body size (i.e. female stature increased disproportionately), and changed body (limb) proportions to levels which places each of the three trait complexes well within the range of modern human variation (Aiello and Wells 2002; Ruff 2002). Crucially, postcanine dental size and the masticatory apparatus were reduced as well (McHenry and Coffing 2000), implying a diminished need for oral processing.[2] To be compatible with an increase in stature and brain expansion relative to body size, this reduced need for oral processing indicates increased access to high quality nutrients (Aiello and Wheeler 1995; Antón 2003), that is, a novel dietary adaptation.

There are reasons to believe that *Homo erectus* pursued a generalist subsistence strategy in response to the climatic changes that framed its emergence (Ungar et al. 2006; Wood and Strait 2004, p. 151). The fact that the strategy promoted reduced sexual dimorphism suggests that the new food sources on which it relied were non-monopolizable. That would have allowed subordinate males to divert effort formerly invested indirectly in reproduction via male dominance contests to a direct reproductive investment in a female and her offspring by provisioning them with high quality nutrients (cf. Chapais 2008). Under the circumstances this would have promoted opportunistic foraging for a variety of dispersed and seasonally fluctuating high-quality nutrients, requiring intensive and extensive search. Curiosity, memory capacity, and strategic planning would have been at a premium, providing a selection pressure for brain expansion.

The lakeshore, riverbank, and other land-water interfaces (and their hinterlands) inhabited by *Homo erectus* contained a variety of such nutrient sources. They include scavenged longbones and skulls that could be split for access to marrow by a blow from a hammerstone (Binford 1981; Brantingham 1998; Stiner 2002, Fig. 3), clams and mussels, easily bashed open with the same primitive tool, and catfish seasonally caught by hand in shallow spawning waters to be eaten raw (Stewart 1994; cf. Asfaw et al. 1999, p. 630; see also Joordens et al. 2009). Exploitable without massive mammalian competition (cf. Brantingham 1998, p. 332; Houston 1979; MacArthur and Levins 1967; see also O'Connell et al. 1988), these are prime

[2] The reduction in masticatory apparatus presumably involved the silencing of a gene encoding a myosin chain specific to the musculature of the mandible, resulting in the loss of its fast-twitch fibres (Stedman et al. 2004). The original dating of this mutation to approximately 2.4 million years ago has been challenged on methodological grounds (Perry et al. 2005), but its alternate dating to approximately 5.3 million years ago is not easily reconciled with the evolution of massive jaw muscles in australopithecines after that date (Hawks 2005). As pointed out by Stedman and colleagues, a diminution of masticatory muscles and their power removed the need for a sagittal crest insertion, and may thus even have released constraints on brain expansion (Stedman et al. 2004).

sources of the long-chain and polyunsaturated fatty acids required for brain growth and maintenance (Broadhurst et al. 1998; Cunnane et al. 1993; Cunnane and Crawford 2003), all consumable with minimal demands on mastication.

A provisioning strategy with bi-parental care is presumably what permitted early *Homo* to increase female stature along with the brain size and protracted development of helpless infants (cf. Alexander 1990). Crucially, it may have made it possible to combine an extended period of postnatal development (which scales with longevity, in turn a correlate of increased body size; see Aiello and Wells 2002; Bird and O'Connell 2006; Charnov 1993) with early weaning by substituting high-fat nutrients for breast milk (Kennedy 2005; from the present point of view, replace the role attributed to 'meat' in that and other accounts with 'marrow and molluscs'). The shortening of inter-birth intervals this would have made possible presents one of the few ways to increase reproductive output in apes, constrained as they are by K-selection. This role of a high-fat diet thus may help explain not only the emergence of *Homo erectus* in Africa, but its rapid spread across vast expanses of Eurasia.

A provisioning strategy is not sustainable without some assurance of paternity on the part of provisioning males. In fact, it is unlikely to occur without it (Buss 2002; Geary 2000, 2005; Møller 2003). Note, however, that increased male investment in female and offspring may supply a path to enhance the level of paternal certainty, rather than being its result (Kvarnemo 2006; Møller and Cuervo 2000). On either direction of causality in this regard, the multi-male multi-female group sociality that presumably prevailed in the transition to *Homo* would have yielded some version of colonial monogamy as the setting for successful pursuit of a provisioning strategy. Such a pattern would be unique among primates, but is well attested in a variety of forms among birds,[3] and is, as it were, 'still with us' in what is called 'ecological monogamy' among present day hunter-gatherers (Marlowe 2000). In their case the ecology, needless to say, is responsible only for the limitation of male

[3] A few examples: Within freely mixing groups of black vultures, stable coalitions of several nuclear families associate at roosts and carcasses, and apparently support one another in competition over food (Rabenold 1986). Consistent associations of acorn woodpecker nuclear families defend year-round joint territories on which they together build and maintain granaries numbering in the tens of thousands of storage sites (Brown 1975). Parrots afford numerous further examples; see for example Pepper (1996).

resources to investment in no more than one woman and her offspring, and not for the male propensity to invest resources in a woman whose children he fathers. That propensity, operative in modern polygyny as well, is a human departure from great ape patterns for which the emergence of *Homo erectus* provides a plausible context.

The need for some form of pair-bond or 'strategic partnership' between male and female *Homo* to underwrite a provisioning strategy makes it likely that these ancestral couples sang, in the form of duets. Though we are the only primate with vocal learning, we are not the only one that sings. Singing is rare in terrestrial mammals generally, but is more common among primates, occuring in some two dozen non-human primate species belonging to the genera *Indri*, *Tarsius*, *Callicebus*, and *Hylobates* (gibbons). Since the clusters of singing species are only distantly related, song is assumed to have evolved independently within each taxonomic group (Geissmann 2000; Haimoff 1986). In each case, the singers exhibit a monogamous social structure, and in each case both males and females sing—usually in the form of duetting though there are monogamous primates that do not sing, the pattern suggests a latent functional coupling between monogamy and pair duets in primates, an association that is particularly strong for the Old World primates to which we belong.

It remains to consider why only one of these paired primates evolved learned song, that is, a reliance on vocal learning, while the rest rely on innately structured calling for their duetting (see, e.g. Geissmann 2000, for gibbons). The answer presumably lies in a difference in the nature of the pair relation. Social monogamy is not a unitary phenomenon (Fuentes 1998; Møller 2003). Specifically, the role of bi-parental care of offspring (including provisioning) that features prominently in the case of bird monogamy is rare among mammals (Kleiman 1977; Komers and Brotherton 1997). The non-human primate instances of monogamy appear to conform to those of other mammals (Komers and Brotherton 1997; Rutberg 1983), while it is exactly a bird-like, substantial, and direct paternal investment in a female and her offspring that was invoked above as the key to increasing reproductive output despite a lengthening developmental period and increased infant dependence in early *Homo*.

With substantial and direct male investment of resources in a bonded female and her offspring, the stakes involved in pair formation and maintenance are raised (for both sexes), bringing into play the entire spectrum of factors considered in the section on 'squandering as asset' above. As noted there, it is only on account of the exacting demands of the sequential stages of vocal learning that the formal properties of a song performance become a veridical and summary indicator of a whole set of developmental and genetic factors relevant to an individual's current capacities. Again, the matter assumes particular significance in K-selected species such as those of the *Homo* line, in which prolonged development and a large brain make individuals into precious repositories of experience (Merker 2004). In such a setting, the ability to advertise the quality of individual developmental history through a compact display is a prime asset, and one uniquely provided by vocal learning. The capacity to do so may accordingly have been attained by our ancestors at their most singular and definitive step to a highly K-selected mode of adaptation, namely as part of the emergence of *Homo erectus*.[4]

A MILLION YEARS OF SONG AND ITS CONVERSION INTO LANGUAGE

If, then, learned song evolved as part of our first major advance in brain size, presumably the second such advance, from roughly 600,000 to 150,000 years ago (Ruff et al. 1997), is associated with the emergence of *Homo sapiens* and the origin of spoken language from this background of an encephalized ape brain devoted to ritualized learned song for display purposes, among its many other tasks. Recent advances in our understanding of the long-term consequences of purely cultural transmission of a repertoire of patterned strings

[4] Note that the conjectural account of the emergence of *Homo erectus* presented in this section of my chapter has many potential ramifications beyond those suggested here. For example, if foraging for mussels, clams, and other aquatic foodstuffs involved more than incidental diving, a number of consequences follow. Voluntary control of respiration would have been enhanced, presumably by a direct axonal connection from the primary motor cortex to the medullary respiratory centres, a neural pathway needed for vocal learning no less than for diving (Okanoya and Merker 2007; Wild 1994). Aquatic environments are the single most conspicuous factor in the ecological background to mammalian vocal learning (cetaceans and pinnipeds, see Janik and Slater 1997), and would suggest a straightforward function not only for the human diving reflex (Schagatay et al. 2000), but for the thickened bones (medullary stenosis) and general skeletal robusticity of *Homo erectus* (Kennedy 1985, 1991), an autapomorphy which on this interpretation would have evolved as ballast for diving. Finally, the slow renewal of exhausted clambeds could have driven *Homo erectus*' migratory movements along the shorelines of Eurasia.

(see Batali 1998, 2002; Hurford 2002; Kirby 1998, 2000, 2001, 2002) raise the possibility that language emerged as an inadvertent consequence of the intergenerational transmission of a rich repertoire of song in an emancipated vocal learner such as the partly hypothetical ancestral *Homo* sketched in the foregoing.

The formal key to that possibility is the functional significance of the so-called 'learner bottleneck' in intergenerational transmission of string-utterances in a population of learning agents over many generations. Any given agent in any given generation is exposed to only a fraction of the utterances being produced in the population as a whole (an aspect of the so-called 'poverty of the stimulus', see Smith 2003; Zuidema 2003). This means that a state of competition for access to the next generation exists among utterances. Any factor, characteristic or circumstance giving an utterance even the slightest advantage in this competition will cause it to last somewhat longer in this repeated filtering through the learner bottleneck over the generations. Computer simulations have shown that in the course of hundreds and thousands of generations of transmission, without differential reinforcement of outcomes or natural selection, this process is capable of transforming an initial state in which nonsense-strings are randomly paired with meanings on an individual basis to a state of semantic and syntactic organization exhibiting compositionality, lexical categories, constituent order, frequency-dependent coexistence of regular and irregular forms, and recursion, all shared by the population as a whole (Batali, 1998, 2002; Kirby, 1998, 2000, 2001, 2002).

In nature, individuals do not go about making random utterances to communicate meanings carried in their heads. In an earlier publication I have provided a detailed account of how an iterated learning model of the kind explored in the computer simulations just cited can become operative in a population of emancipated singers engaged in intergenerational transmission of a rich song repertoire for display purposes (Merker and Okanoya 2007). The interested reader is emphatically referred to that source, since its details will not be repeated here. What follows are supplementary remarks to that analysis, focused on one of its aspects that can be given a fuller exposition in light of the adaptive scenario sketched for early *Homo* in the previous section of this chapter. That aspect is the requirement that reproductive behaviour in the singing population not be confined to a seasonally defined courtship or breeding period or be otherwise constrained to a narrow behavioural context.

That requirement is needed in order to give the learner bottleneck the broadest possible basis for its first and fundamental task in the origin of language as here conceived, namely to differentiate a repertoire of shared, culturally transmitted song-strings such as can be found in traditions of emancipated vocal learners into sub-repertoires associated with different behavioural, motivational, and environmental contexts. The outcome of that process would be a situation in which certain song-strings tended to be sung more often in certain contexts, behavioural situations, or environments than others. Simple instances of such repertoire differentiation in accordance with whether singing serves courtship of females or male rivalry, and even in accordance with specific situations of male rivalry, have been documented in birds with vocal learning (see, e.g. Kroodsma 1988; Trainer 1987).

To allow the learner bottleneck to forge assortative linkages between song-strings and contexts covering the entire behavioural-epistemic space of a species, the scope of such repertoire differentiation needs to be broadened to cover the entire behavioural and ecological space of that species (for details, see Merker and Okanoya 2007, pp. 410–1). This condition is not easy to fulfil because singing as here portrayed is ultimately driven by sexual selection, and thereby has courtship and male rivalry as its natural contexts and occasions. The only way to break the stranglehold of this ultimate causation over the situational uses of singing would be in a species in which the sexual receptivity of females was neither cyclically nor seasonally determined, and in which the motives for courtship therefore potentially extended across all behavioural situations in which the sexes found themselves engaged. But such a species is the human one, with year-round sexual receptivity of women, concealed ovulation, and concealment of copulation (the human tendency to seek the seclusion of privacy for the act of mating) (Alexander 1990; Geary 2000; Geary and Flinn 2001; see also Diamond 1991, pp. 56–71, for an informal summary).

We do not know when we acquired these peculiarities of our sexual behaviour, though again they can be plausibly linked to the adaptive scenario sketched here for early *Homo*. The use of non-monopolizable high-fat nutrients by subordinate males to entice females away from the control of dominant males to themselves would not have gone unopposed by dominant males. Initially, the strategy would therefore have had to be pursued surreptitiously. The exigencies of establishing and sustaining furtive pair associations in a group setting under these circumstances may account for both the concealment of ovulation and our tendency to mate in seclusion. The same behaviours would at a later stage, on the consolidation of the strategy in the form of colonial monogamy, have served to reduce competitive conflict and rivalries in sexual matters (cf. Brown 1978, 1980). A reduction does not, of course, mean elimination, and extra-pair copulation would still exert permanent pressure on the provisioning strategy. Indeed, the intensive and extensive search assumed to be involved in foraging for the kind of non-monopolizable foodstuffs suggested in the foregoing would force males to spend considerable time away from their bonded mates. Perhaps high human copulatory frequencies evolved as a pre-emptive measure in sperm competition in this context, as has been suggested for a similar situation in some species of monogamous birds (Birkhead et al. 1987; Møller 1987; Crowe et al. 2009).

With permanent sexual receptivity on the part of females, the behavioural occasions for courtship singing as well as pair duets (Wachtmeister 2001) would have been extended across the varied behavioural contexts filling the daily, seasonal, and yearly rounds of ancestral *Homo*. Assuming that the repertoire composing the singing tradition of a group of these relatively large-brained ancestors was large, complex, and redundant, the learner bottleneck would ensure its differentiation by behavioural context, as detailed in Merker and Okanoya (2007). Like other memory contents, heard song-strings would be stored in context, making contextual association one of many factors active at the learner bottleneck (see further next paragraph). Over generations the song repertoire would subdivide by preferential usage into a set of sub-repertoires distributed (with considerable overlap) across the varied behavioural contexts and activities of individuals, pairs, and larger groupings. Obvious initial dimensions of such differentiation would be by sex of the partner within the pair, by pairs within the group, by groups with regard to one another, and by the purposes of the song display, such as solo versus duet, affiliative versus agonistic, and so on.

From such a start, repertoire-specificity would eventually infiltrate the entire behavioural-epistemic space of the singers, because any difference in the statistics of string usage between behavioural situations would be perpetuated by the learner bottleneck. Simple associative memory mechanisms would ensure that the context in which certain strings happen to be sung a bit more often than others, whether for motivational or chance reasons, would come to serve as a reminder of those strings when the context was encountered again. Those strings would therefore again be produced with slightly greater likelihood than others, leading the learner bottleneck to perform assortative allocation of strings by context over the generations. The arbitrary nature of this allocation presents no conceptual or functional problem, because the strings are not used to communicate meaning but are purely formal patterns sung for ritual display purposes. The conformal motive intrinsic to vocal learning would eventually make the proper choice of song-strings for each specific context obligatory, adding it to the requirements of an impressive display, and the old sexually selected motives would maintain the song tradition in perpetuity across the generations, as in all vocal learning traditions. Thus the learner bottleneck mechanism is given the long-term leisure it needs to do its slow work in the background.

And that work in the background is not confined to progressively driving the repertoire to differentiate by ever finer distinctions of context. It simultaneously transforms the formal properties of the string-context correspondence itself. Variation and redundancy in the formal patterning of strings will ensure that some behavioural situations that share something in contextual terms also happen to share some substrings. As suggested by Alison Wray on language development grounds (Wray 1998), and as demonstrated by Simon Kirby in the computer simulations already cited, under such circumstances the learner bottleneck will eventually seize on the economy available in such correspondences to segregate out the shared substrings as proper songs for the shared context.

Thus, if song-string 'nnnnonkonknn' (letters here represent song elements) occurs as a substring in the songs proper to 'gathering of allied couples at dusk' and also in the songs proper to 'trekking together', over generations the learner bottleneck will promote its occurrence in strings attached to any context involving 'things done together'. No more than a capacity to segment strings and to generalize is required for this to occur (Hurford 2000), capacities well within the powers of a mammal and a vocal learner (Ghirlanda and Enquist 2003; Okanoya and Merker 2007, pp. 426–7). The repertoire accordingly is launched on a process of progressive string-context assortative and hierarchical decomposition from holistic strings *downwards*. Taking place as an unintended side effect of intergenerational transmission through the learner bottleneck, the process is entirely passive and automatic, and takes place for no reason of instrumental utility whatsoever. Continued long enough, its imperceptible multigeneration dynamic will eventually yield single phrases matched to single generalized contextual abstractions (in effect concepts) amounting to a tacit lexicon with rich, implicit semantic content, along with the formal syntactic efficiency measures demonstrated in the computer simulations of Kirby and colleagues.

None of the conceptually ever richer string-context correspondences are as yet employed to communicate meaning. That is not what drives the process. The right song expertly sung in proper context exhibits the singer's command of the ritual tradition in all its sophisticated intricacy, and indirectly the quality of the singer's developmental background that went into acquiring that command. After pairing, duetting additionally exhibits the pair's powers of joint learning and behavioural coordination. The process of assortative repertoire differentiation and recombinant string matching to context has transformed repertoire use into a sophisticated conceptual match to the singing community's life circumstances with glacial slowness, over many generations, and unbeknownst to any singer taking part in the process in any given generation. To them, the tradition seems fixed, and to consist of a given repertoire to be mastered in all its contextual intricacy.

The individual singer's fidelity of learning and performance is what allows the tradition to maintain the statistical association between song-strings and contexts that gives the repertoire its entirely tacit conceptual content, and places it just one step short of supplying a powerful tool for communicating meaning. Yet that one step is a long one, because, in order to be put to language-like use, a song-string with a given contextual 'meaning' would have to be used in situations other than its proper context. This is the crucial design feature of language called 'displacement' in Hockett's (1960) treatment, broadly construed. But using a song-string outside of its proper context would not only violate a basic performance criterion of the vocal tradition itself, but, more seriously, would undermine the statistical associations that guarantee its string-context linkage in the first place. This circumstance presents an inherent barrier to the conversion of a song tradition as sketched here into a tradition of spoken language as we know it, and would presumably have long delayed such a conversion.

Nevertheless, it is unlikely that the potential instrumental uses of linkages between song-string and context would remain entirely and permanently hidden from the carriers of such a sophisticated song tradition, who are after all apes with brains a good bit bigger than the chimpanzee's. Let us say that a certain song-string is used in a given *Homo erectus* community in the context of discovery of a new clam-bed, perhaps with an obligatory coda indicating the identity of the discoverer. Imagine that a discoverer of one such bed, which happened to feed the entire community for a whole week, some time later was caught pilfering unopened marrow bones from his neighbours. As his irate neighbours descended upon him, he may have been reminded of his clam-bed discovery, and the feasting by all that followed. Having been reminded of the occasion, the clam-bed song-string lay near at hand, and should the hapless fellow have chanted it at that moment, he may have found that it stopped his neighbours in their tracks, and gave him a bit of a reprieve.

Such serendipitous discoveries gave notice that encountered contexts not only remind the singer of the strings that belong to those contexts by ritual tradition, but that the strings can be used to remind hearers (belonging to the same tradition) of the contexts to which they are proper. Thus the strings have potential instrumental uses in being sung 'out of context'. Through the scope and means afforded by such 'situationally decoupled' use of strings for manipulating the attentional deployment of hearers, natural selection is given a foothold to work on the conversion of the sexually selected song tradition to a form compatible with its use for instrumental, communicative purposes, that is, to spoken language proper. Any heritable trait that allowed a singer effectively to further his or her reproductive interests by communicatively exploiting the statistical linkages embedded in the song tradition into which he or she was born would be promoted by this selection pressure. Additional cerebral learning capacity is likely to be the principal such trait, while neural circuitry innately embodying linguistic structure clearly is incompatible with the account of language origins proposed here.

If instrumental utility provides a foothold for the selection pressure, the existence of the prior song tradition provides the leisure for that pressure to do its work. With every false start and failed solution to the decoupling problem, the collapse would return matters back to the state of the song tradition for display purposes from which it had been launched. Even on radical corruption of the statistics of string-context linkages by over-exploitation through decoupled use, new string-context linkages would accrue over the generations by the same glacial learner bottleneck process that had forged them in the first place. With this as a safety net, and instrumental utility as a constant selection pressure, a cerebral capacity expansion allowing the vocal learning tradition to incorporate conventions regarding the use of song-strings outside of the situation underwriting their contextual association would presumably accrue. A left hemisphere specialization for displaced and manipulative string usage drawing on pre-existing hemispheric differences is likely to have been part of this process, as is a neocerebellar to prefrontal pathway involved in cognitive and language functions (Leiner et al. 1993; Middleton and Strick 1994); a pathway that, I conjecture, might have played a crucial early role in this development by allowing potential consequences of displaced string-usage to be anticipated. A good part of the 600,000 year career of *Homo sapiens*, marked by the enlargement of the human brain to its present dimensions, may have been taken up and driven by this conversion process before decoupled string usage became a stable, self-sustaining option in the vocal learning tradition of our ancestors, equipping us with spoken language proper.

This transformation, driven by the instrumental utility of decoupled usage, would have entailed and been facilitated by a number of substantial changes in the vocal learning tradition itself. Communicative intent now came to figure as one of the contextual specifics that presumably would have attracted string markers of its own to differentiate in-context from displaced usage (again by means of the learner bottleneck mechanism). Aesthetic matters of form would have had to give way to instrumental and pragmatic ones. A number of pragmatic and grammatical conventions related to communicative intent presumably facilitated the decoupling process. The pragmatic ones may have included manual gestural signals underscoring communicative intent as a simple convenience without for that reason assuming any particularly central role in language origins. And string markers for communicative intent would have been followed by the full spectrum of devices and processes studied under the heading of grammaticalization and allied topics in historical linguistics (Bybee, 2006; DeLancey, 1993; Lehmann, 1995).

A distinctive mode of delivery would have helped as well, to set decoupled use of strings apart from their use for situation-specific display purposes (a distinction that still survives, if imperfectly, in the differences of style and occasion between song and speech). For decoupled use, performance no longer typically required the loud, broadcasting mode of melodic and rhythmic phrasing proper to song, but could be tailored to circumstances and the purpose of the communication, occasionally reduced to the level of whispers. Note that lowered energetic demands on the part of the new signalling mode by no means rendered it 'cheap' from the perspective of the handicap principle. On the contrary: the greater rather than lesser demands now placed on the developmental process of acquiring and perfecting command of the tradition in fact increased its costliness as a handicap (Merker and Okanoya 2007, p. 414). Now, as before, performance was used to display that background and that command by exhibiting mastery of form and style, often in the form of idle talk; but performance now could be put to an additional use as well, namely that of transmitting meaning in order to inform, deceive, justify, or manipulate, as the occasion and its exigencies demanded or allowed. In short, spoken language had been born.

The passage to a stable spoken language is not likely to have reached completion until a late stage of our descent, presumably no earlier than on the emergence of anatomically fully modern humans, and possibly much later. That, however, is a matter in the study of language proper, and thus belongs to a chapter other than the present one, which has been devoted to the distinctive characteristics and powers of vocal learning that set the stage for such a transition.

CONCLUSION

We began by taking note of the fact that both human song and language are inherently dependent upon vocal learning, a neural mechanism conspicuous for its absence among our primate relatives, even the closest ones. In considering how we might plausibly have come into possession of its powers, we found ourselves grappling with fundamental questions encumbering the interpretation of the fossil record of human descent. In what followed, it was suggested that human language may in fact have been incubated in the medium of vocal learning in the form of ritual traditions of context-specific song. The subsequent conversion to spoken language would have taken place under the selection pressure of a self-interested use of song-strings to manipulate the attentional deployment of hearers.

The account given here perforce has had to rely in part on conjecture, but it should at least have demonstrated this: that no reconstruction of either human origins or the origin of language can be valid without accounting for the way in which vocal learning entered our biology; and what further changes it wrought once it was in our possession. In fact, few issues would seem to provide greater leverage for advancing our self-understanding than to determine how much of our distinctively human nature and capacities we owe to the workings of the vocal learning constellation along our evolutionary trajectory. Should the interpretation sketched here turn out to have any merit, its contribution in that regard would be substantial, and what is more, would have supplied the functional nexus around which our distinctive nature, crowned by human language, was assembled.

REFERENCES

Aiello, L. C. and J. C. K. Wells (2002). 'Energetics and the Evolution of the Genus *Homo*', *Annual Review of Anthropology*, 31: 323–38.

Aiello, L. C. and P. Wheeler (1995). 'The Expensive-Tissue Hypothesis: The Brain and Digestive System in Human and Primate Evolution', *Current Anthropology*, 36: 199–221.

Airey, D. C., J. K. Niederer, A. L. Nelson et al. (1996). 'High Vocal Center and Encephalization: Specialization and Developmental Constraints', *Society for Neuroscience Abstracts*, 22: 1042.

Alexander, R. D. (1990). 'How Did Humans Evolve? Reflections on the Uniquely Unique Species', *University of Michigan Museum of Zoology Special Publication*, 1: 1–38.

Antón, S. C. (2003). 'Natural History of *Homo erectus*', *Yearbook of Physical Anthropology*, 46: 126–70.

Asfaw, B., T. White, O. Lovejoy et al. (1999). '*Australopithecus garhi*: A New Species of Early Hominid from Ethiopia', *Science*, 284: 629–35.

Ball, G. F., K. W. Sockman, D. L. Duffy et al. (2006). 'A Neuroethological Approach to Song Behavior and Perception in European Starlings: Interrelationships among Testosterone, Neuroanatomy, Immediate Early Gene Expression, and Immune Function', *Advances in the Study of Behavior*, 36: 59–121.

Batali, J. (1998). 'Computational Simulations of the Emergence of Grammar'. In: J. R. Hurford, M. Studdert-Kennedy, and C. Knight (eds.), *Approaches to the Evolution of Language: Social and Cognitive Bases*, pp. 405–26. Cambridge: Cambridge University Press.

—— (2002). 'The Negotiation and Acquisition of Recursive Grammars as a Result of Competition among Exemplars'. In: T. Briscoe (ed.), *Linguistic Evolution through Language Acquisition: Formal and Computational Models*, pp. 111–72. Cambridge: Cambridge University Press.

Baylis, J. R. (1982). 'Avian Vocal Mimicry: Its Function and Evolution'. In: D. E. Kroodsma and E. H. Miller (eds.), *Acoustic Communication in Birds*, pp. 51–83. New York: Academic Press.

Beecher, M. D. and E. A. Brenowitz (2005). 'Functional Aspects of Song Learning in Songbirds', *Trends in Ecology and Evolution*, 20: 143–9.

Binford, L. R. (1981). *Bones: Ancient Men and Modern Myths*. New York: Academic Press.

Bird, D. and J. O'Connell (2006). 'Behavioral Ecology and Archaeology', *Journal of Archaeological Research*, 14: 143–88.

Birkhead, T. R., L. Atkin, and A. P. Møller (1987). 'Copulation Behaviour of Birds', *Behaviour*, 101: 101–38.

Boughman, J. W. (1998). 'Vocal Learning by Greater Spear-Nosed Bats', *Proceedings of the Royal Society of London B*, 265: 227–33.

Brantingham, P. J. (1998). 'Hominid–Carnivore Coevolution and Invasion of the Predatory Guild', *Journal of Anthropological Archaeology*, 17: 327–53.

Brenowitz, E. A. (2004). 'Plasticity of the Adult Avian Song Control System'. In: H. P. Ziegler and P. Marler (eds.), 'Behavioral Neurobiology of Birdsong', *Annals of the New York Academy of Sciences*, 1016: 560–85.

Broadhurst, C. L., S. C. Cunnane, and M. A. Crawford (1998). 'Rift Valley Lake Fish and Shellfish Provided Brain-Specific Nutrition for Early Homo', *British Journal of Nutrition*, 79: 3–21.

Brown, J. (1975). *The Evolution of Behavior*. New York: Norton.

Brown, C. R. (1978). 'Sexual Chase in Purple Martins', *Auk*, 95: 588–90.

—— (1980). 'Sleeping Behavior of Purple Martins', *The Condor*, 82: 170–5.

Brown, S., M. J. Martinez, D. A. Hodges et al. (2004). 'The Song System of the Human Brain', *Brain Research: Cognitive Brain Research*, 20: 363–75.

Buchanan, K. L. and C. K. Catchpole (2000). 'Song as an Indicator of Parental Quality in the Sedge Warbler', *Proceedings of the Royal Society of London B*, 267: 321–6.

—— K. A. Spencer, A. R. Goldsmith et al. (2003). 'Song as an Honest Signal of Past Developmental Stress in the European Starling (*Sturnus vulgaris*)', *Proceedings of the Royal Society of London B*, 270: 1149–56.

—— S. Leitner, K. A. Spencer et al. (2004). 'Developmental Stress selectively Affects the Song Control Nucleus HVC in the Zebra Finch', *Proceedings of the Royal Society of London B*, 271: 2381–6.

Burish, M. J., H. Y. Kueh, and S. S.-H. Wang (2004). 'Brain Architecture and Social Complexity in Modern and Ancient Birds', *Brain, Behavior and Evolution*, 63: 107–24.

Buss, D. M. (2002). 'Human Mate Guarding', *Neuroendocrinology Letters, Special Issue*, 23 (Suppl. 4): 23–9.

Bybee, J. (2006). 'Language Change and Universals'. In: R. Mairal and J. Gil (eds.), *Linguistic Universals*, 179–94. Cambridge: Cambridge University Press.

Byrne, R. W. (2000). 'Evolution of Primate Cognition', *Cognitive Science*, 24: 543–70.

—— and J. M. E. Byrne (1993). 'Complex Leaf-Gathering Skills of Mountain Gorillas (*Gorilla g. beringei*): Variability and Standardization', *American Journal of Primatology*, 31: 241–61.

Caldwell, M. C., D. K. Caldwell, and P. L. Tyack (1990). 'Review of the Signature-Whistle Hypothesis for the Atlantic Bottlenose Dolphin'. In: S. Leatherwood and R. R. Reeves (eds.), *The Bottlenose Dolphin*, pp. 199–234. San Diego, CA: Academic Press.

Callan, D. E., V. Tsytsarev, T. Hanakawa et al. (2006). 'Song and Speech: Brain Regions Involved with Perception and Covert Production', *NeuroImage*, 31: 1327–42.

Catchpole, C. K. (1976). 'Temporal and Sequential Organization of Song in the Sedge Warbler (*Acrocephalus schoenobaenus*)', *Behaviour*, 59: 226–46.

—— and P. J. B. Slater (1995). *Bird Song: Biological Themes and Variations*. Cambridge: Cambridge University Press.

Chapais, B. (2008). *Primeval Kinship: How Pair-Bonding Gave Birth to Human Society*. Cambridge, MA: Harvard University Press.

Charnov, E. (1993). *Life History Invariants*. Oxford: Oxford University Press.

Clark, D. A., P. P. Mitra, and S. S.-H. Wang (2001). 'Scalable Architecture in Mammalian Brains', *Nature*, 411: 189–93.

Crockford, C., I. Herbinger, L. Vigilant et al. (2004). 'Wild Chimpanzees Produce Group-Specific Calls: A Case for Vocal Learning?' *Ethology*, 110: 221–43.

Crook, J. H. (1970). 'The Socio-Ecology of Primates'. In: J. H. Crook (ed.), *Social Behavior in Birds and Mammals: Essays on the Social Ethology of Animals and Man*, pp. 103–66. New York: Academic Press.

Crowe, S. A., O. Kleven, K. E. Delmore et al. (2009). 'Paternity Assurance through Frequent Copulations in a Wild Passerine with Intense Sperm Competition', *Animal Behaviour*, 77: 183–7.

Cruickshank, A. J., J.-P. Gauthier, and C. Chappuis (1993). 'Vocal Mimicry in Wild African Grey Parrots *Psittacus erithacus*', *Ibis*, 135: 293–9.

Cunnane, S. C., L. S. Harbige, and M. A. Crawford (1993). 'The Importance of Energy and Nutrient Supply in Human Brain Evolution', *Nutrition and Health*, 9: 219–35.

—— and M. A. Crawford (2003). 'Survival of the Fattest: Fat Babies Were the Key to Evolution of the Large Human Brain', *Comparative Biochemistry and Physiology Part A*, 136: 17–26.

Darwin, C. (1871). *The Descent of Man and Selection in Relation to Sex*. New York: D. Appleton & Company.
DeGusta, D., W. H. Gilbert, and S. P. Turner (1999). 'Hypoglossal Canal Size and Hominid Speech', *Proceedings of the National Academy of Sciences, USA*, 96: 1800–4.
DeLancey, S. (1993). 'Grammaticalization and Linguistic Theory'. In: J. Gomez de Garcia and D. Rood (eds.), *Proceedings of the 1993 Mid-America Linguistics Conference*, pp. 1–22. Colorado: Department of Linguistics, University of Colorado.
Devoogd, T. J. (2004). 'Where Is the Bird?' In: H. P. Ziegler and P. Marler (eds.), 'The Behavioral Neurobiology of Birdsong', *Annals of the New York Academy of Sciences*, 1016: 778–86.
—— J. R. Krebs, S. D. Healy et al. (1993). 'Relations between Song Repertoire Size and the Volume of Brain Nuclei Related to Song: Comparative Evolutionary Analysis amongst Oscine Birds', *Proceedings of the Royal Society of London B*, 254: 75–82.
Diamond, J. (1991). *The Rise and Fall of the Third Chimpanzee*. London: Random Century Group.
Dongen, P. A. M. van (1998). 'Brain Size in Vertebrates'. In: R. Nieuwenhuys, H. J. ten Donkelaar, and C. Nicholson (eds.), *The Central Nervous System of Vertebrates*, iii, pp. 2099–123. Berlin: Springer-Verlag.
Doupe, A. J. and P. K. Kuhl (1999). 'Birdsong and Human Speech: Common Themes and Mechanisms', *Annual Review of Neuroscience*, 22: 567–631.
Dowsett-Lemaire, F. (1979). 'The Imitation Range of the Song of the Marsh Warbler, *Acrocephalus palustris*, with Special Reference to Imitations of African Birds', *Ibis*, 121: 453–68.
Egnor, S. E. R. and M. D. Hauser (2004). 'A Paradox in the Evolution of Primate Vocal Learning', *Trends in Neurosciences*, 27: 649–54.
Emery, N. J. (2004). 'Are Corvids "Feathered Apes"? Cognitive Evolution in Crows, Jays, Rooks and Jackdaws'. In: S. Watanabe (ed.), *Comparative Analysis of Minds*, pp. 181–213. Tokyo: Keio University Press.
—— (2006). 'Cognitive Ornithology: The Evolution of Avian Intelligence', *Philosophical Transactions of the Royal Society of London B*, 361: 23–43.
Enquist, M., A. Arak, S. Ghirlanda et al. (2002). 'Spectacular Phenomena and Limits to Rationality in Genetic and Cultural Evolution', *Philosophical Transactions of the Royal Society of London B*, 357: 1585–94.
Esser, K. H. (1994). 'Audio-Vocal Learning in a Non-Human Mammal: The Lesser Spear-Nosed Bat *Phyllostomus discolor*', *Neuroreport*, 5: 1718–20.
Farabaugh, S. M., A. Linzenbold, and R. J. Dooling (1994). 'Vocal Plasticity in Budgerigars (*Melopsittacus undulatus*): Evidence for Social Factors in the Learning of Contact Calls', *Journal of Comparative Psychology*, 108: 81–92.
Finlay, B. L. and R. B. Darlington (1995). 'Linked Regularities in the Development and Evolution of Mammalian Brains', *Science*, 268: 1578–84.

Fitch, W. T. (2000). 'The Evolution of Speech: A Comparative Review', *Trends in Cognitive Sciences*, 4: 258–67.

—— (2002). 'Comparative Vocal Production and the Evolution of Speech: Reinterpreting the Descent of the Larynx'. In: A. Wray (ed.), *The Transition to Language*, pp. 21–45. Oxford: Oxford University Press.

—— (2006). 'The Biology and Evolution of Music: A Comparative Perspective', *Cognition*, 100: 173–215.

Fragaszy, D. M. and S. Perry (2003). *The Biology of Traditions. Models and Evidence.* Cambridge: Cambridge University Press.

Freeberg, T. M. (2000). 'Culture and Courtship in Vertebrates: A Review of Social Learning and Transmission of Courtship Systems and Mating Patterns', *Behavioural Processes*, 51: 177–92.

Fuentes, A. (1998). 'Re-evaluating Primate Monogamy', *American Anthropologist (New Series)*, 100: 890–907.

Gahr, M. (2000). 'Neural Song Control System of Hummingbirds: Comparison to Swifts, Vocal Learning (Songbirds) and Nonlearning (Suboscines) Passerines, and Vocal Learning (Budgerigars) and Nonlearning (Dove, Owl, Gull, Quail, Chicken) Nonpasserines', *Journal of Comparative Neurology*, 426: 182–96.

Garamszegi, L. Z. and M. Eens (2004). 'Brain Space for a Learned Task: Strong Intraspecific Evidence for Neural Correlates of Singing Behavior in Songbirds', *Brain Research Reviews*, 44: 187–93.

Gardner, T. J., F. Naef and F. Nottebohm (2005). 'Freedom and Rules: The Acquisition and Reprogramming of a Bird's Learned Song', *Science*, 308: 1046–9.

Gathogo, P. N. and F. H. Brown (2006). 'Revised Stratigraphy of Area 123, Koobi Fora, Kenya, and New Age Estimates of its Fossil Mammals, including Hominins', *Journal of Human Evolution*, 51: 471–9.

Geary, D. C. (2000). 'Evolution and Proximate Expression of Human Paternal Investment', *Psychological Bulletin*, 126: 55–77.

—— (2005). 'Evolution of Paternal Investment'. In: D. M. Buss (ed.), *The Evolutionary Psychology Handbook*, pp. 483–505. Hoboken, NJ: John Wiley & Sons.

—— and M. V. Flinn (2001). 'Evolution of Human Parental Behavior and the Human Family', *Parenting: Science and Practice*, 1: 5–61.

Geissmann, T. (2000). 'Gibbon Song and Human Music from an Evolutionary Perspective'. In: N. L. Wallin, B. Merker, and S. Brown (eds.), *The Origins of Music*, pp. 103–23. Cambridge, MA: MIT Press.

Ghirlanda, S. and M. Enquist (2003). 'A Century of Generalization', *Animal Behaviour*, 66: 15–36.

Goldstein, M. H., A. P. King, and M. J. West (2003). 'Social Interaction Shapes Babbling: Testing Parallels between Birdsong and Speech', *Proceedings of the National Academy of Sciences (USA)*, 100: 8050–5.

Gorissen, L., T. Snoeijs, E. Van Duyse et al. (2005). 'Heavy Metal Pollution Affects Dawn Singing Behaviour in a Small Passerine Bird', *Oecologia*, 145: 504–9.

Gramza, A. F. (1970). 'Vocal Mimicry in Captive Budgerigars (*Melopsittacus undulatus*)', *Zeitschrift für Tierpsychologie*, 27: 971–83.

Haimoff, E. H. (1986). 'Convergence in the Duetting of Monogamous Old World Primates', *Journal of Human Evolution*, 15: 51–9.

Harcourt, A. H. (1998). 'Does Primate Socioecology Need Nonprimate Socioecology?' *Evolutionary Anthropology*, 7: 3–7.

Hartshorne, C. (1956). 'The Monotony Threshold in Singing Birds', *Auk*, 73: 176–92.

Hasselquist, D., S. Bensch, and T. von Schantz (1996). 'Correlation between Song Repertoire, Extra-Pair Paternity and Offspring Survival in the Great Reed Warbler', *Nature*, 381: 229–32.

Hawks, J. (2005). 'Hazards of Associating Genetic and Morphological Changes', John Hawks Weblog. <http://www.johnhawks.net/weblog/reviews/genetics/diet/myosin_stedman_stone_2005.html>.

—— (2006). 'A Revised Chronology for Early Homo', John Hawks Weblog. Available: <http://johnhawks.net/weblog/reviews/erectus/brown_2006_chronology_koobi_fora.html>

Hayes, K. J. and C. H. Hayes (1951). 'The Intellectual Development of a Home-Raised Chimpanzee', *Proceedings of the American Philosophical Society*, 95: 105–9.

Hockett, C. F. (1960). 'Logical Considerations in the Study of Animal Communication'. In: W. E. Lanyon and W. N. Tavolga (eds.), *Animal Sounds and Communication*, pp. 392–430. Washington, DC: American Institute of Biological Sciences.

Houston, D. C. (1979). 'The Adaptations of Scavengers'. In: A. R. E. Sinclair and M. Norton-Griffiths (eds.), *Serengeti: Dynamics of an Ecosystem*, pp. 263–86. Chicago, IL: University of Chicago Press.

Humboldt, W. von (1836). *Über die Verschiedenheit des Menschlichen Sprachbaues und ihren Einfluss auf die geistige Entwicklung des Menschengeschlechts*. Berlin: Royal Academy of Sciences. Trans. G. C. Buck and F. Raven (1971), *Linguistic Variability and Intellectual Development*. Miami, FL: University of Miami Press.

Hurford, J. (2000). 'The Emergence of Syntax'. Editorial introduction to the section on syntax. In: C. Knight, M. Studdert-Kennedy, and J. Hurford (eds.), *The Evolutionary Emergence of Language: Social Function and the Origins of Linguistic Form*, pp. 219–30. Cambridge: Cambridge University Press.

—— (2002). 'Expression/Induction Models of Language Evolution: Dimensions and Issues'. In: T. Briscoe (ed.), *Linguistic Evolution through Language Acquisition: Formal and Computational Models*, pp. 301–44. Cambridge: Cambridge University Press.

Imanishi, K. (1957). 'Identification: A Process of Enculturation in the Subhuman Society of *Macaca fuscata*', *Primates*, 1: 1–29.

Itani, J. (1958). 'On the Acquisition and Propagation of a New Food Habit in the Troop of Japanese Monkeys at Takasakiyama'. In: K. Imanishi and S. Altmann (eds.), *Japanese Monkeys: A Collection of Translations*, pp. 52–65. Edmonton: University of Alberta Press.

Iwaniuk, A. N. and J. E. Nelson (2003). 'Developmental Differences Are Correlated with Relative Brain Size in Birds: A Comparative Analysis', *Canadian Journal of Zoology*, 81: 1913–28.

—— K. M. Dean, and J. E. Nelson (2005). 'Interspecific Allometry of the Brain and Brain Regions in Parrots (*Psittaciformes*): Comparisons with Other Birds and Primates', *Brain, Behavior and Evolution*, 65: 40–59.

Iwasa, Y., A. Pomiankowski, and S. Nee (1991). 'The Evolution of Costly Mate Preferences II. The "Handicap" Principle', *Evolution*, 45: 1431–42.

Janik, V. M. and P. J. B. Slater (1997). 'Vocal Learning in Mammals', *Advances in the Study of Behavior*, 26: 59–99.

—— (2000). 'The Different Roles of Social Learning in Vocal Communication', *Animal Behaviour*, 60: 1–11.

Jarvis, E. D., (2004). 'Learned Birdsong and the Neurobiology of Human Language'. In: H. P. Ziegler and P. Marler (eds.), 'Behavioral Neurobiology of Birdsong', *Annals of the New York Academy of Sciences*, 1016: 749–77.

—— S. Ribeiro, M. L. da Silva et al. (2000). 'Behaviourally Driven Gene Expression Reveals Song Nuclei in Hummingbird Brain', *Nature*, 406: 628–32.

Jespersen, O. (1922). *Language, Its Nature, Development and Origin*. London: Allen and Unwin.

Johnson, G. (2003). 'Vocalizations in the Grey Butcherbird *Cracticus torquatus* with Emphasis on Structure in Male Breeding Song: Implications for Function and Evolution of Song from a Study of a Southern Hemisphere Species'. PhD dissertation, Griffith University, Australia.

Johnson, F. and O. Whitney (2005). 'Singing-Driven Gene Expression in the Developing Songbird Brain', *Physiology and Behavior*, 86: 390–8.

Joordens, J. C. A., F. P. Wesselingh, J. de Vos et al. (2009). 'Relevance of Aquatic Environments for Hominins: A Case Study from Trinil (Java, Indonesia)', *Journal of Human Evolution*, 57: 656–71.

Jungers, W. L., A. A. Pokempner, R. F. Kay et al. (2003). 'Hypoglossal Canal Size in Living Hominoids and the Evolution of Human Speech', *Human Biology*, 75: 473–84.

Jürgens, U. (1992). 'On the Neurobiology of Vocal Communication'. In: H. Papousek, U. Jürgens, and M. Papousek (eds.), *Nonverbal Vocal Communication: Comparative and Developmental Approaches*, pp. 31–42. Cambridge: Cambridge University Press.

Kalischer, O. (1905). 'Das Grosshirn der Papageien in anatomischer und physiologischer Beziehung', *Abhandlungen der Preussischen Akademie der Wissenschaften*, 4: 1–105.

Kay, R. F., M. Cartmill, and M. Balow (1998). 'The Hypoglossal Canal and the Origin of Human Vocal Behavior', *Proceedings of the National Academy of Sciences (USA)*, 95: 5417–19.

Kellogg, W. N. and L. A. Kellogg (1933). *The Ape and the Child.* New York: McGraw Hill.

Kennedy, G. E. (1985). 'Bone Thickness in *Homo erectus*', *Journal of Human Evolution*, 14: 699–708.

—— (1991). 'On the Autapomorphic Traits of *Homo erectus*', *Journal of Human Evolution*, 20: 375–412.

—— (2005). 'From the Ape's Dilemma to the Weanling's Dilemma: Early Weaning and its Evolutionary Context', *Journal of Human Evolution*, 48: 123–45.

King, A. P. and M. J. West (2002). 'The Ontogeny of Competence'. In: D. J. Lewkcowicz and R. Lickliter (eds.), *Conceptions of Development*, pp. 77–104. Philadelphia, PA: Psychology Press.

Kirby, S. (1998). 'Language Evolution without Natural Selection: From Vocabulary to Syntax in a Population of Learners'. Technical report, *Edinburgh Occasional Papers in Linguistics*, 98 (1), Department of Linguistics, University of Edinburgh.

—— (2000). 'Syntax without Natural Selection: How Compositionality Emerges from Vocabulary in a Population of Learners'. In: C. Knight, M. Studdert-Kennedy, and J. Hurford (eds.), *The Evolutionary Emergence of Language: Social Function and the Origins of Linguistic Form*, pp. 303–23. Cambridge: Cambridge University Press.

—— (2001). 'Spontaneous Evolution of Linguistic Structure: An Iterated Learning Model of the Emergence of Regularity and Irregularity', *IEEE Transactions on Evolutionary Computation*, 5: 102–10.

—— (2002). 'Learning, Bottlenecks and the Evolution of Recursive Syntax'. In: T. Briscoe (ed.), *Linguistic Evolution through Language Acquisition: Formal and Computational Models*, pp 173–204. Cambridge: Cambridge University Press.

Kleiman, D. G. (1977). 'Monogamy in Mammals', *Quarterly Review of Biology*, 52: 39–69.

Knörnschild, M., O. Behr, and O. von Helversen (2006). 'Babbling Behavior in the Sac-Winged Bat (*Saccopteryx bilineata*)', *Naturwissenschaften*, 93: 451–4.

Komers, P. E. and P. N. Brotherton (1997). 'Female Space Use Is the Best Predictor of Monogamy in Mammals', *Proceedings of the Royal Society of London B*, 264: 1261–70.

Konishi, M. (2004). 'The Role of Auditory Feedback in Birdsong'. In: H. P. Ziegler and P. Marler (eds.), 'The Behavioral Neurobiology of Birdsong', *Annals of the New York Academy of Sciences*, 1016: 463–75.

Koutsos, E. A., K. D. Matson, and K. C. Klasing (2001). 'Nutrition of Birds in the Order *Psittaciformes*: A Review', *Journal of Avian Medicine and Surgery*, 15: 257–75.

Kroodsma, D. E. (1978). 'Continuity and Versatility in Birdsong: Support for the Monotony-Threshold Hypothesis', *Nature*, 274: 681–3.

—— (1988). 'Song Types and their Use: Developmental Flexibility of the Male Blue-Winged Warbler', *Ethology*, 79: 235–47.

—— and H. Momose (1991). 'Songs of the Japanese Population of the Winter Wren (*Troglodytes troglodytes*)', *Condor*, 93: 424–32.

—— and L. D. Parker (1977). 'Vocal Virtuosity in the Brown Thrasher', *Auk*, 94: 783–5.

Kuhl, P. K. (1989). 'On Babies, Birds, Modules, and Mechanisms: A Comparative Approach to the Acquisition of Vocal Communication'. In: R. J. Dooling and S. H. Hulse (eds.), *The Comparative Psychology of Audition: Perceiving Complex Sounds*, pp. 379–419. Hillsdale, NJ: Erlbaum.

Kvarnemo, C. (2006). 'Evolution and Maintenance of Male Care: Is Increased Paternity a Neglected Benefit of Care?' *Behavioral Ecology*, 17: 144–8.

Lehmann, C. (1995). *Thoughts on Grammaticalization*. 2nd rev. edn. Munich: LINCOM Europa.

Leiner, H. C., A. L. Leiner, and R.S. Dow (1993). 'Cognitive and Language Functions of the Human Cerebellum', *Trends in Neurosciences*, 16: 444–7.

Leitner, S., J. Nicholson, B. Leisler et al. (2002). 'Song and the Song Control Pathway in the Brain Can Develop Independently of Exposure to Song in the Sedge Warbler', *Proceedings of the Royal Society of London B*, 269: 2519–24.

Lindstedt, S. L. and M. S. Boyce (1985). 'Seasonality, Fasting Endurance, and Body Size in Mammals', *American Naturalist*, 125: 873–8.

Lockard, J. S. (1980). *The Evolution of Human Social Behavior*. New York: Elsevier North-Holland.

MacArthur, R. H. and Levins, R. (1967). 'The Limiting Similarity, Convergence and Divergence of Coexisting Species', *American Naturalist*, 101: 377–85.

MacDonald, I. F., B. Kempster, L. Zanette et al. (2006). 'Early Nutritional Stress Impairs Development of a Song-Control Brain Region in both Male and Female Juvenile Song Sparrows (*Melospiza melodia*) at the Onset of Song Learning', *Proceedings of the Royal Society B*, 273: 2559–64.

MacLarnon, A. and G. Hewitt (1999). 'The Evolution of Human Speech: The Role of Enhanced Breathing Control', *American Journal of Physical Anthropology*, 109: 341–63.

Margoliash, D. (2003). 'Offline Learning and the Role of Autogenous Speech: New Suggestions from Birdsong Research', *Speech Communication*, 41: 165–78.

Marler, P. (1970). 'Bird Song and Speech Development: Could There Be Parallels?' *American Scientist*, 58: 669–73.

—— (1991). 'The Instinct to Learn'. In: S. Carey and R. Gelman (eds.), *The Epigenesis of Mind: Essays on Biology and Cognition*, pp. 37–66. Hillsdale, NJ: Lawrence Erlbaum Associates.

Marlowe, F. (2000). 'Paternal Investment and the Human Mating System', *Behavioural Processes*, 51: 45–61.

Mayfield, G. R. (1934). 'The Mockingbird's Imitation of Other Birds', *Migrant*, 5: 17–19.

Maynard Smith, J. and E. Szathmáry (1995). *The Major Transitions in Evolution*. Oxford: Oxford University Press.

McHenry, H. M. (1994). 'Tempo and Mode in Human Evolution', *Proceedings of the National Academy of Sciences, USA*, 91: 6780–6.

—— and K. Coffing (2000). '*Australopithecus* to *Homo*: Transformations in Body and Mind', *Annual Review of Anthropology*, 29: 125–46.

Merker, B. (2000). 'Synchronous Chorusing and Human Origins'. In: N. L. Wallin, B. Merker, and S. Brown (eds.), *The Origins of Music*, pp. 315–28. Cambridge, MA: MIT Press.

—— (2004). 'Cortex, Countercurrent Context, and Dimensional Integration of Lifetime Memory', *Cortex*, 40: 559–76.

—— (2005). 'The Conformal Motive in Birdsong, Music and Language: An Introduction'. In: G. Avanzini, L. Lopez, S. Koelsch et al. (eds.), *The Neurosciences and Music II: From Perception to Performance. Annals of the New York Academy of Sciences*, 1060: 17–28.

—— (2008). 'The Ritual Foundations of Human Uniqueness'. In: S. Malloch and C. Trevarthen (eds.), *Communicative Musicality*, pp. 45–59. Oxford: Oxford University Press.

—— and K. Okanoya (2007). 'The Natural History of Human Language: Bridging the Gaps without Magic'. In: C. Lyon, L. Nehaniv, and A. Cangelosi (eds.), *Emergence of Communication and Language*, pp. 403–20. London: Springer-Verlag.

Middleton, F. A. and P. L. Strick (1994). 'Anatomical Evidence for Cerebellar and Basal Ganglia Involvement in Higher Cognitive Function', *Science*, 266: 458–61.

Mitani, J. C. and K. Brandt (1994). 'Social Factors Influence the Acoustic Variability in the Long-Distance Calls of Male Chimpanzees', *Ethology*, 96: 233–52.

—— and J. Gros-Louis (1998). 'Chorusing and Call Convergence in Chimpanzees: Tests of Three Hypotheses', *Behaviour*, 135: 1041–64.

Møller, A. P. (1987). 'Copulation Behaviour in the Goshawk, *Accipiter gentilis*', *Animal Behaviour*, 35: 755–63.

—— (2003). 'The Evolution of Monogamy: Mating Relationships, Parental Care, Sexual Selection'. In: U. H. Reichard and C. Boesch (eds.),

Monogamy: Mating Strategies and Partnerships in Birds, Humans and Other Mammals, pp. 27–40. Cambridge: Cambridge University Press.

—— and M. D. Jennions (2001). 'How Important Are Direct Fitness Benefits of Sexual Selection?' *Naturwissenschaften*, 88: 401–15.

—— and J. J. Cuervo (2000). 'The Evolution of Paternity and Paternal Care in Birds', *Behavioral Ecology*, 11: 472–85.

Nottebohm, F. (1975). 'A Zoologist's View of Some Language Phenomena, with Particular Emphasis on Vocal Learning'. In: E. H. Lenneberg and E. Lenneberg (eds.), *Foundations of Language Development*, pp. 61–103. New York: Academic Press.

—— (1976). 'Discussion Paper. Vocal Tract and Brain: A Search for Evolutionary Bottlenecks'. In: S. R. Harnad, H. D. Steklis, and J. Lancaster (eds.), 'Origins and Evolution of Language and Speech', *Annals of the New York Academy of Sciences*, 280: 643–9.

Nowicki, S., W. A. Searcy, and S. Peters (2002a). 'Brain Development, Song Learning and Mate Choice in Birds: A Review and Experimental Test of the "Nutritional Stress Hypothesis"', *Journal of Comparative Physiology A: Sensory, Neural, and Behavioral Physiology*, 188: 1003–14.

—————— (2002b). 'Quality of Song Learning Affects Female Response to Male Bird Song', *Proceedings of the Royal Society of London B*, 269: 1949–54.

———— (2004). 'Song Function and the Evolution of Female Preferences: Why Birds Sing, Why Brains Matter'. In: H. P. Ziegler and P. Marler (eds.), 'The Behavioral Neurobiology of Birdsong', *Annals of the New York Academy of Sciences*, 1016: 704–23.

O'Connell, J. F., K. Hawkes, and N. Blurton Jones (1988). 'Hadza Scavenging: Implications for Plio-Pleistocene Hominid Subsistence', *Current Anthropology*, 29: 356–63.

Okanoya, K. (2002). 'Sexual Display as a Syntactic Vehicle: The Evolution of Syntax in Birdsong and Human Language through Sexual Selection'. In: A. Wray (ed.), *The Transition to Language*, pp. 44–64. Oxford: Oxford University Press.

—— (2004). 'The Bengalese Finch: A Window on the Behavioral Neurobiology of Birdsong Syntax', *Annals of the New York Academy of Sciences*, 1016: 724–35.

—— and B. Merker (2007). 'Neural Substrates for String-Context Mutual Segmentation: A Path to Human Language'. In: C. Lyon, L. Nehaniv, and A. Cangelosi (eds.), *Emergence of Communication and Language*, pp. 421–34. London: Springer-Verlag.

Ölveczky, B. P., A. S. Andalman, and M. S. Fee (2005). 'Vocal Experimentation in the Juvenile Songbird Requires a Basal Ganglia Circuit', *PLoS Biology*, 3: e153.

Özdemir, E., A. Norton, and G. Schlaug (2006). 'Shared and Distinct Neural Correlates of Singing and Speaking', *NeuroImage*, 33: 628–35.

Parker, G. A. (2006). 'Sexual Conflict over Mating and Fertilization: An Overview', *Philosophical Transactions of the Royal Society of London B*, 361: 235–59.

Payne, K. (2000). 'The Progressively Changing Songs of Humpback Whales: A Window on the Creative Process in a Wild Animal'. In: N. L. Wallin, B. Merker, and S. Brown (eds.), *The Origins of Music,* pp. 135–50. Cambridge, MA: MIT Press.

Pepper, J. W. (1996). 'The Behavioral Ecology of the Glossy Black-Cockatoo, *Calyptorhynchus lathami halmaturinus*'. PhD thesis, University of Michigan.

Pepperberg, I. M. (1994). 'Vocal Learning in Grey Parrots (*Psittacus erithacus*): Effects of Social Interaction, Reference, and Context', *Auk*, 111: 300–13.

Perry, G. H., B. C. Verrelli, and A. C. Stome (2005). 'Comparative Analyses Reveal a Complex History of Molecular Evolution for Human MYH16', *Molecular Biology and Evolution*, 22: 379–82.

Pinker, S. and P. Bloom (1990). 'Natural Language and Natural Selection', *Behavioral and Brain Sciences*, 13: 707–84.

Powys, V. (1995). 'Regional Variation in the Territorial Songs of Superb Lyrebirds in the Central Tablelands of New South Wales', *Emu*, 95: 280–9.

Rabenold, P. P. (1986). 'Family Associations in Communally Roosting Black Vultures', *Auk*, 103: 32–41.

Ricklefs, R. E. (2004). 'The Cognitive Face of Avian Life Histories'. The 2003 Margaret Morse Nice Lecture. *Wilson Bulletin*, 116: 119–33.

Robinson, F. N. and H. S. Curtis (1996). 'The Vocal Displays of the Lyrebirds (*Menuridae*)', *Emu*, 96: 258–75.

Rogoff, B., R. Paradise, R. M. Arauz et al. (2003). 'Firsthand Learning through Intent Participation', *Annual Review of Psychology*, 54: 175–203.

Ruff, C. B. (2002). 'Variation in Human Body Size and Shape', *Annual Review of Anthropology*, 31: 211–32.

—— E. Trinkaus, and T. W. Holliday (1997). 'Body Mass and Encephalization in Pleistocene *Homo*', *Nature*, 387: 173–6.

Rutberg, A. T. (1983). 'The Evolution of Monogamy in Primates', *Journal of Theoretical Biology*, 104: 93–112.

Saito, N. and M. Maekawa (1993). 'Birdsong: The Interface with Human Language', *Brain and Development*, 15: 31–40.

Schagatay, E., M. van Kampen, S. Emanuelsson et al. (2000). 'Effects of Physical and Apnea Training on Apneic Time and Diving Response in Humans', *European Journal of Applied Physiology*, 82: 161–9.

Searcy, W. A., S. Nowicki, and M. Hughes (1997). 'The Response of Male and Female Song Sparrows to Geographic Variation in Song', *Condor*, 99: 651–7.

Seyfarth, R. M. and D. L. Cheney (1980). 'The Ontogeny of Vervet Monkey Alarm-Calling Behavior: A Preliminary Report', *Zeitschrift für Tierpsychologie*, 54: 37–56.

Slater, P. J. B. (1983). 'Bird Song Learning: Theme and Variations'. In: G. H. Bush and G. A. Clark (eds.), *Perspectives in Ornithology*, pp. 475–511. Cambridge: Cambridge University Press.

Smith, K. (2003). 'Learning Biases and Language Evolution'. In: S. Kirby (ed.), *Language Evolution and Computation*, pp. 22–31. Proceedings of the Workshop on Language Evolution and Computation, 15th European Summer School on Logic, Language and Information, Vienna.

Snowdon, C. T. and A. M. Elowson (1999). 'Pygmy Marmosets Modify Call Structure when Paired', *Ethology*, 105: 893–908.

Soma, M., T. Hasegawa, and K. Okanoya (2005). 'The Evolution of Song Learning: A Review from a Biological Perspective', *Cognitive Studies*, 12: 166–76.

Spencer, K. A., K. L. Buchanan, A. R. Goldsmith et al. (2003). 'Song as an Honest Signal of Developmental Stress in the Zebra Finch (*Taeniopygia guttata*)', *Hormones and Behavior*, 44: 132–9.

—— —— S. Leitner et al. (2005). 'Parasites Affect Song Complexity and Neural Development in a Songbird', *Proceedings in Biological Science*, 272: 2037–43.

Staal, F. (1989). *Rules without Meaning: Ritual, Mantras and the Human Sciences*. New York: Peter Lang.

Starck, J. M. and R. E. Ricklefs (1998). 'Patterns of Development: The Altricial-Precocial Spectrum'. In: J. M. Starck and R. E. Ricklefs (eds.), *Avian Growth and Development: Evolution within the Altricial Precocial Spectrum*, pp. 3–30. New York: Oxford University Press.

Stedman, H. H., B. W. Kozyak, A. Nelson et al. (2004). 'Myosin Gene Mutation Correlates with Anatomical Changes in the Human Lineage', *Nature*, 428: 415–18.

Stewart, K. M. (1994). 'Early Hominid Utilisation of Fish Resources and Implications for Seasonality and Behaviour', *Journal of Human Evolution*, 27: 229–45.

Stiner, M. C. (2002). 'Carnivory, Coevolution and the Geographical Spread of the Genus Homo', *Journal of Archaeological Research*, 10: 1–63.

Sugihara, H. (1998). 'Matching of Acoustic Features during the Vocal Exchange of Coo Calls by Japanese Macaques', *Animal Behaviour*, 55: 673–87.

Sundberg, J. (1987). *The Science of the Singing Voice*. Dekalb, IL: Northern Illinois University Press.

Swanson, L. W. (1995). 'Mapping the Human Brain: Past, Present, and Future', *Trends in Neurosciences*, 18: 471–4.

Szekely, T., C. K. Catchpole, A. Devoogd et al. (1996). 'Evolutionary Changes in a Song Control Area of the Brain (HVC) Are Associated with Evolutionary Changes in Song Repertoire among European Warblers (*Sylviidae*)', *Proceedings of the Royal Society of London B*, 263: 607–10.

Tchernichovski, O., T. J. Lints, S. Derégnaucourt et al. (2004). 'Studying the Song Development Process: Rationale and Methods'. In: H. P. Ziegler and

P. Marler (eds.), 'The Behavioral Neurobiology of Birdsong', *Annals of the New York Academy of Sciences*, 1016: 348–63.

Theunissen, F. E., N. Amin, S. S. Shaevitz et al. (2004). 'Song Selectivity in the Song System and in the Auditory Forebrain'. In: H. P. Ziegler and P. Marler (eds.), 'The Behavioral Neurobiology of Birdsong', *Annals of the New York Academy of Sciences*, 1016: 222–45.

Thorpe, W. H. (1961). *Bird Song*. Cambridge: Cambridge University Press.

Titze, I. R. (2003). 'The Voice as a Musical Instrument: Fundamental Differences between Man-Made and Biological Designs'. In: R. Bresin (ed.), *Proceedings of the Stockholm Music Acoustics Conference*, pp. 21–4. Stockholm: SMAC 03.

Trainer, J. M. (1987). 'Behavioral Associations of Song Types during Aggressive Interactions among Male Yellow-Rumped Caciques', *The Condor*, 89: 731–8.

Ungar, P. S., F. E. Grine, and M. F. Teaford (2006). 'Diet in Early *Homo*: A Review of the Evidence and a New Model of Adaptive Versatility', *Annual Review of Anthropology*, 35: 209–8.

Wachtmeister, C.-A. (2001). 'Display in Monogamous Pairs: A Review of Empirical Data and Evolutionary Explanations', *Animal Behaviour*, 61: 861–8.

—— and M. Enquist (2000). 'The Evolution of Courtship Rituals in Monogamous Species', *Behavioral Ecology*, 11: 405–10.

Walker, A. and P. Shipman (1996). *The Wisdom of Bones: In Search of Human Origins*. London: Weidenfeld & Nicolson.

Waxman, D. and J. R. Peck (1998). 'Pleiotropy and the Preservation of Perfection', *Science*, 279: 1210–13.

Whiten, A., J. Goodall, W. McGrew et al. (1999). 'Cultures in Chimpanzees', *Nature*, 399: 682–5.

Wilbrecht, L. and F. Nottebohm (2003). 'Vocal Learning in Birds and Humans', *Mental Retardation and Developmental Disabilities Research Reviews*, 9: 135–48.

Wild, J. M. (1994). 'The Auditory-Vocal-Respiratory Axis in Birds', *Brain Behavior and Evolution*, 44: 192–209.

Williams, H. (2004). 'Birdsong and Singing Behavior'. In: H. P. Ziegler and P. Marler (eds.), 'The Behavioral Neurobiology of Birdsong', *Annals of the New York Academy of Sciences*, 1016: 1–30.

Wood, B. and D. Strait (2004). 'Patterns of Resource Use in Early *Homo* and *Paranthropus*', *Journal of Human Evolution*, 46: 119–62.

Wray, A. (1998). 'Protolanguage as a Holistic System for Social Interaction', *Language and Communication*, 18: 47–67.

Zahavi, A. (1975). 'Mate Selection—A Selection for a Handicap', *Journal of Theoretical Biology*, 53: 205–14.

Zahavi, A. and A. Zahavi (1997). *The Handicap Principle: A Missing Piece of Darwin's Puzzle*. Oxford: Oxford University Press.

Ziegler, H. P. and P. Marler (eds.) (2004). 'Behavioral Neurobiology of Birdsong', *Annals of the New York Academy of Sciences*, 1016.

Zuidema, W. (2003). 'How the Poverty of the Stimulus Solves the Poverty of the Stimulus'. In: S. Becker, S. Thrun, and K. Obermayer (eds.), *Advances in Neural Information Processing Systems 15 (Proceedings of NIPS 2002)*, pp. 51–8. Cambridge, MA: MIT Press.

Part V

Perspectives from musicology

10

Music as an Emergent Exaptation

Ian Cross

INTRODUCTION

On the surface, music and language share many features. Both appear to consist of patterned sequences of sounds in time; both present complex sequential structures which may involve some degree of recursion or hierarchy in their structure; and both can be conceived of as media in which information is transmitted from a producer and received and decoded by a listener.

From Darwin onwards, theorists have interpreted this apparently close relationship in different ways. For Darwin, music preceded and may have given rise to language; he hypothesized that music and, later, language both arose through processes of sexual selection, suggesting that (1871, p. 880) 'it appears probable that the progenitors of man, either the males or females or both sexes, before acquiring the power of expressing their mutual love in articulate language, endeavoured to charm each other with musical notes and rhythm.' Myers (1913) subsequently proposed that '[w]e may be disposed to conclude that the beginnings of music have been derived from speech. It would be safer, however, to conclude that both have been evolved from a mechanism designed for the vocal "expression of meaning".'

For both Darwin and Myers, language and music were part of the human communicative toolkit, and human powers of communication, exemplified in the flexibility and efficacy of language, have been taken to be perhaps the most significant factor that differentiates us from the rest of the animal world. While other species are self-evidently able to communicate with other members of the same species by sound, sight, touch, or smell, only humans have the capacity to communicate to each other information about complex states of affairs in the material and social worlds. Despite the best efforts of researchers over the last thirty-odd years, neither chimpanzees nor bonobos, our closest evolutionary relatives, seem capable of employing communicative media to express their perceptions, desires, intentions, or attitudes with the level of syntactic and semantic sophistication that seems to be naturally acquired by an average five-year-old human child (see Conway and Christiansen 2001).

Communication through language, enabling complex and useful information to be represented and exchanged, seems to be the principal guarantor of our intellectual powers and of our cultural complexities, features that are likely to have had considerable potency in enabling humans to have emerged as a highly successful species over the 200,000 years or so in which modern humans have been in existence. In other words, it is highly probable that language, in enabling and sustaining our cognitive and social proficiency, was an adaptive factor in human evolution (see Pinker 1994); it enabled humans, individually and in groups, to communicate useful and accurate information to each other and hence to survive and reproduce in situations where other species could not.

Like language, music can also be conceived of as a communicative medium; indeed, both seem to fit equally well within a widely used theoretical model of communication, the 'information theory' model of Shannon and Weaver (1949). In this model a sender (a speaker or performer) makes use of a channel (the air) to encode and to send information to a receiver (an interlocutor or listener) who receives and decodes it. The model seems to provide a basis for considering music and language to be analogous systems of communication, setting aside for the moment the nature of the information that is communicated.

But while this model seems to fit most instances of language use quite well (and perhaps fits all instances of animal communication: see Seyfarth and Cheney 2003), it is not clear that it is so generally applicable to music. Although in, perhaps, most musical situations in contemporary Western global culture it is easy to categorize participants as sender or receiver, performer or listener, if we look to non-Western musical practices, many seem to have as their *raison d'être* not the transmission of musical information from active performer to passive listener, but collective engagement in the synchronous production and perception of complex patterns of sounds and movement (see Arom

1991; Blacking 1976; Espi-Sanchis and Bannan, this volume). Music in these guises does not seem as easily assimilable as is language into the model of communication provided by information theory.

MUSIC

So perhaps music is not a communicative medium; or it may be that it simply mimics aspects of the communicative medium of language, being an opportunistic and non-adaptive exploitation for solely pleasurable purposes of capacities to communicate that have arisen in humans through evolutionary processes for the serious business of survival, as Pinker (1997) suggests. However, Pinker's proposal that music constitutes nothing more than 'auditory cheesecake' is predicated on the notion that music is nothing more nor less than sound organized into patterns in time, albeit pretty patterns. This view of music is simply wrong, ignoring the multiplicity of ways in which human musicality may be expressed, and the superabundance of significances that may be accorded to music across times and within cultures.

For most musicologists, and for almost all ethnomusicologists, music is not just sonic pattern. Music involves action and interaction. Indeed the structure of its sonic patterns may be as much determined by the actions that produce them as by any abstract considerations of sonic design, as both Baily (1985) and Nelson (2002) have shown is the case for, respectively, the music of the Afghan *dutar* and the improvisation of blues guitar solos. In music, sound and action may be intrinsically interlinked, and action in music is, more often than not, a form of interaction that is typically expressed in terms of entrainment to some common temporal framework or pulse. Indeed, in many non-Western contexts, active and collective entrainment with the sonic structure of the musical sound seems to be as much a part of the music as is the structure of the sound itself, perhaps evidenced in the lack of distinction made in certain societies between what in the West might be separately categorized as music and as dance (Gourlay 1984).

And finally, the issue of what—if anything—music is communicating must be addressed. Whereas in language it is usually possible to specify the subject of an utterance with some precision, this is almost never the case for music. Music appears to be a strangely malleable and flexible phenomenon of which the meaning or significance can rarely be pinned down unambiguously. As Langer (1942, p. 195) puts it, in music '[t]he actual function of meaning, which calls for permanent contents, is not fulfilled; for the *assignment* of one rather than another possible meaning to each form is never explicitly made'. Devereux and LaBarre (1961, p. 369) propose that '[i]n addition to viewing art as a harmless safety valve, society and the artist alike consider the artistic utterance as *unrepudiable* in form but *repudiable* as to content'. In effect, one and the same piece of music can bear quite different meanings for performer and listener, or for two different listeners; it might even bear multiple disparate meanings for a single listener or participant at a particular time. Music has a sort of 'floating intentionality' (the word 'intentionality' here simply means 'aboutness'); it can be thought of as gathering meaning from the contexts within which it happens; and in turn contributing meaning to those contexts.

Music seems to be inherently ambiguous, yet it is intuitively plausible that music is a communicative medium. How, and what, then, is it communicating? In order to explore this, it is necessary to consider whether or not the model of communication provided by information theory is in fact adequate to account for all aspects of human communication. This model postulates that information is sent to a receiver who decodes the information encoded in the message received. When applied to human linguistic communication, this requires a very close match between the capacities of the sender and of the receiver. The capacities of the sender and receiver must be similar enough for the receiver to be capable of all the processes that are necessary to decode the message accurately. In effect, this model presupposes that sender and receiver both possess a body of shared knowledge that will render the information that is communicated unambiguously intelligible. This seems to be a reasonable proposition in respect of language; one might expect that a speaker and a hearer belonging to the same language community might be bound by the same set of rules and conventions that they can both bring to bear so that the speaker's message is intelligibly decoded by the hearer.

MUSIC AND MEANING

However, this model can be questioned, at the very least on the grounds that, while a pool of mutual knowledge seems plausible, it can never be guaranteed; hence, neither can the intelligibility of a linguistic message. Without the certainty of mutual knowledge, there will always be a degree of ambiguity in linguistic communication. The exchange of a verbal message requires that inferences are made by the hearer as to the speaker's intended referent, and that inferences are made by the speaker in respect of the hearer's capacity to identify the speaker's intended referent. Sperber and Wilson (1986) have developed a theory of *ostensive-inferential communication* in which a speaker is characterized as not simply broadcasting a message to a listener, but as producing a stimulus that is intended to change the cognitive environment of a listener by seeking to achieve some degree of resemblance between a speaker's and hearer's thoughts (Sperber 1996). This theory claims that human cognition is geared to the maximization of *relevance*, defined by Sperber (ibid.) as a human 'tendency to optimize the effect-effort ratio' in any particular situation and hence minimize the cognitive load involved in processing the information that an individual encounters. For Sperber and Wilson, a presumption of relevance by both speaker and hearer is what frames and guides processes of inference in human linguistic communication and reduces the ambiguity inherent in a communicative act.

Accepting that a degree of ambiguity seems to be inherent in all (even linguistic) acts of human communication, then music's apparent ambiguity does not debar it from being considered to be a communicative medium. However, language has referents—it is *about* something—and hence language can be said to communicate information about states of affairs. Indeed, aboutness or *intentionality* can be considered intrinsic to any act of communication. But what is it, if anything, that music can be said to be *about*? What information is music communicating? Language allows the articulation of complex propositions that can be interpreted as referring exclusively to particular states of affairs. While music appears to possess a similar potential to that of language for the articulation of complex syntactic structures (including hierarchical structures), its reference can never be so direct or unequivocally interpretable. Music can be interpreted as referring both to itself and beyond itself, and it is only in its reference to itself that ambiguity may be minimized or even resolved.

Music unfolds in time. As it does so, it presents patterns that may give rise to expectations for listeners or participants as to how it will continue. In the ongoing flow of the music, those expectations may be realized or rescinded. As it flows in time, music thus generates, for participants, reference to future possibilities of unfolding; when those future possibilities become actualities, the significance of those earlier musical events may become clear, their reference (at least partially) disambiguated, giving rise to what Leonard Meyer (1956) has called music's 'evident meanings'.

Those patterns of evident meaning, together with the music's sonic and gestural qualities as it unfolds, may also yield a degree of reference—this time beyond the music itself—and to some extent sense, in the elicitation of emotion or the evocation of specific conceptual-intentional complexes with which aspects of the music have become associated through individual experience or cultural convention (or possibly because of bio-social predispositions). But while those conceptual-intentional complexes may themselves be complex, they are neither propositional in form nor are they likely to be decomposable; they are also more likely than not to vary from participant to participant, taking the form of what Leonard Meyer referred to as connotative complexes. Their sense and reference are not bound to a specific situation or set of circumstances but rather to a range of situations, very much in the way that a particular emotional or affective mind-brain-body state may be relevant to a range of circumstances for any one individual (Oatley and Johnson-Laird 1998). Hence while music's reference may (retrospectively) be disambiguated, being largely rooted in the temporal structures of the music itself, its sense cannot.

Given this intrinsic ambiguity, when adopting an evolutionary view it seems that Pinker may have been right; the human faculty for music may indeed be parasitic on the human faculty for language, simply exploiting capacities that have arisen for evolutionarily adaptive ends. However, this inherent ambiguity together with the quality of the actions and interactions that appear to be integral to music suffices to differentiate music from language. Music's attributes of *embodying*, *entraining*, and *transposably intentionalizing* time in sound and action (see Cross 2003a) enable it to be efficacious in contexts where language, precisely because of its capacity to be interpreted unambiguously, may be unproductive or impotent; and it can therefore be suggested that the emergence of musicality is likely to have been crucially adaptive in processes of human evolution.

Music can be efficacious for groups, for individuals within groups, and for individuals. If one imagines a group of people involved in a collective musical behaviour, their individual behaviours are likely to be coordinated within a temporal framework and thus stand in more-or-less predictable relationships in respect of each other. This endows the collective activity with a high degree of coherence which is more than likely to help establish a strong sense of group identity in this directed and synchronized modulation of action, attention, and affect (see McNeill 1995). The reinforcement of group identity can function in collective musical behaviours not only because of the music's capacity to entrain but also because music allows each participant to interpret its significances individually and independently without the integrity of the collective musical behaviour being undermined. Music's inexplicitness, its ambiguity or floating intentionality, may thus be regarded as a highly advantageous characteristic of its function for groups; music, then, might serve as a medium for the maintenance of human social flexibility.

In addition to this efficacy at the level of the group, music may be efficacious for individuals within groups which are engaged in collective musical behaviours. This is perhaps most evident if we consider a group of children interacting musically. Here it may be that social flexibility is not just being maintained but formed. Music's powers of entrainment, together with its ambiguity, may allow each participating child to explore forms of interaction with others while minimizing the risk that such exploration might give rise to conflict, effectively underlying the gestation of a social flexibility (see Cross 2003a; Bannan and Woodward 2009). One only has to envisage a group of children interacting verbally and unambiguously rather than musically to see (and hear) how quickly conflict is likely to emerge in linguistic rather than musical interaction!

For individuals, any efficacy for music beyond the purely hedonic seems harder to articulate, but a clue might be found in Meyer's notion of 'connotative complexes', where he implies that music does not so much embody metaphors as constitute a metaphorizing medium, one through which seemingly disparate concepts may be experienced as interlinked. While it is more than feasible that music fulfils this role for mature members of a culture, it seems more viable to suggest that music's efficacy at the level of the individual may be greatest and most universal in infancy and in childhood.

Over the last twenty years it has become evident that infants and children, though hugely flexible in that their neural systems are immensely plastic, are not general-purpose learning machines (see, for example, Spelke 1999). Rather, they seem to be predisposed to pick up certain types of information and to deal with them in particular and distinct ways. These capacities emerge too rapidly to be explained on the basis of the operation of a general-purpose learning mechanism. Moreover, they seem to be specific to the particular domains within which they are displayed. An ability to deal with information in one domain, for example, the social world, is unlikely to be transferable to another domain, say, the physical; a child may show a highly developed capacity to reason about the social world which may not be manifested in their reasoning about physical objects (Cummins 1998). Infants seem to be predisposed to pick up and deal with information in these distinct domains rapidly and effectively.

Yet the hallmark of the human species is a generalized ability to deal with information that is not specific to any particular domain. It is in the emergence of this domain-general intellectual flexibility that music is likely to play a role. Music's floating intentionality, its potential for its meaning or aboutness to be transposed from one situation to another, allows that one and the same musical act might be co-opted by an infant or child in dealing with information in two quite different domains. This could help in the emergence of the capacity to relate or to integrate information across domains, and assist in the emergence of a domain-general competence. In effect, early musical, or rather, proto-musical, behaviours may be functional in individual development in giving rise to a *metaphorizing* capacity. The attributes of music that may facilitate this transposition of its significances and hence allow the *redescription* of information across domains (see Karmiloff-Smith 1992) may lie in its capacity to mirror forms of emotional and cognitive dynamics, or it may be that positive emotions evoked by engagement with music afford the conditions for rich integration of representations and enhanced exploratory behaviour (see Damasio 1995).

Indeed, individual musical behaviours in childhood have been characterized as fundamentally exploratory, and children seem to be predisposed to engage in music-like activities from birth. Even very young infants can engage in music-like or *proto-musical* behaviours—which involve not only sound perception and production but also movement (see Papousek 1996)—and they are highly motivated to do so. While early proto-musical and proto-linguistic behaviours appear to be indissociable, linguistic behaviours become increasingly bound by considerations of relevance (after Sperber and Wilson 1986) so as to constrain the extent to which they can substitute one for another in the linguistic contexts in which they are deployed. However, proto-musical and musical behaviours are likely to retain a degree of 'floating intentionality', continuing to be appropriate in a wide range of dissimilar situations and types of information, their individual and social functionality being closely tied to their effective ambiguity.

MUSIC IN EVOLUTION

Overall, it appears that music plays crucial roles for humans in individual and social development, and that a predisposition to engage in music-like activities seems to be part of our biological heritage. That biological heritage is, by and large, a consequence of the operation of evolutionary processes, and it can be suggested that music may have played a significant role in human evolution.

The intellectual and social flexibilities that mark out modern humans seem to have emerged in the hominin lineage which arose within the last seven million years. The hominin line can be traced through the australopithecines (small-brained bipeds) through *Homo habilis* (appearing two and a half million years ago, with a somewhat increased brain capacity and first evidence of stone tools). Subsequently *Homo ergaster* appears about two million years ago, with a considerably increased brain capacity, and a more robust physique with a barrel-shaped rib-cage, a leap in the sophistication of stone tools, and dispersal beyond Africa into Eurasia. Between seven and five hundred thousand years ago, *Homo heidelbergensis* appears, the predecessor of both the *Neanderthals*, who arise some three hundred and fifty to two hundred and fifty thousand years ago, and ourselves, modern *Homo sapiens*, who emerged as an African species some two hundred thousand years ago (see White et al. 2003).

Successive hominin species had ever larger brains, and left traces of progressively more sophisticated life-ways. However, while each successive species developed increasingly complex skills in dealing with the problems of survival, those skills were restricted to relatively narrow domains; while the cognitive capacities of each successive species were increasingly highly developed they were also somewhat inflexible. Only with the advent of *Homo sapiens sapiens*, modern humans, do we find unambiguous evidence for a capacity to transfer skills flexibly from one domain to another—a generalized cross-domain intellectual capacity—together with the ability to make use of symbols, the earliest evidence for symbolic behaviour consisting of engraved pieces of ochre found in South Africa and dated to 77,000 BP (Henshilwood et al. 2002). This intellectual and social flexibility (or, as Henshilwood and Marean (2003) put it, a capacity for 'fully symbolic *sapiens* behaviour') marks us off from our hominid predecessors. And it can be suggested that music (and I include dance in this) is likely to have played a significant role in the emergence of these capacities, by acting so as to enhance cognitive flexibility, and so as to acquire and maintain competence in managing social relationships.

And it is, after all, only with modern humans that we find unambiguous evidence for musical behaviours in the form of musical instruments in the ancient archaeological record; the earliest yet found being a bone pipe from Geissenklösterle in southern Germany (from 36,000 BP), and a large assemblage of musical bone pipes having been found in Isturitz in southern France covering a time span of some fifteen to twenty thousand years, first appearing at around 30,000 BP (see d'Errico et al. 2003). It is notable that the date of the earliest is around the time of the earliest appearance of modern humans in Europe; effectively, almost as soon as modern humans reach Europe they are leaving traces of sophisticated musical behaviours, which strongly suggests that humans brought music with them out of Africa; and to me, and to many others, strongly suggests that musicality constitutes a specific and unique attribute of modern humans.

This is not to suggest that music arose *ab initio*, full-blown, with the emergence of modern humans. It is more likely that components of musicality were possessed to some degree by our predecessor species, but that only with modern humans did an integrated capacity for music appear. It is likely that human musicality is built from a number of disparate capacities that arose in response to a variety of evolutionarily selective pressures at different times and over different timescales in the hominid lineage, some of which may be tentatively identified. The evolutionarily adaptive value of social flexibility might have underpinned the probable use by *Homo ergaster* of complex vocal signals (control of which would have been enabled by the barrel-shaped chest) to communicate affect or emotional state in order to regulate social interactions; the same factor may well have led, with *Homo heidelbergensis*, to the modern human vocal tract, which would allow the articulation of the full range of vocal sounds (including musical sounds) of which modern humans are capable (see Morley 2002 and this volume).

The selection pressures of sociality may also have impacted on the rate of individual maturation within the hominids; each successive species appears to be progressively more *altricial* than its predecessors, consecutive species spending a progressively longer proportion of their total lifespan in progessively differentiated juvenile states (see Bogin 1999). Joffe (1997) has shown that primate species which have complex social organizations are more likely to be altricial, proposing that complex social organization requires an extension of the learning period in order to enhance the flexibility with which members of a species manage their social interactions. A significant feature of the behaviour of juvenile animals in that learning period—particularly in predatory or social species—is play, which can be thought of as consisting of action and interaction that '*appears* to be purposeless' (Bekoff 1998) in a world largely constructed by the participants. It usually involves behaviours that are likely to occur in functional contexts being employed in modified forms, and (when interactive) can be said to require the 'negotiation of cooperative agreement' (ibid.). It may take stereotyped forms or require stereotyped eliciting behaviours (the 'play bow' in canids, the 'play-face' in many primates), and can be thought of as enabling juveniles to learn to deal with their environment (in terms of manipulating features of it), and to acquire the skills necessary to deal with conspecifics (in terms of rehearsing and exercising the skills necessary for the management of social interaction), as well as being self-stimulating—'fun'—in its own right.

It can be suggested that, in an increasingly altricial lineage, the need to accommodate to population structures with an increasing proportion of members with access to juvenile modes of cognition and behaviour (other things being equal) may have favoured the emergence of something like musicality as a means of assimilating the exploratory value of those juvenile modes of cognition into the adult behavioural repertoire (see Cross 2003b). Given that play is particularly a feature of the behaviour of juveniles in social mammals, and given that it is likely to have positive survival value for members of those species who engage in it, it is probable that group behaviours that enable yet regulate it so as to co-opt its utility into the adult repertoire are likely to have some adaptive—or exaptive—value (see Belyaev 1979). And it can be hypothesized that the absolute duration of, as well as the proportion of total species lifespan represented by, the period in which juvenile modes of thought and behaviour are available will correlate with the requirement for such enabling and regulating mechanisms. Music can be interpreted as one such mechanism, emerging as a consequence of the progressive extension and stage-differentiation of the juvenile period in the later hominin lineage.

To conclude: music takes so many forms and fulfils such diverse functions in the present day that any attempt to understand it in evolutionary terms must only be partial; music's significance in the contemporary world cannot be 'reduced' to its putative role in human evolution. Nevertheless, music's powers in the present are likely to be underwritten by the features that appear to have rendered music functional in evolution: its potential to keep people together in time, and to clear a social and mental space for the unhindered exploration of the capacity to mean.

REFERENCES

Arom, S. (1991). *African Polyphony and Polyrhythm.* Cambridge: Cambridge University Press.

Baily, J. (1985). 'Music Structure and Human Movement'. In: P. Howell, I. Cross, and R. West (eds.), *Musical Structure and Cognition* (pp. 237–58). London: Academic Press.

Bannan, N. and S. Woodward (2009). 'Spontaneity in the Musicality and Music Learning of Children'. In: C. Trevarthen and S. Malloch (eds.), *Communicative Musicality,* pp. 465–94. Oxford: Oxford University Press.

Bekoff, M. (1998). 'Playing with Play: What Can We Learn about Cognition, Negotiation and Evolution? In: D. D. Cummins and C. Allen (eds.), *The Evolution of Mind,* pp. 162–82. Oxford: Oxford University Press.

Belyaev, D. K. (1979). 'Destabilizing Selection as a Factor in Domestication', *Journal of Heredity,* 70: 301–8.

Blacking, J. (1976). *How Musical is Man?* London: Faber.

Bogin, B. (1999). *Patterns of Human Growth,* 2nd edn. Cambridge: Cambridge University Press.

Conway, C. M. and M. H. Christiansen (2001). 'Sequential Learning in Non-Human Primates', *Trends in Cognitive Science,* 5: 539–46.

Cross, I. (2003a). 'Music and Biocultural Evolution'. In: M. Clayton, T. Herbert, and R. Middleton (eds.), *The Cultural Study of Music: A Critical Introduction,* pp. 19–30. London: Routledge.

—— (2003b). 'Music and Evolution: Causes and Consequences', *Contemporary Music Review*, 22, 3: 79–89.

Cummins, D. D. (1998). 'Social Norms and Other Minds: The Evolutionary Roots of Higher Cognition'. In: D. D. Cummins and C. Allen (eds.), *The Evolution of Mind*, pp. 30–50. Oxford: Oxford University Press.

Darwin, C. (1871). *The Descent of Man and Selection in Relation to Sex*. London: John Murray.

d'Errico, F., C. Henshilwood, G. Lawson et al. (2003). 'Archaeological Evidence for the Emergence of Language, Symbolism, and Music—an Alternative Multidisciplinary Perspective', *Journal of World Prehistory*, 17, 1: 1–70.

Damasio, A. (1995). *Descartes' Error: Emotion, Reason and the Human Brain*. London: Picador.

Devereux, G. and W. LaBarre (1961). 'Art and Mythology'. In: B. Kaplan (ed.), *Studying Personality Cross-Culturally*, pp. 361–403. Evanston, IL: Row, Peterson.

Gourlay, K. A. (1984). 'The Non-Universality of Music and the Universality of Non-Music', *The World of Music*, 26, 2: 25–36.

Henshilwood, C. S., F. d'Errico, R. Yates et al. (2002). 'Emergence of Modern Human Behavior: Middle Stone Age Engravings from South Africa', *Science*, 295, 15: 1278–80.

—— and C. W. Marean (2003). 'The Origin of Modern Human Behavior: Critique of the Models and their Test Implications', *Current Anthropology*, 44, 5: 627–51.

Joffe, T. H. (1997). 'Social Pressures Have Selected for an Extended Juvenile Period in Primates', *Journal of Human Evolution*, 32, 6: 593–605.

Karmiloff-Smith, A. (1992). *Beyond Modularity*. Cambridge, MA: MIT Press.

Langer, S. (1942). *Philosophy in a New Key*. Cambridge, MA: Harvard University Press.

McNeill, W. H. (1995). *Keeping Together in Time*. Cambridge, MA: Harvard University Press.

Meyer, L. B. (1956). *Emotion and Meaning in Music*. Chicago, IL: University of Chicago Press.

Morley, I. (2002). 'Evolution of the Physiological and Neurological Capacities for Music', *Cambridge Archaeological Journal*, 12, 2: 195–216.

Myers, C. (1913). *The Beginnings of Music: Essays and Studies Presented to William Ridgway*. Cambridge: Cambridge University Press.

Nelson, S. (2002). 'Melodic Improvisation on a Twelve-Bar Blues Model: An Investigation of Physical and Historical Aspects, and their Contribution to Performance'. PhD thesis, City University, London.

Oatley, K. and P. N. Johnson-Laird (1998). 'The Communicative Theory of the Emotions: Empirical Tests, Mental Models and Implications for Social Interactions'. In: J. M. Jenkins, K. Oatley, and N. L. Stein (eds.), *Human Emotions: A Reader*, pp. 84–97. Oxford: Blackwell.

Papousek, H. (1996). 'Musicality in Infancy Research: Biological and Cultural Origins of Early Musicality'. In: I. Deliège and J. A. Sloboda (eds.), *Musical Beginnings*, pp. 37–55. Oxford: Oxford University Press.

Pinker, S. (1997). *How the Mind Works*. London: Allen Lane.

—— (1994). *The Language Instinct*. London: Allen Lane.

Seyfarth, R. M. and D. L. Cheney (2003). 'Signalers and Receivers in Animal Communication', *Annual Review of Psychology*, 54: 145–73.

Shannon, C. and W. Weaver (1949). *The Mathematical Theory of Communication*. Urbana, IL: University of Illinois Press.

Spelke, E. (1999). 'Infant Cognition'. In: R. A. Wilson and F. C. Keil (eds.), *The MIT Encyclopedia of Cognitive Sciences*, pp. 402–4. Cambridge, MA: MIT Press.

Sperber, D. (1996). *Explaining Culture*. Oxford: Blackwell.

—— and D. Wilson (1986). *Relevance: Communication and Cognition*. Oxford: Blackwell.

White, T. D., B. Asfaw, D. DeGusta et al. (2003). 'Pleistocene *Homo Sapiens* from Middle Awash, Ethiopia', *Nature*, 423: 742–7.

11

Musicians' Performance Prosody

Johan Sundberg

INTRODUCTION

Why does almost every person in the world actively seek possibilities to listen to music? This question is probably the most important challenge that music science faces.

Music is communication. It may communicate many things, but one important component is information on its structure, and this information can be wrapped in various expressive colours. One strategy to gain an improved insight into the puzzle as to why music seems attractive and available to anyone is to examine music performance. The two basic questions would be:

1. What principles do musicians apply when they convert scores into a musical performance?
2. What is the acoustic code that they use thereby?

The act of performance is quite relevant to how we perceive a piece of music. In our time we have many opportunities to hear music played by machines rather than by living musicians. Such examples mostly sound quite pathologic, particularly for music from the classical Western repertoire.

For many years our research group at the department of Speech Music Hearing, KTH, has studied the reasons for the computer's shortcomings as a musician. Our method has mainly been analysis-by-synthesis. This implies that we use a computer to generate performances of music files on a synthesizer. A professional musician, the late violinist Lars Frydén, assessed the emerging performances and recommended how they could be improved. We implemented his recommendations as performance rules in the computer's control program, and then tested these rules on various music examples. After many years of such experiments we had formulated about two dozen performance rules. Anders Friberg combined these rules into a generative grammar of music performance, *Director Musices* (Friberg 1995). It adds some of the departures that musicians typically make from the nominal score, thus making the performance musically more interesting or appealing.

Our performance rules can be divided into three types. The *grouping rules* represent one type. They serve the purpose of grouping, that is, showing where the structural boundaries are in the composition and which tones belong together. The *differentiation rules* constitute another type. These rules enhance the difference between musical categories such as note values or scale tones or intervals, for example, by increasing the dissimilarities between them. The *emphasis rules* represent the third type. They add emphasis to unexpected tones and de-emphasize expected tones.

EXAMPLES OF PERFORMANCE RULES

1 Grouping

One of the strongest principles in the performance of Western classical music is phrasing. Todd (1985) demonstrated that many of the deviations from the nominal inter-onset-intervals in a performance of a Chopin *Mazurka* could be accounted for in terms of one single phrasing pattern, characterized by a speeding up in the beginning of a phrase and the slowing down towards the end of the phrase. Figure 11.1 shows the phrasing pattern used for the deviations from nominal inter-onset-intervals in the *Director Musices* program. By varying the different parameters of this model, Friberg (1995) found that this pattern could be used to explain most of the timing of three pianists' performances of Robert Schumann's *Träumerei*.

It is interesting that a similar principle applies also to music composition. Sundberg and Lindblom (1976) found that the frequency of notes decreased, that is, the nominal duration of notes tended to increase, towards the end of phrases in nursery tunes. In

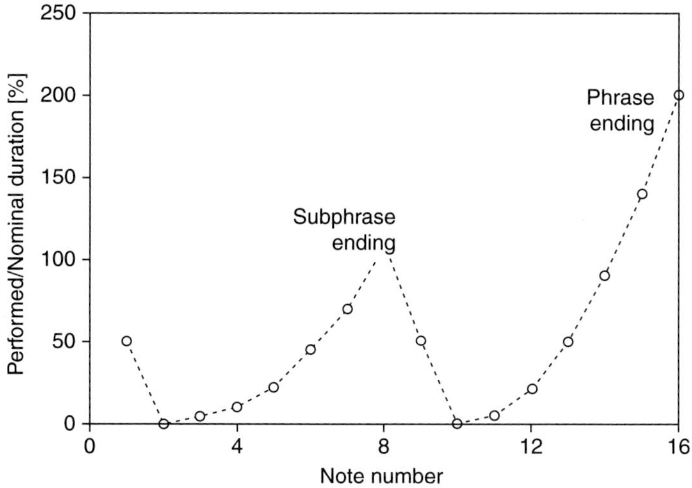

Figure 11.1. Illustration of the effect of the phrase arch rule on the tempo within a phrase consisting of two sub-phrases.

other words, it is rare that the end of a phrase in such songs contains a sequence of short notes.

The slowing down of the tempo towards the end of phrases is an important principle also in speech prosody, where it is referred to as final lengthening. Thus, when we speak we generally lengthen the last syllable of the phrase. This parallel suggests that the code used in music performance for signalling the end of a phrase is well chosen in the sense that its meaning should be immediately understandable to all who understand speech.

This is not to say that speech is the origin of the code for phrase marking in music. The anatomy of the final *ritardando* suggested another possibility. Vi Verrillo and I measured the inter-onset-interval pattern in about twenty pieces from the baroque era and computed the mean *ritardando* curve, normalizing tempo and the duration of the *ritardando* (Sundberg and Verrillo 1980). Figure 11.2 shows this mean *ritardando* curve. It seemed an attractive hypothesis that the slowing down of the tempo towards the end of a piece is an icon, such that final *ritardando* alludes to the slowing down of the footsteps of a stopping runner. Anders Friberg and I (1999) tested this hypothesis that the *ritardando* curve was similar to the footstep frequency of stopping runners. Thus we hired four dancers and asked them to run at different footstep frequencies and to stop at a given place. Their movements were recorded on video and audio. The videos were then presented to a jury of choreographers who rated the aesthetic quality of the stopping. The stoppings that received high ratings were then analysed. We found to our great surprise (and disappointment) that there was no systematic change of footstep frequency during the stopping. The slowing down towards the end of the running was realized in terms of a decrease of the step length. Thus there was no support for the hypothesis that there was an analogy between note and footstep frequency. Instead, analysis of the material revealed that the runner's velocity changed in the way illustrated in Figure 11.2, that is, in almost exactly the same way as the tempo changed during the final *ritardando*. In this sense, the *ritardando* seemed to lend some support for the assumption that the *ritardando* is an icon where the analogy is between tempo and velocity. Provided that we experience some kind of movement when we listen to music, this finding is not surprising. It suggests that we understand the meaning of the final *ritardando* because we recognize its similarity to the experience we have of stopping locomotion.

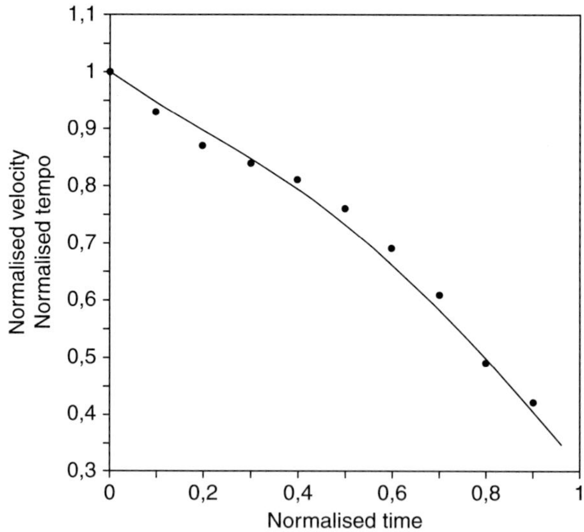

Figure 11.2. Comparison of the mean tempo decrease during a final ritardando (dots, according to Sundberg and Verrillo 1980) and the average velocity decrease during a runner's stopping (solid curve, according to Friberg and Sundberg 1999).

2 Differentiation

Another principle emerging from the performance rules is the differentiation of categories. It seems clear that we perceive pitch, pitch intervals, and note values in terms of categories. Thus, pitches are perceived categorically, such that a number of fundamental frequencies are perceived as one particular scale tone. For example, not only the frequency of 440 Hz is perceived as the pitch of A4, but also the frequencies in a narrow band surrounding 440 Hz. Likewise, fundamental frequency ratios are perceived as musical intervals; not only the ratio $1{:}2^{4/12}$ is perceived as a major third but also all ratios between 64:81 and 4:5. By and large, the interval category width is about a quartertone. For note values, a similar situation exists, even though the categories actually overlap, musical context being a factor of major influence.

Musicians tend to enhance the difference between such categories when they play. For example, solo instrumentalists playing instruments with adjustable intonation tend to play major thirds wider and minor thirds narrower than what is referred to as just intonation (Fyk 1995). Also, the individual scale tones are complemented by various aggregate properties, so the tritone is often played louder, longer, and with more vibrato than adjacent scale tones. And short notes are sometimes played shorter and long notes longer than nominal.

Is the principle of enhancing and complementing differences between categories specific to music? Randy Diehl (1991) has observed a similar principle in speech, and proposed the term *acoustic enhancement* for it.

But the differentiation principle is applied not only in acoustic communication such as music and speech. If we assume that the various parts of a house are perceived in terms of categories such as foundation, walls, windows, and roof, architecture also offers an example, these categories typically being differentiated by visual enhancement in terms of different colours. Moreover, in nature the categorical perception of gender is typically enhanced by means of appearance. Female birds in polygamous species have a different feather cover to male birds. Thus, it seems fair to conclude that differentiation or category enhancement is a rather widespread principle.

Such enhancement can be achieved by increasing the difference along the dimension that separates the categories, as in the case of intonation where major thirds are played widened and minor thirds are played narrowed. But differentiation may also be facilitated by adding another distinguishing property, such as lengthening some scale tones and shortening others. Speech offers some more examples of the same principle. For example, the difference between long and short vowels is sometimes enhanced by diphthongization of the long variant or by a difference in formant frequencies.

3 Emphasis

Expectancies have been shown to play a major role in music listening (Krumhansl 1990). This would be the background of a third principle that musicians apply when they play. Thus, the *Director Musices* performance grammar contains a rule that responds to remarkable scale tones by adding duration and increasing loudness and, if the instrument allows, width of vibrato. In the *Director Musices* performance grammar, the remarkableness of a tone is determined by the tone's distance along the circle of fifths from the root of the prevailing chord. The grammar quantifies this property of distance to the root of the chord in terms of *melodic charge*. Tones with a high melodic charge are lengthened and played louder and, if possible, with a wider vibrato than other tones. A related rule creates changes of loudness and tempo according to remarkableness of chords. The quantitative measure used for this type of remarkableness is a derivative of the melodic charge and is referred to as *harmonic charge*. These rules obviously apply only to traditional Western tonal music. For atonal music a cousin of the melodic charge is used, the *chromatic charge*.

In vocal music, the lyrics add another kind of remarkableness to the music, reflecting another kind of expectancy, the semantic. We have not developed any rule predicting this type of remarkableness. Rather, analysis of sung performance has demonstrated this performance principle. Figure 11.3 analyses a professional singer's performance (of lines from a Schumann song) in terms of the comparison between actual and nominal note durations, measured as inter-vowel-onset intervals, which can be regarded as the equivalent of the syllable in vocal music (Sundberg and Bauer-Huppmann 2007). There are four cases of lengthening that amount to values greater than 1.5 times the nominal value. Of these, that of the first note, *Es schw*, may depend on the articulatorily complex sequence of consonants at the end of the note. The other three notes that were considerably lengthened, *e l* before *ieb*, *ie L* before *ipp*, and *en* before *au*, are all unstressed syllables, and the singer's aim seemed to be to

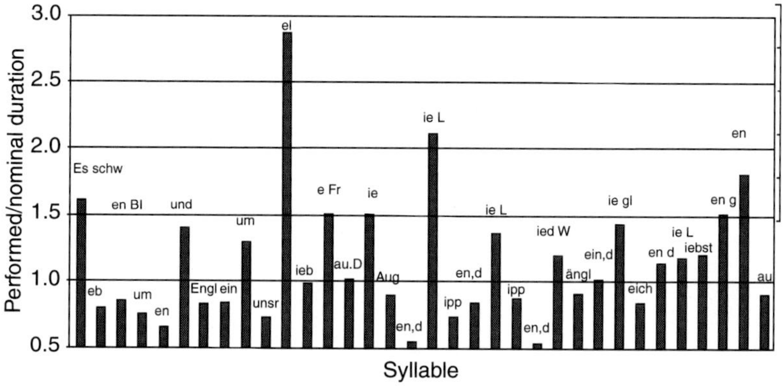

Figure 11.3. Ratio between nominal and performed note durations in a professional singer's performance of Robert Schumann's *Lied* 'Im Rhein' from *Dichterliebe*, op. 48 (words by Heinrich Heine). The chart shows the syllables for the lines '*Es Schweben Blumen und Englein / Um unsre Liebe Frau, / Die Augen, Die Lippen, die Wänglein, / Die gleichen der Liebsten genau.*' Note durations were measured as vowel-onset-to-vowel-onset. (From Sundberg et al. 1995).

emphasize the subsequent stressed syllable. Many similar examples were found in the material. This emphasis code could be referred to as *emphasis by delayed arrival.*

It may appear odd that syllables preceding the stressed syllable of semantically important words be lengthened. However, emphasis by delayed arrival is often used also in speech. This is illustrated in Figure 11.4, which shows the variation in syllable duration when a sentence was read three times a by a professional actor who deliberately emphasized a different word each time. Note that here also the principle was to lengthen the syllable preceding the stressed syllable of the semantically enhanced word. This indicates that in this case the same code for emphasis is used both in speech and music.

SO WHAT?

Performances of a given piece of music may vary considerably. Intuitively, then, it seems unlikely that a variety of differing performances can be described by a limited set of rules. It is important to

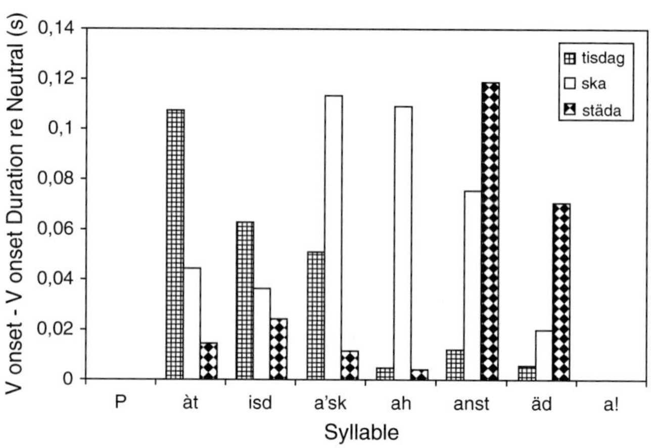

Figure 11.4. Syllable duration ratios between the neutral and emphasized versions for a sentence pronounced by a professional actor in a neutral manner and when she emphasized one of the words in the sentence: '*På tisdag ska han städa*' (On Tuesday he should tidy). Black, grey, and white columns refer to the versions where the words 'tisdag', 'ska', and 'städa', respectively, were emphasized. (From Sundberg 2000).

note, however, that few rules, if any, are compulsory. Rather, it seems more appropriate to regard the rules as optional. For example, there would be a number of acoustic cues available to a performer for expressing emphasis. Likewise, the differentiation of note value categories can probably be achieved not only by lengthening long notes but also by shortening short notes. Also the opposite occurs, though with a different effect on the performance. A performance rule seems more like a word in a dictionary. There are many codes available to convey the same information. The interesting thing is that the performer adds information by applying a set of principles, and here we have identified three such principles: grouping, differentiation, and emphasis.

It should be stressed that the classification of rules into one of these principles is sometimes ambiguous. For example, the melodic charge rule may serve the purpose of differentiation of the scale tone categories; but it is equally possible to regard it as an application of the emphasis principle. However, this ambiguity does not seem problematic. The interesting thing is that the expressivity of a performance seems to emerge from the application of a set of performance principles.

Some examples of parallels between musical and non-musical communication have been discussed above. The principles of category differentiation, grouping, and emphasis are applied in music as well as in speech and also in some other extra-musical communication systems. This suggests that they serve some relevant purposes, and thus that they represent important components in a communication system. In other words, sorting perceptions into categories, dividing a signal sequence into chunks, and taking into account the receiver's expectancies are probably important in a communication system.

It is also clear from the above that the acoustic code used for enhancing differences between categories, for grouping and for emphasis are not specific to music. Rather, the same or similar codes are used for the same respective purposes in speech and/or some other contexts. This suggests that music is an application of communication skills acquired in the extra-musical domain. Also, it implies that it should be possible to understand some of the information conveyed by the performance of a piece of music by making use of knowledge assembled in non-musical communication.

These findings support the assumption that music perception is an application of skills that are acquired and used elsewhere. It then does not seem all that surprising that music is a human universal; its acoustic code should not be difficult to understand. In other words, it should be easy to convert the acoustic signal flow in a musical performance into meaningful information.

This certainly fails to provide an answer to the question why man possesses and needs music. On the other hand, it shows that music is not special. It is actually somewhat similar to speech, even though instrumental music lacks an equivalent to semantics. Instead, both the piece of music itself and the musicians performing it strive to convey information on the structure of the piece, for example by marking the structural boundaries in the composition. Moreover, as speakers do in verbal communication, musicians help the listener to process the information by enhancing category differences and by stressing unexpected/remarkable events. It seems possible to argue that the existence of music would have been much more difficult to explain if music had been special, demanding specific perceptual and cognitive skills of the listener.

Also probably of relevance to man's need for music are the emotional aspects in which music is typically embedded. It is quite interesting that the *Director Musices* grammar can be used to colour performance in what listeners perceive as different emotional clothes (Bresin 2000). As parallels between music and speech with regard to emotional expression have recently been reviewed by Juslin and Laukka (2003) we refrain from discussing them here.

In summary, the performance principles applied in music performance are similar to those found in other communication systems. The acoustic codes used for these principles are familiar to us from our experience of speech and other forms of communication and from the experience of motion. There seem to be good reasons to believe that this belongs to the explanation why music is a universal companion to man.

REFERENCES

Bresin, R. (2000). 'Virtual Virtuosity. Studies in Automatic Music Performance', doctoral dissertation, Department of Speech Music Hearing, KTH, Stockholm.

Diehl, R. (1991). 'The Role of Phonetics within the Study of Language', *Phonetica*, 48: 120–34.

Friberg, A. (1995). 'A Quantitative Rule System for Musical Expression', doctoral dissertation, Department of Speech Music Hearing, KTH, Stockholm.

—— and J. Sundberg (1999). 'Does Music Performance Allude to Locomotion? A Model of Final *Ritardandi* Derived from Measurements of Stopping Runners', *Journal of the Acoustical Society of America*, 105: 1469–84.

—— (1995). 'Matching the Rule Parameters of Phrase Arch to Performances of "Träumerei": A Preliminary Study'. In: A. Friberg and J. Sundberg (eds.), *Proceedings of the KTH Symposium on Grammars for Music Performance*, pp. 37–44. Stockholm: TMH, Speech, Music, and Hearing.

Fyk, J. (1995). *Melodic Intonation, Psychoacoustics, and the Violin*. Zielona Góra: Organon Publishing House.

Juslin, P. N. and P. Laukka (2003). 'Communication of Emotions in Vocal Expression and Music Performance: Different Channels, Same Code?' *Psychological Bulletin*, 129: 770–814.

Krumhansl, C. L. (1990). *Cognitive Foundations of Musical Pitch*. New York: Oxford University Press.

Sundberg, J. (2000). 'Emotive Transforms', *Phonetica*, 57: 95–112.

——and J. Bauer-Huppmann (2007). 'When Does a Sung Tone Start?' *Journal of Voice*, 21, 3: 285–93.

—— J. Iwarsson, and H. Hagegård (1995). 'A Singer's Expression of Emotions in Sung Performance'. In: O. Fujimura and M. Hirano (eds.), *Vocal*

Fold Physiology: Voice Quality Control, pp. 217–29. San Diego, CA: Singular Press.

—— and B. Lindblom (1976). 'Generative Theories in Language and Music Descriptions', *Cognition*, 4: 99–122.

—— and V. Verrillo (1980). 'On the Anatomy of the Ritard: A Study of Timing in Music', *Journal of the Acoustical Society of America*, 68: 772–9.

Todd, N. (1985). 'A Model of Expressive Timing in Tonal Music', *Music Perception: An Interdisciplinary Journal*, 3, 1: 33–57.

12

Harmony and its Role in Human Evolution

Nicholas Bannan

Rien de plus cher que la chanson grise
Où l'Indécis au Précis se joint.
[Nothing is dearer than the grey song
Where the undefined combines with the precise.]

Paul Verlaine (1884)

INTRODUCTION

Musical aspects of human vocalization received comparatively little attention in the twentieth-century research literature on language acquisition and evolution. However, universal elements of human musical vocalization, such as unison singing and the capacity for harmonizing, are as unique to our species as speech. These abilities require investigation, both in their own right, as examples of special attributes of significance to the nature of human culture, and in terms of the features that they share with the neural and anatomical structures on which language depends. Some time between *Homo*'s last common ancestor with the African apes and the emergence of the modern traits all members of our species inherit, an evolutionary trajectory shaped vocal behaviour which exploits the innate ability to duplicate with absolute precision the pitch contour of acoustic models, allowing both serial imitation and simultaneous unison performance. A possible link between such aspects of singing and the production of language may exist in the function of vocal timbre, in which the categorical differences that give rise to vowel sounds in language also define intervallic relationships consistent with the generation of melody. This chapter presents a commentary on the historical and interdisciplinary emergence of this view, and outlines a framework for future research in the field.

BACKGROUND

A sequence of significant publications devoted to theories of language use and acquisition (Chomsky 1957, 2001; Jenkins 2000; Juscyk, 1997; Johanson and Edgar 1996; Pinker, 1994, 1997) has, perhaps surprisingly, made no reference to any role for song or musical response. Definitions of singing have tended to view it as an extension or addition to communication in words:

> Singing is a fundamental mode of musical expression. It is especially suited to the expression of specific ideas, since it is almost always linked to a text; even without words, the voice is capable of personal and identifiable utterances. It is arguably the most subtle and flexible of musical instruments, and therein lies much of the fascination of the art of singing. (Jander and Harris 2001)
>
> In recitation and declamation, as in singing, the voice is sustained beyond the requirements of speech. But in song, the voice is sustained at greater length than in the others, and its sound distributed over a wider range of pitch. (Lomax 1982, p. 789)

Does it help to view the voice as a 'musical instrument', albeit one that developed biologically, like the syrinx of the singing birds?

> The voice is a musical instrument, like any other, but with the additional power of framing the sounds called words and allying them with its tones. (Scholes 1950, p. 991)
>
> The human voice may not exactly be an instrument, but it is certainly the oldest means of music-making, and it remains the most natural and most expressive . . . the voice is actually a part of the body of the performer. (Sadie 1985, p. 33)
>
> The voice organ can be regarded as a wind instrument consisting of an air pressure supply driving an oscillator, the output of which is fed into a resonator from which the sound is radiated to the air outside the instrument. (Sundberg 2001)

Sundberg writes with the authority of a researcher who has systematically observed and tested the anatomical operations he describes, employing all the computer-aided technology of the modern laboratory. With more basic if ingenious tools, many of his own devising, Helmholtz (1863) was able to give almost the same description a century before:

> In the larynx, the elastic vocal chords act as membranous tongues . . . They have the advantage over all artificially constructed tongues of allowing the width of their slit, their tension, and even their form to be altered at pleasure with extraordinary rapidity and certainty, at the same time that the resonant tube formed by the opening of the mouth admits of much variety of form, so that many more qualities of tone can be thus produced than on any instrument of artificial construction. (Helmholtz 1863, p. 98)

The influence of Helmholtz's pioneering work in the field of acoustics and its consequences for human communication pervades subsequent research. In particular, two independent paths have been taken in disciplines that arose in the wake of his discoveries: speech science, and its relation to audiology; and musicology. In speech science, *source-filter theory* (Fant 1970; Fry 1979; Assmann and Nearey 1987; Alcántra and Moore 1994) has established the means by which vowels and voiced consonants are produced through the modification by the vocal tract, tongue, and lips of fundamental frequencies formed by the vocal folds. Source-filter theory provides an insight into the uniquely human capacities it describes, represented by the evolved structure of the *Homo* larynx and respiratory tract, the facial musculature, and, above all, the neural adaptations involved in their fine control. The theory extends in most respects also to accounting for the production of singing (Titze 1994; Sundberg 1987; but see also Titze 2001 for special exceptions). The capacity of source-filter theory to explain the process of vocalization, the means by which this is refined, and its intentional employment acquired through vocal learning, depends on aural processing that allows the accurate imitation of models in all four of their parameters: pitch, duration, amplitude, and timbre. While response to sound stimuli that exploits sensitivity to these parameters is widespread across a variety of unrelated species, a few of which also possess the capacity for imitation and vocal learning, the relationship between perception and production in humans that permits unison singing and precisely organized rhythmic-prosodic interaction is found in no other species. Above all, human exploitation of controlled exhalation in articulate vocalization contrasts markedly with the relatively limited vocal repertoire of chimpanzees and gorillas.

A separate discipline with firm roots in the implications of Helmholtz's work has developed in musicology, with consequences for methods of musical analysis, for composition and orchestration, and for vocal and instrumental pedagogy. The theory and practice of music demand precise sensitivity to the tuning of pitches, to the combination of tones and timbres, and to the way these interact over time. Training in music in widely separated cultures has been found to attach considerable significance to transmission of the capacity for exact and faithful reproduction of models, as well as to the development of understanding and experience that refines and accelerates this ability. Following Helmholtz, pedagogies that systematize this process for early learning within the Western tradition have been devised by, amongst others, Kodály (1974) and Jaques-Dalcroze (1921). Concurrently, academic music analysis has viewed the relationship between the part of a musical work and the whole in terms of hierarchical tonal relationships expounded in the work of Schenker (1906/1954: see also Krumhansl 1990; Parncutt 1989).

These two traditions in speech science and music theory share in common both the legacy of Helmholtz and an early association with ideas in the new science of evolution that emerged concurrently in the late nineteenth century.

HELMHOLTZ AND THE ORIGIN OF 'EVOLUTIONARY MUSICOLOGY'

In both anecdotal observation and systematic research, Helmholtz stressed the potential for musical resonance of the speaking voice, and, by implication, the musical features of speech:

> It is interesting in calm weather to listen to the voices of men who are descending from hills to the plain. Words can no longer be recognized, or at most only such as are composed of M, N, and vowels, as *Mamma, No, Noon*. But the vowels contained in the spoken words are easily distinguished. Wanting the thread which connects them into words and sentences, they form a strange series of alternations of quality and singular inflections of tone. (Helmholtz 1863, p. 68)

Reports of the musicality of spoken utterances under certain conditions had interested earlier authors, especially the French encylopaedists (Thomas 1995); while von Humboldt in his major work on human language (1836) called us a 'singing species'. With the aid of the resonating devices he invented, Helmholtz was able to explain these phenomena and reveal the harmonic nature of human oral communication:

> The upper partials of the human voice have nevertheless been heard at times by attentive observers. Rameau had heard them at the beginning of last century. And at a later period Seiler of Leipzig relates that while listening to the chant of the watchman during a sleepless night, he occasionally heard at first, when the watchman was at a distance, the Twelfth of the melody, and afterwards the prime tone. The reason of this difficulty is most probably that we have all our lives remarked and observed the tones of the human voice more than any other, and always with the sole object of grasping it as a whole and obtaining a clear knowledge and perception of its manifold changes of quality. (Helmholtz 1863 p. 104)

In reporting the dual nature of voice perception identified here, of verbal meaning and quality, Helmholtz acknowledges as significant the way that what the ear pays attention to is defined by context. Under normal circumstances, we listen to speech to discern meaning. But meaning is not the only component of vocalization that we are able to perceive.

Helmholtz went on to speculate on the evolutionary sequence that may have given rise to modern human vocality:

> It is not my intention to deny that music in its initial state and simplest forms may have been originally an artistic imitation of the instinctive modulations of the voice that correspond to various conditions of the feelings.... An endeavour to imitate the involuntary modulations of the voice and make its recitation richer and more expressive, may therefore very possibly have led our ancestors to the discovery of the first means of musical expression, just as the imitation of weeping, shouting, or sobbing, and other musical delineations may play a part in even cultivated music, (as in operas), although such modifications of the voice are not confined to the action of free mental motives, but embrace really mechanical and even involuntary muscular contractions. (Helmholtz 1863, pp. 370–1)

Significantly, Helmholtz traces the origins of such vocalizations in the emotional motivations that gave rise to them, suggesting correlates between these and the forms of expression found in art music. As we shall see, singing teachers were quick to elaborate on these speculations, initiating a tradition that appears to have developed quite independently of speech science and linguistics. Nevertheless, a modern neurologically informed account of physical and psychological aspects of vocal learning such as that of Deacon (1997) is largely consistent with Helmholtz's view:

> Not only do we exhibit a mix of arousal-independent learned vocalizations (speech and song) and highly stereotypic vocalizations that are innately linked to certain emotions (laughter and sobbing), but our visceral-emotional and skeletal-muscle-skill-learning systems often compete for access over vocalization, as well as often complementing one another in the same vocal behaviour. It is as though we haven't so much shifted control from visceral to voluntary means but superimposed one upon the other. (Deacon 1997, pp. 243–4)

Darwin, an admirer of Helmholtz who corresponded with him, took the more radical step of proposing that an evolutionary sequence that led to language commenced at the earlier stage of animal communication:

> I cannot doubt that language owes its origin to the imitation and modification of various natural sounds, the voices of other animals, and man's own instinctive cries, aided by signs and gestures. When we treat of sexual selection we shall see that primeval man, or rather some early progenitor of man, probably first used his voice in producing true musical cadences, that is in singing, as do some of the gibbon-apes of the present day; and we may conclude from a widely spread analogy, that this power would have been especially exerted during the courtship of the sexes—would have expressed various emotions, such as love, jealousy, triumph—and would have served as a challenge to rivals. It is, therefore, probable that the imitation of musical cries by articulate sounds may have given rise to words expressive of various complex emotions. (Darwin 1871, p. 132–3)

The implications for human origins of Darwin's evolutionary theory were distilled more starkly in this statement than in anything he had published previously. For the first time, the direct link between animal behaviour and emergent human capacities such as language, thereto seen as an unbridgeable and determining difference, were set out explicitly. Interpretations of Darwin's theory followed, including a contribution from Nietzsche that again focuses on the assumed emotional underpinning of primitive vocalizations and their similarity to music:

> All degrees of pleasure and displeasure—expressions of *one* primeval ground that we cannot see through—find symbolic expression in the *tone of the speaker*, while all other representations are designated by the *gesture symbolism* of the speaker. Insofar as this primeval background is the same in all human beings, the *tonal background* is also universal and intelligible despite the differences between languages. Out of this develops the more arbitrary gesture symbolism that is not wholly adequate to its foundation, and with this begins the multiplicity of languages whose variety we may consider metaphorically as a strophic text for this primeval melody of the language of pleasure and displeasure. (Nietzsche 1871, p. 107)

Deacon again provides a modern summary of how some components of vocal learning exploit the instinctive responses of evolved human anatomy:

> Accentuating the symptoms of respiration by making a hissing sound—produced as air passes through the narrow constriction of the throat and mouth—may have provided some of the earliest forms of vocal communication. The subsequent differentiation and specialization of sound-breathing patterns has likely built on this otherwise primitive index of arousal. Because of this ancient pattern, vocal communication has inherited many of the organizational features of a partly automatic, partly controllable motor system. These include the automatic, reflex-like production of sounds when in an appropriate arousal state; stereotypic vocalizations; and a minimal role for learning. (Deacon 1997, p. 234)

Darwin's study of *The Expression of Emotion in Man and Animals* deals with both visual and acoustic signals, and the relationship between the two, and he initiates an incomplete attempt to account for the connection between emotion, expression, and vocalization:

> When the next full expiration follows, the mouth is slightly closed, and the lips, from causes hereafter to be discussed, are somewhat protruded; and this form of the mouth, if the voice be at all exerted, produces, according to Helmholtz, the sound of the vowel _O_. Certainly a deep sound of a prolonged _Oh!_ may be heard from a whole crowd of people immediately after witnessing any astonishing spectacle. If, together with surprise, pain be felt, there is a tendency to contract all the muscles of the body, including those of the face, and the lips will then be drawn back; and this will perhaps account for the sound becoming higher and assuming the character of _Ah!_ or _Ach!_ As fear causes all the muscles of the body to tremble, the voice naturally becomes tremulous, and at the same time husky from the dryness of the mouth, owing to the salivary glands failing to act. Why the laughter of man and the

> tittering of monkeys should be a rapidly reiterated sound, cannot be explained. During the utterance of these sounds, the mouth is transversely elongated by the corners being drawn backwards and upwards.... But the whole subject of the differences of the sounds produced under different states of the mind is so obscure, that I have succeeded in throwing hardly any light on it; and the remarks which I have made, have but little significance. (Darwin 1872, p. 38)

Darwin's admission of defeat in addressing the complexity of the relationship between emotion, instinctive expression, and learned vocalization throws down a challenge to evolutionary theory that has yet to be fully answered in speech science, music psychology, or neurology. The voice scientist and singing teacher Lucie Manén re-formulated the challenge a century after Darwin, and her account prefaces an examination of what researchers focusing on the development of the trained singing voice have brought to this debate:

> Singing is the artistic expression of human emotion in sound. The expression of emotion is a reaction to internal and external stimuli in which the whole system participates: facial expression, gesture and attitude and vocal exclamations, together with heartbeat and breathing, reflecting the internal state of the individual. These reactions occur as reflexes. From the sensory organs—eyes and ears—impulses are carried via the nerves to the muscles which produce attitudes and gestures, to the muscles controlling facial expression, to the muscles of the larynx governing exclamations, to the centre of breathing and to the centres controlling the heartbeat. Chemical substances are poured into the bloodstream to supply the system with all those resources on which the human being unconsciously calls in any given state of emotion, be it hatred or love, joy or pain, fear or aggression. (Manén 1974, p. 11)

VOICE EVOLUTION AND THE PERSPECTIVE OF SINGING TEACHERS

Working in nineteenth-century London in a milieu excited by the ideas of Darwin and Helmholtz, the celebrated voice teacher Manuel Garcia framed his method in allusions to the latest discoveries in the physics of sound and the biology of the voice:

> Q. What is the object of vocal study?
> A. To make the voice irreproachable in its intonation, firm, strong, flexible, extended, and to correct its faults.
> Q. Is that all?
> A. To teach the student the art of phrasing, to familiarize him with the different styles, and to develop his expression. (Garcia 1894, p. 1)

So begins a distillation of his life's work of the first published singing teacher to have clinical experience as a voice scientist, expressed in the dialogue form of Classical philosophy. It later continues:

> Q. How many vowels are there?
> A. Though *grammarians* admit generally nine (in Italian), the number is, in fact, unlimited.
> Q. How is this?
> A. The mouth being formed of elastic and movable organs, has an unlimited power of modifying its shape and capacity, and each change is a mould whereby a particular vowel is formed.
> Q. Is there any analogy between vowels and timbres?
> A. The most intimate. We know that the changes of form in the vocal tube determine analogous changes in timbre. Vowel-shades and timbres are but two different names to express the same modifications of sonority in the voice. The result of this mutual dependence between Pharynx, Timbres, and Vowels is that a change in one produces changes in both the others.
> Q. Is the great variety of timbres of any practical use?
> A. They are the physiognomy of the voice. They tell the involuntary emotions which affect us, and assume a more clear or covered tint, a timbre more brilliant or more obscure, according to the nature of those feelings. (Garcia, 1894, p. 45)

Working in Germany in the generation following Garcia, Husler wove evolutionary theory into the rationale for his highly regarded vocal teaching:

> The singing voice is evoked by a special psychological disposition and a long and complex series of physical functions. The necessary psychological disposition is common to all mankind for, though it cannot as a rule be satisfied other than indirectly—as in the making of music or verse—the urge to sing is inborn. It is one of the attributes of Man, and for its fulfilment requires no extraneous aids or adaptations. In highly civilized races this inner urge is slowly fading... The physical prerequisites for singing, the anatomical constituents, are not chance endowments limited

> to a few individuals. Every human being normally possesses by nature the physical means of singing—even if he 'has no voice', or to be accurate, even if he cannot sing. In the normal person, inability to sing is due either to an impairment of the vocal organ or, in the majority of cases, to the natural inborn faculty being hampered and obstructed: it needs to be released, to be 'unlocked'... Most people believe... that the singer possesses a special physical equipment and also that, in some miraculous way, he himself fashions his own instrument. This, however, is to place singers outside the laws of nature, because so complex and significant a faculty as singing must necessarily be given to a whole species. (Husler and Rodd-Marling 1965, p. 1)

Here is an expert singing teacher, whose clients included some of Europe's most acclaimed soloists and ensembles, arguing for the universality of vocal ability. Husler refers to the need to 'release' the voice, implying that vocal learning may take different paths according to the motivation or circumstances of the individual. This would be consistent with the issue we will later address of people who speak but believe themselves unable to sing.

Manén also alludes to an evolutionary approach in suggesting links, on lines proposed by Darwin (1872) and Helmholtz (1863), between specific sounds and emotions. While research to investigate this claim has yet to be carried out, it has proved useful in singing teaching as a means of developing optimal resonance and voice quality, as well as providing a template for singers to call upon in the pursuit of precise timbral blend and tuning in duets, ensembles, and choirs:

> In any... state of emotion the exclamation of the human voice follows laws common to human beings of all races. For describing events with little emotional content and in a neutral mood, the human voice uses exclamations in its lower range. As soon as the individual changes the emphasis of what he speaks or sings, the pitch of the voice changes. It rises with increasing emotional content. For the expression of pleasure the natural exclamation is 'ah'. The exclamation 'ee' denotes disgust and hatred, while 'oo' denotes fear and horror. (Manén 1974, p. 11)

What evidence exists of this holistic, corporeal experience of the production of song, and in what ways does this differ from the process of speaking in which speech science tells us that the production of sounds selects from arbitrary categories (Hockett 1960; Chomsky 1957)? Molly Adkins, an American singer trained in the European *bel canto* tradition, recorded and analysed the consequences for her physical production of a singing voice of mastering, after eight years of training, the Japanese solo folk song repertoire. Adkins's analysis (2000) of acoustic data derived from her singing in these two distinct styles reveals the different 'settings' (Laver 1980) required to achieve the target timbres. The formant peak averaged across vowels in Japanese folk song was higher (3.5–4.5 kHz) than for Italian art song (3–4 kHz), arising from a comparatively raised larynx and pharyngeal narrowing.

What Adkins's experience illustrates is that the authenticity required of expert singing if it is to convince the expert listener is not only a costly affair but also requires adaptation to culturally derived characteristics that make different demands on the anatomy. Singing so as to achieve acceptable results depends therefore on *quality of transmission* that involves immersion in the physical, linguistic, musical, and cultural determinants of a repertoire. It also involves control of volume, both with respect to the 'broadcast' phenomenon of high-energy communication over distance, and the more intimate features of laments and lullabies. Rehearsal to meet such expectations exists wherever song learning demands it. Nattiez observed this amongst the Inuit of Northern Canada:

> we were wholly justified in studying the sound component of *katajjaq* as a musical fact, especially since the Inuit were themselves quite conscious of what linguists call well-formedness in certain sound-parameters of the genre. Studying the beginners' lessons in *katajjaq* makes this clear. The old woman who teaches the children corrects sloppy intonation of contours, poorly meshed phase displacements, and vague rhythms exactly like a Western vocal coach. (Nattiez 1990, pp. 56–7)

Similarly, John Blacking studied the learning of song amongst the Venda of Southern Africa. An important feature in children to which Blacking draws attention is their attraction to song for its own sake. He quotes Kodály in support of this: 'the child delights in the play of musical forms at a stage when he does not care about the meaning of the text' (Kodály 1974, p. 142); and compares this to subjects of his own research:

> I found this to be true of the music-making of Venda children, and it reminds us that we can enjoy music from different parts of the world without having long lectures about the cultural background. That can

come later, when children have been allowed to use their 'direct intuition'. (Blacking 1987, p. 142)

Moyle (1997) captures a sense of the importance attached by Aboriginal adults to accurate learning in his observation of song transmission in Balgo, Western Australia:

> Because singing is a social event, demonstrable executant ability is a status marker, but it is a two-edged marker. Because performance errors are usually immediately apparent and are always adversely regarded, singers join in audibly only when they feel confident about the song. Less experienced singers usually start after the others, and tend to sing more softly. (Moyle 1997, p. 43)

The capacity for specialized, rehearsed vocalization pervades every known culture (Trân Quang and Bannan, this volume). Yet singers who master the style of more than one culture, such as Trân Quang Hai, Molly Adkins, and Patricia Shehan Campbell (1991), demonstrate the flexibility with which, given time and effort, the patterns of vocal habituation that we acquire through enculturation in a mother tongue and native culture can be modified. With similar application, adult non-singers can develop a singing voice (Knight 2000; Buttolph 1993; see also Mithen 2008). Vocal learning is open to lifelong extension and adaptation. In relation to this, target repertoires and stylistic attributes are themselves subject to change.

The singer and voice scientist John Potter (1998) has made a particular study of the emergence of the operatic singing voice as it developed dramatically in the nineteenth century. This tradition was subject in turn to rapid and irreversible changes in canons of taste that can be traced in the legacy of sound recordings that emerged from the turn of the twentieth century onwards, such as expressive rhythmic fluctuation, portamento and glissando, and the nature of vibrato (Potter 2006; Leech-Wilkinson 2006). Potter's analyses convey that what audiences perceive to be 'good singing'—the target model of the vocal learner—is itself fluid. The consequences define the socially mediated acceptability of the performance: the recordings Potter has analysed of great singers of the period 1895–1910 often proving grating or risible to modern audiences.

As Potter (1998) describes it, the transition from the *bel canto* technique associated with the performance in their time of Handel and Mozart to the requirements of Verdi, Wagner, and Puccini represents a sophisticated acoustic solution to combining the delivery of text with the need to surmount the high-intensity accompaniment of the Romantic period orchestra: irrespective of target vowel sound, the singer seeks to activate continuously spectral reinforcement in the audible range of the singer's formant (Sundberg 1987; see also Walker 1997) in precisely the bandwidth in which the competing instrumental sound is less strongly present. It would be Potter's view that this conscious, artificial adaptation of 'natural' vocal technique was introduced for this specific purpose as a new phenomenon that allowed singers to cope with the demands of this generation of composers. While this view may be historically accurate, the anatomical properties it depends on, allowing the emission of high-intensity sound in which fundamental pitch is filtered to broadcast a timbre made up of harmonic peaks at advantageous frequencies, are very old in our species.

MONOTONY, MONOPHONY AND HARMONY

A significant feature of the human development of group singing has been the capacity for participants' complementary ranges to give rise to effects which transcend mere amplification:

> Why is a consonant union of voices pleasanter than a single voice? Is the singing of voice against voice, a consonant union of voices in Octaves? This singing of voice against voice occurs when young boys and men sing together, and their tones differ as the highest from the lowest of the scale. (Aristotle, *Problems*, xix: 39, quoted in Helmholtz 1863, p. 237n.)

Aristotle describes here an outcome of evolved human anatomy: the acoustic dimorphism of the adult male and female of the species, linked to the fact that the infant and child voice of both genders relates to the female adult range up to puberty. In widespread cultures, men and women, and/or men and children, sing melodies in parallel octaves. In some, the interval of parallel movement between voices may be others that are recognizably consonant: the fifth, or the third. Dargie (1988) transcribed Xhosa group vocalization in which the melody proliferated into parallel reinforcement employing full three-part chords.

James Burney, a lieutenant on Cook's second Pacific expedition (1772–75), notated a Maori melody and reported (Burney 1975, p. 57) that it was sometimes sung in parallel thirds 'except the two last notes which are the same'. This encounter with expressive harmony and closure achieved by cadential reinforcement in a people whose practices can have had no contact with European musical conventions was evidently a surprise to a young man who knew his European music history: his father authored the first *General History of Music* in English. Burney records that his captain, James Cook, recognized the singing to be 'harmonious enough but very dolefull (sic) to a European ear'. They were to meet other examples of convergent evolution of part-singing. In Tonga (ibid. p. 84), Burney notated the singing of a four-note melody which he ascribed to the Aeolian mode, and illustrated the way that some singers maintained the lowest (tonic) note as a drone: 'it puts me in mind of the Church Singing among Roman Catholics'. Cook found Tongan singing 'musical and harmonious, noways harsh or disagreeable'. Despite more than 20,000 years of genetic, geographical, cultural, and linguistic separation, Polynesians and Europeans proved susceptible to the same elements of musical organization in devising group song. What aspects of a universal human musical ability might account for this?

Sachs (1962; see also Trân Quang and Bannan, this volume) proposed a threefold taxonomy of human singing: *pathogenic*, from the Greek *pathos* and characterized by emotionally motivated utterances, often of wide range; *logogenic*, where singing arises in the performance of words, and is characterized by monotone chant; and *melogenic*, in which musical conventions emerge that generate melody. These can be observed in a wide range of musical practices, sometimes consecutively, and even concurrently in a single vocal work composed of multiple voices performing together. The developmental puzzle is to consider how harmony singing may have arisen. Clearly, monophonic unison arose out of coordinated behaviour: the more identical the performance of participants, the better (this can be equally true of dance). But how do we account for parallel harmony, or for the widespread practice of performing melody over a sustained drone? What is the significance of monotone performance, whether in accompanying an independent melody, or as the means of chanting text?

For Darwin, harmony represented a technology developed through culture, and, by implication, dependent on existing powers of speech:

> Helmholtz has explained on physiological principles why concords are agreeable to the human ear; but we are little concerned with these, as music in harmony is a late invention. (Darwin 1871, p. 866)

But as we have seen, Helmholtz was also interested in the harmonic properties he revealed in speech itself. Again, this suggests that Helmholtz's legacy has been subjected to branching interpretations: on the one hand, harmony as an invention or technology; on the other, harmony as an instinctive response, acquired alongside the anatomy of the modern human vocal tract, to the anatomy and neurology of mammalian aural perception. As ethnomusicology has revealed the divergent musical practices of the world's cultures, this duality could be seen as representing *explicit* and *implicit* 'theories of harmony' (Bannan 2008). Schenker portrayed the consequences of the former for the composition of European music in consciously Darwinist language:

> I have repeatedly had occasion to show the truly biologic characteristics displayed by tones in various respects. Thus the phenomenon of the partials could be derived from a kind of procreative urge of the tones; and the tonal system, particularly the natural [major] one, could be seen as a sort of collective order, similar to a state, based on its own social contracts by which the individual tones are bound to abide. (Schenker 1906, p. 84).

Schenker argues that response to the harmonic series arises from the very nature of music, but that the great achievements of the European tradition were made possible by the development of a sophisticated theoretical framework arising from these principles, which he viewed as absent from other musics. Of course, explicit theories of music had long existed in philosophical and practical documentation in ancient Greek, Chinese, and classical Arabic and Indian sources. However, *implicit* theories of harmony are clearly evident in the many vocal and instrumental practices worldwide that spring from exploration of the harmonic series (Trân Quang and Bannan, this volume; Espi-Sanchis and Bannan, this volume).

As Fitch (2006) has pointed out, ethnomusicology has proved less ready than linguistics to establish universals within the medium it represents. However, the following could be claimed as musical universals more-or-less present (Brown 1991) in the world's cultures:

1. the capacity for chant on a monotone (*monotony*)
2. the capacity for unaccompanied melody (*monophony*)
3. the capacity for harmony arising from the combination of different pitches.

Of these, (1) and (2) can be performed by individuals while also being open to group unison participation. However, (3) depends on the simultaneous contribution of different pitches and/or vocal contours by more than one singer. Notably, (3) can arise from a simultaneous combination of (1) and (2) (*melody over a drone*); or from the combination of two or more complementary (*polyphony*) or parallel lines (*heterophony*) which have the properties of (2). Essential to this formulation is the significance of (1), *monotony*, which as a term in everyday speech has had something of a bad press, suggesting limitation or boredom, but which in fact represents a significant physiological achievement with a specific role in human culture.

One can only speculate on how and when such abilities may have evolved. A further factor that must not be overlooked is the response of singers not merely to co-participants, but also to the environment. The phenomena of echo and reverberation can allow an individual performer to achieve both amplification and a sense of interaction with 'another': small wonder that this may have appeared to arise from magical forces, and have acquired associations with ritual behaviour and cultural observance. In the fledgling field of acoustic archaeology, Watson and Keating (1999) have produced acoustic evidence which suggests that prehistoric sites such as burial mounds and stone circles may have defined the nature of the rituals which exploited their properties. They found resonant responses that amplified easily accessible frequencies of the human voice, as well as periods of echo that can elicit rhythmic interaction. Pedagogical experiments that replicate these acoustic stimuli electronically (Bannan 1998) illustrate the fascination and motivation they can arouse in young singers. A modern composition that exploits a similar phenomenon is the sound installation *I am sitting in a room* by Alvin Lucier (1970) in which a recording of spoken text describing the process the listener will perceive is broadcast into a room and repeatedly re-recorded and replayed until the meaningful language of the original script is replaced by sounds that arise from the resonant responses of the room itself. Over a period of the minutes this process takes, language melts into a kind of primal music in which the speech can no longer be discerned. Such examples blur the distinction between linguistic and musical experience.

MUSICALITY IN VOCAL LEARNING AND CULTURE

Hockett (1960, pp. 89–96) identified thirteen design features of vocal communication:

1 Vocal-auditory channel
(an acoustic system which leaves the limbs free for other purposes)
2 Broadcast transmission/directional reception
(signal audible at distance/binaural direction-finding traces signal source)
3 Rapid fading (transitoriness)
(leaves no trace: may explain fascination for echo/mimicry. Contrasts explicitly with communication by *olfactory signs*)
4 Interchangeability
(user can both transmit and receive)
5 Total feedback
(user 'hears what he says': internalisation leads to consciousness)
6 Specialization
(speech is a dedicated system, not a by-product of other behaviour)
7 Semanticity
(capability of communication to be specific and interpretable)
8 Arbitrariness
('whale'=small word, big thing: 'micro-organism'=vice versa)
9 Discreteness
(significance of small variations in the code: e.g. 'pin' not same as 'bin')
10 Displacement
(the capacity to refer to absent objects, the past/ future, etc.)
11 Productivity
(the capacity of the system to allow meaningful coinage)
12 Traditional transmission
(capacity for the code to be used to develop/acquire the code)
13 Duality of patterning
(economical generation: e.g. 'cat'/'tack'/'act' share the same phonemes)

His aim was to demonstrate that, while many unrelated species possess component communicative abilities similar to those of humans, ours is the only species to synthesize the full range represented in his model into an integrated, variable behaviour capable of infinite yet meaningful productivity.

He also proposed using these design features as a means of differentiating between human language and music, though his analysis of music in terms of these features is based on a category of 'music' that is, by implication, instrumental, and comprises 'only Western music since the time of Bach' (Hockett 1960, p. 94):

Instrumental music

1	Is auditory but *not* vocal.
2	It involves broadcast transmission and directional reception.
3	It involves transitoriness.
4	It *does not* involve interchangeability.
5	It involves total feedback.
6	It involves specialization.
7	It does not (in general) involve semanticity.
8	(The box for arbitrariness is left blank.)
9	It *partly* involves discreteness.
10	(The box for displacement is left blank.)
11	It involves productivity.
12	It involves traditional transmission.
13	(The box for duality of patterning is left blank.)

While Hockett's design features represent a useful basis for examining what happens when animals communicate vocally, providing an insight into the different contexts in which musical vocalization may occur, his own application of them to music is limited and internally inconsistent. Usefully, though, it provides a framework for the further examination of distinctions between language and music, specifically where both are expressed vocally: that is, when the music is sung. Indeed, Fitch (2006, p. 178) has extended Hockett's model, proposing the need to add a fourteenth feature, *complexity*, and suggests an interpretation of *interchangeability*: although someone unable to play an instrument cannot therefore respond on that instrument to an instrumental model, this can be achieved by singing. Indeed, this idea plays an important part in Cox's (2001) theory of participatory ('mimetic') listening.

Brown (2007) also complements Hockett's approach by proposing design features for the way different species sing. For instance, humans share with several avian species (such as parrots) the ability to control timbre and duration in producing the target acoustic objects of speech; are able, like fireflies and certain crickets and frogs (Merker 2000) to entrain precisely to a common pulse; and can, like humpback whales (Payne 2000) generate and exchange new repertoires. But we are the only species to combine these abilities so as to sing in unison (both real and, even more significantly, at the octave), let alone to contribute differing but acoustically complementary pitches to collective performance through singing in harmony: a musical component of the *complexity* whose inclusion Fitch (2006) argues is essential in any comprehensive definition of specifically human communicative attributes. While the group performance features outlined above depend on the matching of rhythm and fundamental frequency, they also convey information arising from conformity of timbre, ranging from coordinating the production of matched consonants and vowels to the specialized capacity to reinforce a given timbre by performing it simultaneously at different specifically related frequencies: in harmony. This 'third dimension' to human exploitation of pitch relationships that relate the 'vertical' (simultaneous) to the 'horizontal' (sequential) is open to explanation on evolutionary lines.

Further elaboration of the implications of Hockett's claims for the differences between language and music is beyond the scope of this chapter, but for present purposes the following can be stated:

- To the extent that musical behaviour is an evolved characteristic, vocal music, which arises directly from the potential of evolved anatomy, is far older than instrumental, which employs extra-somatic tools.
- If music is to be examined as a human universal, theorization about its nature and structural properties cannot be limited to consideration of one culture or historical period.
- The infinite productivity of human language is arguably matched by our species' capacity for variety of song, in terms of mimicry, imitation, polyphony, and styles of vocal production (Trân Quang and Bannan, this volume ch. 6). Merker (2002) characterizes the human capacity for music as a 'Humboldt system' in which the finite media principally represented by discrete pitches and rhythmic durations give rise to combinations and syntheses of potentially infinite complexity and uniqueness.
- Developments in the understanding of music cognition, both theoretical (Cox 2001) and clinical (Hodges 2000; Parsons 2001) illustrate that *displacement* has a function in musical behaviour, and that *duality of patterning* also represents a feature of musical structure evident in both productivity and perception whereby

elements can be embedded within musical discourse at different levels with variable functions (see Hofstadter 1979; see also Jan 2007).

Hockett's model was intended to account for language as a set of distinct but overlapping traits with an evolutionary history shared with ancestor species (1960, p. 93). He did not set out to consider, as Darwin had proposed, that song may have played a part in this process. A full account of the evolution of human song intended to complement and extend Hockett's model will need to recognize the different conditions in which song is practised: for instance, as simultaneous chorus, or wordless emotional expression; as well as how these may have conferred survival value in specific social/environmental contexts (Dunbar, Foley, Gamble, Merker, Morley, this volume).

A robust account of the nature of human capacity for song therefore needs to focus on features that have hitherto been treated as peripheral to language, but which this chapter argues are central to all vocal behaviour, and to its evolution. Principal amongst these are the fixed-pitch relationships that link together vocal timbre, resonance, melodic contour, and the capacity for simultaneous performance; as well as the neural and limbic features of response and memory that assign and retrieve emotional and motivational affect inseparable from musical perception and production.

THE HARMONIC SERIES AND HUMAN VOCALIZATION

Anatomical evidence examined in relation to its acoustic potential by Sundberg (1987), specifically the relationship between lung capacity and the efficiency of the vocal folds to turn sub-glottal pressure into broadcast sound, illustrates that the human vocal instrument evolved for high-energy communication over distances. This clearly applies equally to the neonates of the species (Locke 2006; Soltis 2004), which cry loudly, unlike those of chimpanzees (Falk 2004). This anatomical arrangement in our upright early ancestors preserved the capacity for instinctive, call-like production; but critical adaptations such as the development of voluntary breathing (Hunley 1998; Walker and Shipman 1996) and a more sophisticated vocal tract (Frayer and Nicolay 2000; Koppe et al. 1999; Altmeier and Ward 1998; Kay et al. 1998; Falk 1988; Franciscus and Trinkhaus 1988; Lieberman 1975; Morley, Clegg, this volume) endowed the physical capacity for imitation of a wide range of target sounds on which human vocal learning depends. The implication of increased neoteny in modern humans compared to our common ancestor with chimpanzees should not be overlooked, permitting a more direct mapping of movements (Meltzoff and Moore 1992) and phonation (Minami and Nito 1998; Kessen et al. 1979) due to the adult human's greater preservation of the anatomical proportions of its infant form. While fossil evidence (Johanson and Edgar 1996; Gould 1977) allows a glimpse of this anatomical morphology, it represents poorly the crucial evidence for changes in neural organization that determine the relationship between perception and production which allows vocal learning to take place.

The performance advantages acquired during hominin evolution exploited with increasing precision and intentionality an existing mammalian aural system (Braun 1999; Wright et al. 2000). This has the capacity in our species to endow the temporary window in which infants appear instinctively to match pitch in the first months of life (Minami and Nito 1998; Kessen et al. 1979), though this ability seems not to be retained towards the end of the first year (Siegel et al. 1990). Nevertheless, it illustrates that pitch-matching forms part of the suite of abilities that relate to human vocal learning. Neonates also process intervals according to consonance and dissonance (Ross et al. 2007; Sethares 1993). Indeed, human response to organized sound commences several months prior to birth, as the first of our senses to become active initiates the laying-down of musical memories in the womb (Lecanuet 1996; Woodward 1992; Lester and Boukydis 1992; Shetler 1990); it then proceeds actively, through play, mimicry, and rehearsal (Bannan and Woodward 2009; Eckerdal and Merker 2009; Stefanics et al. 2009; McMullen and Saffran 2004; Street 2003; Ilari et al. 2003; Trehub 2001, 1990; Malloch 1999; Minami and Nito 1998; Trevarthen 1994; Kessen et al. 1979) that develop and exploit control of the four parameters of production: fundamental frequency, duration, volume, and timbre.

Both technical competence and pragmatic motivation (understanding of when and in what way it may be appropriate to vocalize) emerge from creative play behaviour, solitary as much as social, and from intersubjective responses that include simultaneous participation (Bannan and Woodward 2009). Communication becomes possible as instinctive acoustic production is harnessed intentionally (Deacon 1997). A key feature of this process is the achievement of intentional sustained *monotony*, whereby fluctuation in fundamental frequency, that is in neonates solely subject to emotional motivation or state of health (Soltis 2004), is gradually brought under control in the processes that permit proto-song in the second year, and production of the first articulate words. The capacity to match, maintain, and recall individual pitches permits relations between pitches to become organized both 'vertically', commencing with the achievement of unison that could be taken as the default position, and 'horizontally'. Were singing to have evolved prior to language (Darwin 1871), these represent the properties of sophisticated group participation that enculturation into vocal behaviour could endow without dependence on linguistic exchange. While music in modern humans continues to exploit these abilities, language would appear not to, except in rhythmically organized speech such as oaths and prayers. Nevertheless, as Helmholtz and Garcia speculated, traces of this relationship between the 'horizontal' and the 'vertical' in musical perception and production parallel the acoustic properties of language that are represented in the source-filter theory of speech.

However the productive and perceptual abilities that permit these musical behaviours originally evolved, they have been harnessed in quite different ways in the two media through which humans communicate vocally:

1. In language perception and production, it is the variation of formants (and their transitions) in relation to fundamental frequency that provides vowels, the phonetic engines of efficient vocal continuity and significant carriers of distinct meaning. While fundamental frequency (F_0) is not, and need not be, fixed, the production of vowels in specific acoustic relationship to F_0 relates to harmonicity, whereby the categorical differences between the formants of vowels map onto frequencies generated by the harmonic series of the voiced sound (Ross et al. 2007).

2. In music, control of fundamental frequency allows the establishment of *tonality*, the means by which pitch relationships operate on short-term memory to achieve musical coherence in relation to the passing of time. Tonality of one kind or another is universal: it emerges as the pitch-centricity that gives rise to the scales and modes in which

music is organized; the hypnotic properties of the reciting-note of chant; the phenomenon of combining melody with a drone that represents arguably the simplest means of exploiting the 'three-dimensionality' of music whereby melodic contour and rhythmic entrainment are complemented by harmonic depth-of-field; and more complex polyphonies based on independent voices, and so on.

Critical to both these productive features of language and music is their perceptual underpinning in the phenomenon known as the 'harmonic series'.

This acoustic principle is well known to musicians, who have for millennia exploited its properties in myriad ways to construct and master a range of musical instruments from available materials (see Espi-Sanchis and Bannan, this volume ch. 7). The extent to which vocalists have learned to control the relationship between F_0 and higher harmonics so as to produce spectral melodies is increasingly known and understood (Trân Quang and Bannan, this volume ch. 6). This conscious exploitation of overtone production that exists both in singing and in the use of the vocal tract as a resonator for or component of instrumental technique, such as in Xhosa bow music (Dargie 1988; see also Espi-Sanchis and Bannan, this volume), or performance on instruments as unrelated as the Jew's harp (Ledang 1972) and didjeridu (Tarnopolsky et al. 2006), illustrates knowledge, whether implicit or explicit, of the properties of the harmonic series. What requires examination is what role implicit knowledge of the harmonic series, reinforced socially in group performance, may have played in the vocal prerequisites for the development of language.

Skilled instrumentalists can isolate harmonics as distinct pitches, on stringed instruments by lightly touching the node of a string at the precise point that suppresses other vibrations, and on wind instruments by varying the pressure and direction of the air supply. When they do this, overtones that may have contributed to the sound of the fundamental, or remained latent in the instrument, are brought into relief as distinct pitches. In the same way, practised singers in cultures as widely separated as southern Africa and Mongolia can achieve the same effect in a stule known as 'overtone singing'.

Meanwhile, the aggregate sound of performing at the same time notes related to one another in this way results in a chord: for example, A (440 Hz) + E (1320 Hz/3rd harmonic) + c# (2200 Hz/5th harmonic) produces the euphonious chord known as A major. Differences between spoken vowels depend on similar features of resonance, suggesting that vowels employed in speech may represent an adaptation to language of harmonically differentiated properties of high-energy singing.

If speech emerged from an existing capacity for shared song, then unison singing, in addition to imitation, would have required vowels to be 'tuned' to consonance between participants, as occurs in group singing today. The low-energy acoustics of normal speech do not depend on this phenomenon, though it may still be vestigially apparent. Evidence for this would include: that resonant singing, especially that which depends on blend between several singers performing simultaneously, requires attunement to collective timbre as well as relative pitch; and that infant-directed speech by adults and older siblings tends towards greater clarity of vowel production as well as heightened pitch contour.

The link between the otherwise distinct functions in language and music represented by vowel production and tonality arises due to the evolved perception of categorical boundaries that are themselves consequences of the neuro-anatomy of aural discrimination (Braun 1999; Langner et al. 1998).

FORMANTS, VOWELS, HARMONICS, AND INTERVALS

As Braun points out (2009), all mammals are mobile wind instruments whose sound-making capacities evolved with their respiratory tracts, and the relationship between the sounds we are thus capable of making and perceiving has shaped the neural and anatomical features of the mammalian auditory system. This involves not merely the processing of humanly produced sound—whether our own or that of others—but its perception within a complex acoustic environment made up of natural sounds and the productions of other species, as well as discrimination between them. Acutely selective hearing provides one of the most vital defence mechanisms on which our survival depends. In humans, this has conferred the different ways in which we pay attention to sounds. For instance, we efficiently discriminate between human vocalizations and sounds produced by non-human agents (Vouloumanos and Werker 2004; Belin et al. 2000). We also listen to and produce speech and music as having differently organized acoustic properties. For instance, when someone shouts 'Fire!' we do not stop to admire the tone of their voice, but respond instantly to the information provided because it aids self-preservation. By contrast, in listening to or performing the Elizabethan madrigal 'Fyer, Fyer' by Thomas Morley (1595), we do not rush from the concert hall in response to the meaning of the words: indeed, their presentation as music invites us instead to appreciate their performance (see Deutsch et al. 2008 for an analogous experiment). Barthes captured something of the differential processing of the communicative systems of music and language in his illustration of the dominance of language: 'Language . . . is the only semiotic system capable of *interpreting* another semiotic system' (Barthes 1977, p. 179).

While language and song are produced employing the same anatomical structure (Sundberg's 'wind instrument' (2001); Braun's 'air tube like . . . a flute or bassoon' (2009)), this difference in how they are perceived has had consequences for the ways in which their products are studied. Where musicians conceive of timbre as arising from the combination of notes and its dependence on harmonic content in terms of the relationship between the 'vertical' and the 'horizontal' as these are sequenced through rhythmic organization over time (Schenker 1906; Wishart 1996; Sethares 1993; Carlos 1987), speech science analyses vowels in terms of the categorical differences in *formants*. While the capacity to perceive and match vowel sounds relates to bandwidths and timbral characteristics that derive from the properties of the harmonic series (Braun 1999), perception and production of vowels does not involve consciousness of harmonic series relationships. The relationship in speech between prosodic contour and the 'vertical' characteristics of vocal timbre (including vowels) would not normally be viewed as musically organized.

De Boer (2000) approached linguistic discrimination of vowels and its development through computer modelling that set out to replicate how a vowel system may have evolved. Statistical surveys of languages and the speech sounds they employ (Maddieson 1984) illustrated that '[t]he sound systems of human languages show remarkable regularities' (de Boer 2000, p. 441). De Boer set out to investigate whether 'the structure of vowel systems is determined by self-organization in a population under constraints of perception and production' (p. 445), stating:

> Explanations of phonetic phenomena that do not depend on innate properties of the brain have been based on independent evidence from physics and communication theory. They start from the assumption that speech is meant to provide communication over an unreliable channel using the human vocal tract and perceptual system (p. 442).

Addressing the ontogenetic parallels to the phylogenetic organization of vowels, de Boer observes:

> When children learn the sound system of their native language, they do not explicitly optimize it. Rather they learn to produce sounds that will eventually become extremely similar to those of their parents and peers (p. 443).

It is this process of convergence on 'target' vowels that de Boer's work examines. But what defines the nature of the 'target', and what is the mechanism that permits discriminate perception to develop?

The reliability with which children learn vowels relates to whether they can reproduce them sufficiently distinctly to avoid confusing listeners. The 'raw material' that vowel production embraces can be set out as a sequence in which progressive shaping involving the lips and tongue position moves from narrow/suppressed via open/enriched to narrow/enriched—referred to by singing teachers as the 'vowel circle'. Sung performance of this sequence on a monotone will generate distinct harmonics:

oo – oa – or – o – ar – a – e – é – i – ee – ü

The following box shows why a variety of distinct vowel classes is essential to distinguishing correctly between otherwise similar words. In this presentation, the left-hand column is set out vertically so that each subsequent vowel maps onto that generated by the next harmonic, beginning with the dull sound, of 'boot' in which harmonic activation is suppressed by the narrowing of the lips, and gradually 'opening out' to the harmonic-rich front vowels. The words in the right-hand column represent those whose vowel colour does not fall within this harmonic series sequence, but which is achieved by alternative means.

The main vowels in English, illustrated by words and names beginning with a B *sound and ending with a* T *sound*

Boot (English)	
Bʊat	
Bought	
Bott	
Bart	Burt
Bat	Beart
Bet	But
Bate	
Bit	
Beet	
Boot	Bute
Büt (Scottish)	

In experiments that explored the two dimensions of pitch perception, the 'vertical' features of timbre and the 'horizontal' features of melody, Sethares (1993) tested the relationship between the timbre of synthesized tones and the relative consonance or dissonance of their combination according to the precise tuning of scales designed to match their properties. His results confirmed Carlos's (1987) observation that 'the timbre of an instrument strongly affects what tuning and scale sound best on that instrument'. Since mankind's use of his own voice co-evolved with the perceptual system responsible for its tuning, it would seem likely that Carlos's dictum applies especially to the human voice and both its phylogenetic and ontogenetic acquisition. Might the intervals of F_0 contour (prosody) thus be derived from those present in the harmonic series, perceived both as timbre and specific dyad (two-note chord)? And might the condition of people who acquire language but are less confident singers arise due to their sufficient and reliable grasp of phonetic production for speech in their acoustic and social environment excluding, or even suppressing, retention of vocal production with song-like characteristics? We will consider this question as one of pragmatics relating to the individual's socially and culturally influenced motivation to produce song-like resonance in vocalization.

The composer Karlheinz Stockhausen exploited this relationship between vowel timbre and musical harmonics in his vocal composition *Stimmung* (1968). The piece requires six singers trained in the production of vocal harmonics, which Stockhausen relates to specific vowels, to improvise sung responses to spoken names and texts, all underpinned by the intonation of a six-note chord (one pitch per singer) that is itself derived from the 2nd, 3rd, 4th, 5th, 7th and 9th harmonics of a low, unheard B flat. *Stimmung* has proved influential on the subsequent generation of composers, not least as overtone singing in a variety of cultures has become more widely known (Trân Quang and Bannan, this volume).

In research that replicated some aspects of de Boer's on vowel acquisition, the composer and music scientist Miranda (1999a) explored the relationship between vowel acquisition and more general features of timbre discrimination and productivity. The focus of Miranda's work is to illustrate the potential of artificial life modelling (*Alife*) for musicological research. Building on the example of de Boer (1997), Miranda's account of the acoustics on which spoken language depends asserts: 'Speech sounds normally have four or five prominent formants whose central frequencies are crucial for the ear to distinguish between different sounds'. It is these 'central frequencies' that interest Miranda in the generative *Alife* modelling he devised: for, unlike de Boer's imitation games, what Miranda constructs is a silicon tool for organizing musical sounds. In previous work (Miranda 1999b), he had 'demonstrated how a pitch system for musical composition can be naturally derived from the formant frequencies of a given vowel set' (Miranda 1999a, p. 16). In preparing for this, he had synthesized melodies for the words of the Latin Mass using randomly generated notes, evaluating the way that (in a manner consistent with the predictions of Sethares (1993) and Carlos (1987)) the melodies in which F_0 more regularly matched the associated vowel tended to sound more natural: 'those melodies containing specific configurations of matching vowels with non-matching ones tended to form coherent musical prosodies' (p. 16).

The phenomenon revealed in Miranda's *Alife* model is, though, extremely old, and can be discerned where chant-like music is performed under suitable acoustic conditions that permit the transmission of the full richness of the resonant human voice. The performance of chant, unsupported by instrumental accompaniment, exemplifies the capacity for sustained and resonant singing in which the qualities of vowels reveal themselves as both verbal and musical phenomena. Singing of this kind depends on an 'intentional resonance' (see Bannan

2000) whereby the performer presents language in a different manner to the production of everyday speech.

Fig. 12.1 sets out to convey in music notation the relationship between chanted text (what the singer performs, notated conventionally in the bass clef) and the spectral melody (what may be heard, notated without consonants in the treble clef) that can be discerned as arising from its performance. If listeners clearly hear the English text, even from a distance within the reverberant acoustic of a cathedral in which most of the consonants are suppressed leaving only the 'melody of vowels' intoned on a single repeated pitch, they do so through perceiving categorical differences in formants over the sustained and repeated monotone. A capable cantor presents these sounds so as to conform to a singing style that both communicates information—the text of the prayer—and conveys a calm and serene atmosphere in which the voice is perceived as stress-free and consoling. In many cases where expert singers intone such chants in resonant acoustics, distinct harmonics can clearly be heard as a spectral melody accompanying the earth-bound singer. This 'vowel melody' will not represent a fixed phenomenon: the voices of different singers will resonate in a manner related to personal voice quality; and the same singer will generate variant performances on different occasions. While these factors may, then, give rise to different 'melodies', in all cases the pitches revealed will be those derived from the harmonic series of the reciting-note. Fig. 12.1 represents a snapshot of the author's performance on a single occasion. The precise musical properties presented in the figure were arrived at by meticulous and repeated listening to the phenomenon in performance subsequent to observing harmonics in live, on-screen representation of a voice analysis programme [the CD has recordings of this prayer being sung].

In such phenomena, music and language meet, providing a tantalizing glimpse into their complementary functions as much as their shared developmental pathways.

ELEMENTS OF MUSIC 'FOSSILIZED' IN SPEECH, BEHAVIOUR, AND CULTURE

A variety of behaviours in language and oral culture indeed suggest a musical underpinning for their production and perception. These

Evening Collect

Figure 12.1. The 'melody of vowels' discernible over monophonic chant. The prayer is intoned on the note A (220 Hz), noted here rhythmically in the bass clef as a representation of what the singer performs. The upper part, notated in the treble clef, illustrates the 'melody of vowels' that can be perceived.

include: the ways in which adults modify their speech and song in performing to children; the role of musical behaviour in early infant socialization; the rhythmic nature of children's vocal play; the categorical shaping of aural preferences that relate to the way one learns language; the relationship between speech and melody in whistled languages; the involvement of music in a variety of therapies, especially where language use is inhibited; timbral variation in vocalization that expresses or conveys emotion; and features of timbral significance in wordplay, rhymes, and mnemonics.

Acoustic analysis of the universal practice of infant-directed speech (IDS) illustrates that vowel sounds that map the varied formants of adult speakers employ a greater categorical clarity than is the case in everyday speech (Eckerdal and Merker 2009; Trehub 2001; Trainor et al. 2000; Trevarthen 1999; Fernald 1992). This is as much a feature of the perceived 'musicality' of IDS as raised pitch levels, expressive contour, continuity of phonation, and slowed tempo. Liu, Kuhl, and Tsao (2003) hypothesized that 'the degree to which mothers expanded their vowel-space during infant-directed speech would be strongly associated with infants' performance in tests of speech perception' (p. F2). De Boer and Kuhl (2003) employed a computational learning model to test whether vowels with IDS characteristics were more easily learned than those in normal adult speech. The fact that they proved to be complements the research of Golinkoff and Alioto (1995) which illustrated that adults presented with IDS examples rather than normal speech learned new vocabulary more efficiently in a second language.

Further evidence for this transference to language of the non-verbally expressive and emotional properties of vocalization that we might term 'musical' occurs in the nonsense-phrases and onomatopoeia of children's speech and verbal games, especially in mnemonic pairs and sequences. No horse goes 'clop clip': in such expressions across various languages, the vowel with the 'high' harmonic always precedes the lower, as in a musical cadence. The following are examples of this:

ding dong
tick tock
clip clop
spicks and specks
riff-raff
pring prang

The latter, *pring prang*, is a category of children's singing game from Javanese Sunda (Romet, 1992). 'Riff-raff' was first recorded in the Glossary of Anglo-Indian (1886) edited by Yule and Burnell, entitled *Hobson-Jobson*. Such acoustic pairings would seem to be widespread, and have their origins in the relationship between production and perception. Words with entirely unrelated meanings (or no meaning at all) are linked in the mind as a consequence of their articulatory demands. Proximate and related neural systems give rise, often unconsciously, to a range of verbal features such as puns, malapropisms, spoonerisms, and Freudian slips.

As children develop, their response to music varies as they respond to age-related stimuli in their social environments (Bannan and

Woodward 2009). Romet (1992) illustrated the difference in style between a lullaby sung by a Javanese mother and a play song in which a child aged three or four from the same culture might participate. Her transcriptions and commentary illustrate the changes of emphasis evident in the child's weaning from lullaby onto play song that are open to a variety of interpretations (see box below). The contrast between the two repertoires would seem to conform to criteria for analysing the organization and perception of music, and its social and evolutionary purpose, in the work of several authors: Grainger 1934 (in Blacking 1987); Sachs 1962; Wishart 1996; Merker 2000; Deacon 1997:

	Neng Neleng Kung	***Pring Prang***
	Sung to an infant by an individual adult	Sung by groups of children
	Timeless,	Measured
	Melismatic	Syllabic
	Organic	Repetitive
(Grainger 1934)	Melodic	Rhythmic
(Sachs 1962)	Melogenic	Logogenic
(Wishart 1996)	Continuum	Lattice
(Merker 2000)	Individual	Group
(Deacon 1997)	Serial	Simultaneous

The criteria adopted here from Wishart (1996), the opposition and complementarity between 'continuum' and 'lattice' processing of sound, represent an especially useful tool for discerning varied kinds of musical behaviour. Where *continuum* maps onto sustained, 'effortless' vocalization (e.g. singing for the length of an extended breath), *lattice* comprises sequences of distinct sounds organized rhythmically. Both conditions inform music and language, and provide means of expressing and communicating different emotions. They also define different ways in which group musical performance can be built on the features of *seriality* and *simultaneity*, conforming to the performance properties of monophony, polyphony, and heterophony.

An example of the relationship between speech phonetics and musical pitch is conveyed in the way that the timbral features of vowel sounds in language are replicated by the distinct intervals traced in the contour of melodies in whistled languages. This phenomenon was first demonstrated for the *silbo* of the Spanish island of Gomera by Classe (1957), and has parallels in similar surrogate-verbal language production in other parts of the world (Meyer 2004). Meyer concluded 'The study of whistled languages might be a means to better understand the role and status of prosody in the language faculty and some elements of its evolution linked to musical aspects'.

The relationship in development and language acquisition between the oral and aural domains can be traced in the correspondence found between early exposure to the local acoustic environment represented by human vocalization and the characteristics of speech (dialect) acquired. Raggozine and Deutsch (1994) discovered that subjects' preferences in resolving the ambiguous interval of the tritone presented in aural perception tests mapped onto existing test data that correlated, with geographical precision, with the location of the aural environment in which they had learned their mother tongue. Their test procedure required subjects to select how they 'resolved' the acoustic phenomenon of the tritone, an interval that inverts as itself and thus gives an ambiguous sense of which note is top and which bottom. The profile derived from iterations of the test procedure suggested that exposure to language and its imitation shapes the way we process pitch. The consequences for accurate vocal learning, including target accent, prosody, rhythm, and timbre, of what Merker (2005) refers to as the 'conformal motive' represent a compelling holistic process, and perhaps this confers sensitivities that are not themselves expressed in vocal production but which leave a trace on aural perception.

A field that provides rich evidence of a role for music as a paralinguistic medium of considerable significance is that represented in music therapy and music medicine. El Mogharbel et al. (2003) traced the development of an autistic-spectrum child who never acquired nor understood speech yet was able expressively and creatively to generate extended musical structures through a form of scat-singing. Where language has been lost through disease or trauma, musical communication can often remain available as a medium of self-expression and participation, especially where subjects who are unable to speak prove able to sing or whistle (Bannan and Montgomery-Smith 2008; Sacks 2007, 1985; Aldridge 1996; Morgan and Tilluckdharry 1982). Specific musical therapies have been devised for certain conditions, such as melodic intonation therapy (Belin et al. 1996; Seki & Sugishita 1983), which can prove effective in the treatment of aphasia. Tamplin (2008) reports the treatment of subjects with dysarthric speech resulting from stroke or trauma who responded well to exercises involving singing. New digital aids support the fluent language production of stutterers by providing a 'chorus' represented by a feedback loop which replays their own speech as if performed in unison with others (Kalinowski et al. 2004). The invention of this technology was preceded by research into the phenomena of choral speech (Kalinowski and Saltuklaroglu 2003) and the presentation of simultaneous silent mouthing (Saltuklaroglu et al. 2004).

The link here would seem to be the specific features of human vocal timbre, and their capacity to elicit emotion, memory, and categorical response both in 'passive' perception and in imitation. As Schellenberg et al. (1999) argue, while timbre may be difficult to define, its effects can be both instantaneous and powerful. They played excerpts of songs only 0.1 of a second long, which subjects were nevertheless able to recognize and match to waveforms presented visually. Traube and Depalle (2004) discovered that analogies between vowel colour and guitar-plucking techniques can be clearly determined such that they can be agreed upon by performers and listeners and thus play a meaningful part in pedagogy. Perception of emotion in the voice itself is also demonstrably related to timbre. Waarammaa et al. (2006) found that perception of emotional valence in synthesized vowel samples could be manipulated through boosting or suppressing acoustic energy in the *F*3 region. The subtlety and diversity of human vocal communication exploits this capacity to assign meaning to timbral variation.

HOW SUCCESSFUL AN ADAPTATION IS THE HUMAN VOICE?

> And we know what Helmholtz, the highest authority in Europe on the subject, has said about the human eye; that if an optician had sold him an instrument so carelessly made, he would have thought himself fully justified in returning it. (Darwin 1872, p. 464)

Darwin's placement in the evolutionary context of *The Descent of Man* of Helmholtz's evaluation of the engineering properties of the human eye parallels attempts to reverse-engineer our species' capacity for vocalization. This is important when considering the developmental pathway, involving a series of anatomical and neural adaptations and exaptations (Bannan 2008, 2003; Morley 2002 and this volume), that must have bridged the gap between the animal communication of our nearest common ancestor with the great apes and modern human vocality. In particular, to what extent does the human voice represent, anatomically and acoustically speaking, an effective sound-producing mechanism?

> Although the singing voice resembles string and wind instruments in many ways, there is a major size difference that forces the biological system to take advantage of nonlinearities. Both the length of the vocal fold and the length of the vocal tract are well below standard (in terms of musical instruments) . . . Even in comparison to the tubes and horns of brass and woodwind instruments, we are woefully undersized. (Titze 2003, p. 21)

But the evolved proportions of the human vocal tract provide the constraints under which neural mechanisms have developed to make the most of what they provide.

> The result of this undersizing of our airways is that we cannot resonate many harmonics of our sound source at any one time. By musical instrumental standards, this should result in a rather dull and uninteresting instrument. (Titze 2003, p. 22)

The modern human brain has turned this deficiency into the defining capacity of our species.

As Darwin could have explained to Helmholtz, evolution does not need to design perfection: just to respond sufficiently to selective forces such that the organism survives and reproduces. The difficulty in tracing the evolution of the capacity for speech is to get from animals calls to language in one adaptation (see Merker, this volume). As language-users, we need to think beyond the 'continuity paradox' (Bickerton 1991) whereby it is impossible to conceive or express within the medium of language what meaningful vocal communication without language may have been like. A combinatorial system able to utilize the increased articulatory capacity of gracile early *Homo* to vary meaningfully and intentionally the timbre of high-energy vocalization represents a potential pre-linguistic stage consistent with the fossil and comparative anatomical evidence.

A developmental Rubicon must have been crossed when hominins acquired the ability to modify timbre over an unchanging F_0:

> if a singer wants to sing a crescendo on a constant pitch, she has to make complicated adjustments to vocal fold tension to overcome the effects of a greater blowing pressure. (Titze 2003, p. 22)

Like the achievement of monotony, we take this capacity for granted. Yet the engineering involved in control of these apparently simple actions is highly complex.

In retaining lifelong the capacity to match vocal models, we continually and subtly reinvent our vocal performance in response to a variety of conditions as voices alter in range in relation to maturation and other biological features: adolescent voice change in males (Cooksey 1997); variation in range and timbre due to menarche, pregnancy and menopause in women (Lã and Davidson 2005; Abitbol et al. 1989), and to physical states such as dehydration that affect voice quality (Verdolini et al. 2002):

> We can tell when someone is eating while they are talking . . . we know when they breathe hard, when they are nervous, when their throat is dry or swollen, when they are sad, happy, or disgusted—all from the voice signal. (Titze 2003, p. 23)

WHY DO MANY PEOPLE WHO EASILY ACQUIRE LANGUAGE SEEM UNABLE TO SING?

This question has to be addressed because it is often raised as an objection to the view that singing may have played a role in human evolution. The point is made: 'if singing preceded language and is viewed as a human universal, then why am I "tone-deaf"?' How can a capacity have had survival value that modern humans can fail to express?

In a recent self-experiment in collaboration with the neurologist Lawrence Parsons, Steven Mithen has set out to overcome his own self-declared 'tone-deafness', tracing the effect on his capacity to learn and its consequences for the organization of his brain (Mithen 2008). The initial outcomes of this project are consistent with existing research in music education and community singing (Knight 2000; Bannan 2000; Buttolph 1993), as well as with the implications of perceptual testing (Sloboda et al. 2005): very few people are genuinely unable to respond to changes in pitch; and, vitally, the skill of doing so improves with active participation. This is consistent with the implications of Hockett's (1960) 4th and 5th design features: *interchangeability* (whereby the communicator can both transmit and receive) and *total feedback* (we 'hear what we say'). But for some people there seems to be a developmental interruption whereby their production in singing does not seem to keep pace with their perception (Dalla Bella et al. 2009). However, such 'false' *amusics* (Sloboda et al. 2005) can prove capable of making considerable progress, as Mithen has found in his own case (see also Welch 2001).

This phenomenon has yet to be adequately explained. Welch (2001) suggests that it may relate to the interruption of singing development by overloading involved in the rapid development of speech and vocabulary. Knight's (2000) subjects reported a sense of rejection associated with their singing by peers, carers, or teachers that silenced them until her approach proved able to release their singing voices (see also McKinney 1994). Bannan (2000) modelled a differential response, however caused, whereby some individuals retain a sense of fascination and pleasure in the sheer experience of resonant voice-use, whereas for others this can be inhibited. To overcome this, inexperienced singers need to re-educate the relationship between vocalization, motivation, and emotion, preferably in a supportive social environment:

> Resonant voice is defined as a voice production that is both easy and vibrant in the facial tissues. The perception of 'ease' and 'vibrancy' belong primarily to the person producing the sound, but listeners can have similar perceptions. (Titze 2001, p. 519)

While these explanations for 'tone-deafness' and its remediation are proving effective in music education and therapy, they provide only a glimpse of the evolutionary factors that may be involved. They do, though, illustrate the powerful effect on behaviour of what one might term the pragmatics of singing. How might this have come about?

Combining words with music in singing (Dunsby 2004; see also Welch 2001; McKinney 1994) is a cognitively and anatomically complex task, albeit one that, in most cultures, humans perform with comparative ease. But language (Deacon 1997; Barthes 1977) demands that it be acquired as the dominant medium: a process that may be underlined by schooling and literacy (Welch 2001; Bjørkvold 1992). It may be that, in falling short, inexperienced singers have to 'drop' that which the allocation of limited perceptual or productive resources defines as the inferior, or less useful, medium. But there is clearly an emotional aspect to this, which would suggest a pragmatic component to the ability to sing related to self-esteem (Knight 2000) and self-identification as having a singing voice acceptable to others (Bannan 2000).

Drawing on a proposition outlined by Deacon (1997, pp. 111–13) that has been forcefully voiced by Blackmore (2007), one explanation would be that both music and language have developed as memetic parasites that have adapted to their hosts in order to exploit their resources. On this basis, whatever the sequence of adaptations that research may reveal, language has emerged as the dominant medium, able to inhibit musical responses. This may prove consistent with the architecture of 'perceptual rivalry' (Pettigrew 2001; Ngo et al. 2008) whereby the different 'switching' rates between the hemispheres of the brain correlate with cognitive preferences and personality traits. It may begin to explain why some children accelerate in their musical processing during key developmental windows while others appear not to (Welch 2001). Pedagogical experiments designed to illuminate how musical learning may be possible in the absence of language provide an insight into this issue (Bannan 2005, 2004, 1998).

CONCLUSIONS: HARMONY, PERCEPTION, SONG, AND CULTURE

The human voice has developed to exploit in complex ways the capacity for variation over time, both in terms of short-term contrast and, through the recursive properties of music (Hofstadter 1979) and of grammar (Chomsky 2001), over longer and more potentially meaningful structural spans. But in what ways has the human voice remained a musical instrument?

If Carlos (1987) is right that 'the timbre of an instrument strongly affects what tuning and scale sound best on that instrument', then the human voice is, by comparison with manufactured musical tools employed around the world and in recorded history, a very odd musical instrument indeed, in that its commonest characteristic is ever-changing timbre consequent both on the production of words and in the communication of varied emotional states (Scherer 1992;

Salgado 2000). Analysing the singing voice from the perspective of de Cheveigné and Kawahara (1999) illustrates that, unlike the costly achievement of fixity with which musical instruments are crafted, wherein performers depend in their engagement with their instrument on predictable physical dimensions (string length and tautness, length of tube, position of finger-holes, etc.), the sheer number of variables that impinge on the production of the human voice (see Fig. 12.2) represents a considerable contrast. This is captured in the

When we perceive the vocalizations of others, we can assign to the acoustic signal values that allow us to 'read' into it information about any—or all—of the following:

Culture

Health Gender Homeostasis Maturity

Emotion Pragmatics

Distance between participants Acoustic environment

Figure 12.2. 'Input' features that determine vocal timbre.

mathematical complexity with which acousticians have to deal when accounting for how sounds are produced and modified in the human vocal tract: the 'output' features of vocal performance within a flexible two-tube system (Sundberg 2001; Lieberman 1975) endowed with variable resonators, filters and inhibitors. In terms of any analysis that approaches the adaptation for singing of the human voice from the angle 'form before function' (i.e. that accepts the potential for physical form to have arisen by random adaptation prior to neural changes that harnessed its latent capabilities), it is precisely within the feature of timbre that we locate those properties of human communication conferred by evolution that rendered our sound-producing mechanisms the perfect complement to the configuration of our auditory capacity.

Through looking at song in addition to language, we can place the development of human vocalization in a broader context that conforms well to the demands of Tinbergen's (1951) framework for studying evolved behaviour. 'Reverse-engineering' the full range of human vocal activity cross-culturally reveals properties of song in comparison with speech that could be taken to confer survival advantages. These represent candidate adaptations on natural selection lines where specific anatomical features have been exploited:

1. Cost-benefit: expenditure of energy is minimized where respiration can be slowed without harm to the organism in support of variable and resonantly amplified utterance. This may have a role in empathetic vocal affect, such as songs for calming both listeners and/or the singer himself—singing can evoke a sense of well-being (Dunbar, this volume).
2. A wide variety of responses to social and environmental stimuli influence the repertoire of human vocalization, both emotionally motivated and under full voluntary control, conferring flexibility in the exploitation of learned patterns of vocal production and interaction.
3. Simultaneous vocalization can prove effective in myriad ways: as collective aggressive display, as entrained celebration or ritual, and as a means of eliciting a variety of shared emotions.

The behaviour central to these developments, which depends on adaptations which separate our species from the chimpanzee would seem to be chant:

> As an adaptive extension of primate vocalization systems and primate display acts, proto-chanting may represent a missing link in the transition from non-human primate to hominid communication, and in the shift from early to later varieties of hominid communication. (Oubré 1997, p. 187)

Three critical adaptations could thus be proposed on which the acquisition of song depended, and in which phylogeny is recapitulated in ontogeny:

1 Voluntary breathing
(from pant-hoot to controlled exhalation)

2 Monotony
(vocal control of F_0 transcends emotional motivation)

3 Tonality
(F_0 encoded in short -term memory permits melodic patterning and group choral song)

These adaptations would have made possible the capacity for chant that exploited the capacity to reinforce sustained monotone phonation while 'colouring' the song through precise formant variation is a candidate for the kind of proto-song that Merker (this volume), Brown (2007, 2000), Fitch (2006) and Mithen (2005) have argued as a stage preceding the development of language. While chant could be either solitary or gregarious, its resonant production could have been consistent with sensitivity to the properties of the harmonic series. An *implicit* 'theory of harmony' could thus shape musical interaction prior to, and in the absence of, language.

Support for this speculation is provided by a study carried out by Walker (1997) of the reliability of F_0 maintenance in monotone chant across a variety of cultures, including Australian Aborigine song and Western opera. Walker's data illustrates that, while F_0 wanders, the element of song that remains stable is timbre:

> Greater concentrations of energy and formant activity at F1 and F2 in the Aboriginal data suggest greater importance being placed on vocal articulators than on vocal fold movement. The former yields vowel pitch and the latter musical pitch information... And in Western opera singing, the specially focused high spectral energy reflected in formant data appears to be as important as musical pitch... It cannot be supported, therefore, that musical pitch is a 'natural' a priori component in all the singing studied in this paper. (Walker 1997, p. 341)

These aspects of human vocal capacity shaped by sensitivity to the harmonic series are so important that it needs stressing further: while several species of bird and cetacean sing expressively, and some have extraordinary capacities for the imitation of other species and sound sources, human beings, who can also parallel such feats of imitation, are the only species to have evolved the ability to sing in unison in both the dimensions of rhythmic coordination and precise pitch attunement. The further widespread ('almost universal', Brown 1991) feature of singing different notes that aggregate to make chords is equally something that only the later stages of the hominin line appear to have attained. Not only is this as astonishing an achievement as language itself: one explanation for the evolution of this ability is that it may have been a necessary condition for the development of the anatomical, aural, and neural substrates on which language depends.

ACKNOWLEDGEMENTS

This chapter is based on presentations given at the Music, Language and Human Evolution workshop at Reading University, 2004, supported by the European Science Foundation; and at the Cradle of

Language conference, Stellenbosch University, 2006. It incorporates material from an article 'Language out of Music; the Four Dimensions of Vocal Learning' published in *The Australian Journal of Anthropology*, December 2008. I thank Jonathan Dunsby, Robert Walker, Christopher Tonkin, and, especially, Björn Merker for valuable suggestions on previous versions of this chapter.

REFERENCES

Abitbol, J., J. de Brux, G. Millot et al. (1989). 'Does a Hormonal Vocal Cord Cycle Exist in Women? Study of Vocal Premenstrual Syndrome in Voice Performers by Videostroboscopy-Glottography and Cytology on 38 women', *Journal of Voice*, 3: 157–62.

Adkins, M. (2000). 'Bi-musicality in Japanese Folk Song and Italian Art Song: A Study on Vocal Timbre'. In: B. A. Roberts and A. Rose (eds.), *The Phenomenon of Singing 2*, pp. 1–6. St John's, NF: Memorial University of Newfoundland.

Alcántra, J. and B. Moore (1994). 'Identification of Vowel-Like Harmonic Complexes: Component Phase, Level, and Fundamental Frequency', *Journal of the Acoustical Society of America*, 97, 6: 3813–24.

Aldridge, D. (1996). *Music Therapy, Research and Practice in Medicine: From Out of the Silence.* London: Jessica Kingsley.

Altmeier, W. A. III and C. V. Ward (1998). 'A Pediatrician's View: Why Do We Have Paranasal Sinuses?' *Pediatric Annals*, 27: 784–5.

Assmann, P. and T. Nearey (1987). 'Perception of Front Vowels: The Role of Harmonics in the First Formant Region', *Journal of the Acoustical Society of America*, 81, 2: 520–34.

Bannan, N. (2008). 'Language out of Music: The Four Dimensions of Vocal Learning', *The Australian Journal of Anthropology*, 19, 3: 272–93.

—— (2005). 'Music Teaching without Words'. In: M. Dottori, B. Ilari, and R. Coelho de Souza (eds.), *Proceedings of the 1st International Symposium on Cognition and the Musical Arts*, pp. 400–5. Curitiba, Brazil: Universidade Federal do Paraná.

—— (2004). 'A Role for Action Research Projects in Developing New Pedagogical Approaches to Aural and Musicianship Education'. In: J. Davidson (ed.), *The Music Practitioner: Exploring Practices and Research in the Development of the Expert Music Performer, Teacher and Listener*, pp. 295–308. Aldershot: Ashgate.

—— (2003). '"Reverse-Engineering" the Human Voice: Examining the Adaptive Prerequisites for Song and Language'. In: R. Kopiez, A. Lehmann, and I. Wolther (eds.), *Proceedings of the Fifth Triennial Conference of the European Society for the Cognitive Sciences of Music.* [CD-ROM]. Hochschule für Musik und Theater, Institute for Research in Music Education Monograph No. 6, Hanover.

—— (2000). 'Instinctive Singing: Lifelong Development of "the Child within"', *The British Journal of Music Education*, 17, 3: 295–301.

—— (1998). 'Aural Feedback, Vocal Technique, and Creativity'. In: B. A. Roberts (ed.), 'Sharing the Voices: The Phenomenon of Singing', *Proceedings of the International Symposium, St. John's, Newfoundland, Canada, June 1997*, pp. 11–19. St John's, NF: Memorial University Press.

—— and C. Montgomery-Smith (2008). 'Singing for the Brain: Reflections on the Human Capacity for Music Arising from a Pilot Study of Group Singing with Alzheimer's Patients and Carers', *Journal of the Royal Society for the Promotion of Health*, 128, 2: 73–8.

—— and S. Woodward (2009). 'Spontaneity in the Musicality and Music Learning of Children'. In: S. Malloch and C. Trevarthen (eds.), *Communicative Musicality*, pp. 465–94. Oxford: Oxford University Press.

Barthes, R. (1977). *Image, Music, Text,* trans. S. Heath. London: Fontana.

Belin, P., R. J. Zatorre, P. Lafaille et al. (2000). 'Voice-Selective Areas in Human Auditory Cortex', *Nature*, 403: 309–12.

—— P. Van Eeckhout, M. Zilbovicius et al. (1996). 'Recovery from Nonfluent Aphasia after Melodic Intonation Therapy: A PET study', *Neurology*, 47: 1504–11.

Bickerton, D. (1991). *Language and Species.* Chicago, IL: University of Chicago Press.

Bjørkvold, J.-R. (1992). *The Muse within: Creativity and Communication, Song and Play from Childhood through Maturity*, trans. W. Halverson. New York: HarperCollins.

Blacking, J. (1987). *'A Commonsense View of All Music'*. Cambridge: Cambridge University Press.

Blackmore, S. (2007). 'Did Musical Memes Change our Minds?' Unpublished keynote address, Music and Evolutionary Thought conference, University of Durham School of Music and Institute of Advanced Studies, Durham, UK.

Braun, M. (2009). 'On the Mechanism of Major-Minor Perception'. <http://www.neuroscience-of-music.se/mode perception.htm>.

—— (1999). 'Auditory Midbrain Laminar Structure Appears Adapted to F_0 Extraction: Further Evidence and Implications of the Double Critical Bandwidth', *Hearing Research*, 129: 71–82.

Brown, D. E. (1991). *Human Universals.* New York: McGraw-Hill.

Brown, S. (2007). 'Contagious Heterophony: A New Theory about the Origins of Music', *Musica Scientiae*, 11, 1: 3–26.

—— (2000). 'The "Musilanguage" Model of Music Evolution'. In: N. Wallin, B Merker, and S. Brown (eds.), *The Origins of Music*, pp. 271–301. Cambridge, MA: MIT Press.

Burney, J. (1975). *With Captain James Cook in the Antarctic and Pacific*, ed. B. Hooper. Canberra: National Library of Australia.

Buttolph, K. (1993). 'Tone-Deafness: The Myth Explored', *Sounding Board*, Spring: 10–13.

Campbell, P. S. (1991). *Lessons from the World*. New York: Schirmer.

Carlos, W. (1987). 'Tuning: At the Crossroads', *Computer Music Journal*, 11, 1, Microtonality: 29–43.

Chomsky, N. (2001). *New Horizons in the Study of Language and Mind*. Cambridge: Cambridge University Press.

—— (1957). *Syntactic Structures*. The Hague: Mouton.

Classe, A. (1957). 'Phonetics of the Silbo Gomero', *Archivum Linguisticum*, 9: 43–61.

Cooksey, J. (1997). 'Voice Transformation in Male Adolescents'. In: L. Thurman and G. Welch (eds.), *Body, Mind and Voice: Foundations of Voice Education*, pp. 405–515. Minneapolis, MN: The VoiceCare Network.

Cox, A. (2001). 'The Mimetic Hypothesis and Embodied Musical Meaning', *Musica Scientiae*, 5, 2: 195–212.

Dalla Bella, S., J.-F. Giguère, and I. Peretz (2009). 'Singing in Congenital Amusia', *Journal of the Acoustic Society of America*, 126: 414–24.

Dargie, D. (1988). *Xhosa Music: Its Techniques and Instruments, with a Collection of Songs*. Cape Town: David Philip.

Darwin, C. R. (1871). *The Descent of Man, and Selection in Relation to Sex*, 2nd edn. London: John Murray.

—— (1872). *The Expression of the Emotions in Man and Animals*. London: John Murray.

de Boer, B. (2000). 'Self-Organisation in Vowel Systems', *Journal of Phonetics*, 28: 441–65.

—— (1997). 'Generating Vowel Systems in a Population of Agents'. In: P. Husbands and I. Harvey (eds.), *Fourth European Conference on Artificial Life*, pp. 503–10. Cambridge, MA: MIT Press.

—— and P. Kuhl (2003). 'Investigating the Role of Infant-Directed Speech with a Computer Model', *Acoustics Research Letters Online*, 4, 4: 129–14.

de Cheveigné, A. and H. Kawahara (1999). 'Missing-Data Model of Vowel Identification', *Journal of the Acoustical Society of America*, 105, 6: 3497–508.

Deacon, T. (1997). *The Symbolic Species*. London: Allen Lane.

Deutsch, D., R. Lapidis, and T. Henthorn (2008). 'The Speech-to-Song Illusion'. *156th Meeting of the Acoustical Society of America*, Paper 2aMU6. <http://www.acoustics.org/press/156th/deutsch.html>.

Dunsby, J. (2004). *Making Words Sing: Nineteenth- and Twentieth-Century Song*. Cambridge: Cambridge University Press.

Eckerdal, P. and B. Merker (2009). '"Music" and the "Action Song" in Infant Development: An Interpretation'. In: S. Malloch and C. Trevarthen (eds.), *Communicative Musicality*, pp. 241–62. Oxford: Oxford University Press.

El Mogharbel, C., I. Laufs, M. Wenglorz et al. (2003). 'The Sounds of Songs without Words'. In: R. Kopiez, A. Lehmann, I. Wolther et al. (eds.), *Proceedings of the Fifth Triennial Conference of the European Society for the Cognitive Sciences of Music*. Institute for Research in Music Education Monograph No. 6, Hanover, Germany.

Falk, D. (2004). 'Prelinguistic Evolution in Early Hominins: Whence Motherese?' *Behavioral and Brain Sciences*, 27: 491–503.

—— (1988). 'Enlarged Occipital/Marginal Sinuses and Emissary Foramina: Their Significance in Hominid Evolution'. In: F. E. Grine (ed.), *Evolutionary History of the 'Robust' Australopithecines*, pp. 85–96. New York: Aldine de Gruyter.

Fant, G. (1970). *Acoustic Theory of Speech Production*. The Hague: Mouton.

Fernald, A. (1992). 'Meaningful Melodies in Mothers' Speech to Infants'. In: H. Papousek, U. Jürgens, and M. Papousek (eds.), *Nonverbal Vocal Communication: Comparative and Developmental Approaches*, pp. 262–82. Paris and Cambridge: Editions de la Maison des Sciences de l'Homme and Cambridge University Press.

Fitch, W. T. (2006). 'The Biology and Evolution of Music: A Comparative Perspective', *Cognition*, 100: 173–215.

Franciscus, R. G. and E. Trinkhaus (1988). 'Nasal Morphology and the Emergence of *Homo erectus*', *American Journal of Physical Anthropology*, 75: 517–28.

Frayer, D.W. and C. Nicolay (2000). 'Fossil Evidence for the Origin of Speech Sounds'. In: N. Wallin, B. Merker, and S. Brown (eds.), *The Origins of Music*, pp. 217–34. Cambridge, MA: MIT Press.

Fry, D. B. (1979). *The Physics of Speech*. Cambridge: Cambridge University Press.

Garcia, M. (1894). *Hints on Singing*. London: Ascherberg, Hopwood and Crew.

Golinkoff, R. M. and A. Alioto (1995). 'Infant-Directed Speech Facilitates Lexical Learning in Adults Hearing Chinese: Implications for Language Acquisition', *Journal of Child Language*, 22: 703–26.

Gould, S. J. (1977). *Ontogeny and Phylogeny*. Cambridge, MA: Harvard University Press.

Grainger P (1934). 'A Commonsense View of All Music'. Synopsis of 12 illustrated lectures delivered from Melbourne by the Australian Broadcasting

Commission. repr. as Appendix A of J. Blacking (1987) '*A Commonsense View of All Music*', pp. 151–80. Cambridge: Cambridge University Press.

Helmholtz, H. von (1863). *Die Lehre den Tonemptfindugen als psychologische Grundlage für die Theorie der Musik*. Braunschweig: F. Vieweg und Sohn. Trans. A. J. Ellis, *On the Sensations of Tone* (1885 repr. 1954). New York: Dover.

Hockett, C. F. (1960). 'The Origin of Speech', *Scientific American*, 203: 89–96.

Hodges, D. A. (2000). 'Why Are We Musical? Support for an Evolutionary Theory of Human Musicality'. In: C. Woods, G. Luck, R. Brochard et al. (eds.), *Proceedings of the Sixth International Conference on Musical Perception and Cognition*. [CD-ROM]. Keele: Department of Psychology, University of Keele, UK.

Hofstadter, D. (1979). *Gödel, Escher, Bach: An Eternal Gold Braid*. London: Harvester.

Humboldt, W. von (1836). *Über die Verschiedenheit des Menschlichen Sprachbaues und ihren Einfluss auf die geistige Entwicklung des Menschengeschlechts*. Berlin: Royal Academy of Sciences. Trans. G. C. Buck and F. Raven (1971), *Linguistic Variability and Intellectual Development*. Miami, FL: University of Miami Press.

Hunley, K. L. (1998). 'Vertebral Canal Size and Function: A Comparison of Extant and Fossil Hominoids', *Journal of Human Evolution*, 34: A10.

Husler, F. and Y. Rodd-Marling (1965). *Singing: The Physical Nature of the Vocal Organ. A Guide to the Unlocking of the Singing Voice*. London: Hutchinson.

Ilari, B., L. Polka, and M. Sundara (2003). 'Preferences for "A Cappella" and Accompanied Songs: A Study with Infant Listeners'. In: R. Kopiez, A. C. Lehmann, I. Wolther et al. (eds.), *Proceedings of the Fifth Triennial Conference of the European Society for the Cognitive Sciences of Music*. [CD-ROM]. Institute for Research in Music Education Monograph No. 6, Hanover.

Jan, S. (2007). *The Memetics of Music: A Neo-Darwinian View of Musical Structure and Culture*. Aldershot: Ashgate.

Jander, O. and E. Harris (2001). 'Singing'. *Grove Music Online*. <http://www.oxfordmusiconline.com/subscriber/article/grove/music/25869>.

Jaques-Dalcroze, É. (1921). *Rhythm, Music and Education*, trans. H. Rubinstein (1980). London: Dalcroze Society.

Jenkins, L. (2000). *Biolinguistics*. Cambridge: Cambridge University Press.

Johanson, D. C. and B. E. Edgar (1996). *From Lucy to Language*. New York: Simon & Schuster.

Juscyk, P. W. (1997). *The Discovery of Spoken Language*. Cambridge, MA: MIT Press.

Kalinowski, J., A. Stuart and M. Rastatter (2004). US Patent Application for 'Methods and Devices for Delivering Exogonously Generated Speech Signals to Enhance Fluency in Persons who Stutter'. <http://www.freepatentsonline.com/6754632.html>.

Kalinowski, J. and T. Saltuklaroglu (2003). 'Choral Speech: The Amelioration of Stuttering via Imitation and the Mirror Neuronal System', *Neuroscience and Biobehavioural Reviews*, 27: 339–47.

Kay, R. F., M. Cartmill, and M. Barlow (1998). 'The Hypoglossal Canal and the Origin of Human Vocal Behaviour', *Proceedings of the National Academy of Sciences, USA*, 95: 5417–19.

Kessen, W., J. Levine, and K. Wendrich (1979). 'The Imitation of Pitch in Infants', *Infant Behaviour and Development*, 2: 93–9.

Kodály, Z. (1974). *The Selected Writings of Zoltán Kodály*, ed. F. Bónis, trans. L. Halápy and F. Macnicol. London: Boosey and Hawkes.

Knight, S. (2000). 'Exploring a Cultural Myth: What Adult Non-Singers May Reveal about the Nature of Singing'. In: B. A. Roberts and A. Rose (eds.), *The Phenomenon of Singing 2*, pp. 144–54. St John's: Memorial University of Newfoundland.

Koppe, T., H. Nagai, and K. W. Alt (1999). *The Paranasal Sinuses of Higher Primates: Development, Function, and Evolution*. Berlin: Quintessence.

Krumhansl, C. (1990). *The Cognitive Foundations of Musical Pitch.* Oxford: Oxford University Press.

Lã, F. and J. W. Davidson (2005). 'Investigating the Relationship between Sexual Hormones and Female Western Classical Singing', *Research Studies in Music Education*, 24, 1: 75–87.

Langner, G., C. E. Schreiner, and U. W. Biebel (1998). 'Functional Implications of Frequency and Periodicity Coding in Auditory Midbrain'. In: A. R. Palmer, A. Rees, A. Q. Summerfield et al. (eds.), *Psychophysical and Physiological Advances in Hearing*, pp. 277–85. London: Whurr.

Laver, J. (1980). *The Phonetic Description of Voice Quality.* Cambridge: Cambridge University Press.

Lecanuet, J.-P. (1996). 'Prenatal Auditory Experience'. In: I. Deliège and J. Sloboda (eds.), *Musical Beginnings: Origins and Development of Musical Competence,* pp. 3–34. Oxford: Oxford University Press.

Ledang, O. (1972). 'On the Acoustics and the Systematic Classification of the Jew's Harp', *Yearbook of the International Folk Music Council*, 4: 95–103.

Leech-Wilkinson, D. (2006). 'Portamento and Musical Meaning', *Journal of Musicological Research*, 25: 233–61.

Lester, B. M. and C. F. Z. Boukydis (1992). 'No Language but a Cry'. In: H. Papousek, U. Jürgens, and M. Papousek (eds.), *Nonverbal Vocal Communication,* pp. 145–73. Cambridge: Cambridge University Press.

Lieberman, P. (1975). *On the Origins of Language*. New York: Macmillan.

Liu, H.-M., P. Kuhl, and F.-M. Tsao (2003). 'An Association between Mothers' Speech Clarity and Infants' Speech Discrimination Skills', *Developmental Science*, 6, 3: 1-10.

Locke, J. L. (2006). 'Parental Selection of Vocal Behaviour', *Human Nature*, 17, 2: 155–68.

Lomax, A. (1982). 'Singing'. In: *Encyclopaedia Britannica, Macropaedia*, xvi: 789. Chicago, IL: Encyclopaedia Britannica.

Lucier, A. (1970). 'I am sitting in a room', *Source: Music of the Avant-Garde*, 7–8. <http://www.ubu.com/sound/lucier.html>.

Maddieson, I. (1984). *Patterns of Sounds.* Cambridge: Cambridge University Press.

Malloch, S. (1999). 'Mother and Infants and Communicative Musicality'. In: I. Deliège (ed.), 'Rhythms, Musical Narrative, and the Origins of Human Communication', *Musicae Scientiae*, special issue, 1999–2000: 29–57.

Manén, L. (1974). *The Art of Singing: A Manual of Bel Canto*. London: Faber Music.

McKinney, J. (1994). *The Diagnosis and Correction of Vocal Faults: A Manual for Teachers of Singing and for Choir Trainers*, rev. repr. Nashville, TN: Genevox.

McMullen, E. and J. Saffran (2004). 'Music and Language: A Developmental Comparison', *Music Perception*, 21, 3: 289–311.

Meltzoff, A. N. and M. K. Moore (1992). 'Early Imitation within a Functional Framework: The Importance of Person Identity, Movement, and Development', *Infant Behaviour and Development*, 15: 479–505.

Merker, B. (2005). 'The Conformal Motive in Birdsong, Music, and Language: An Introduction', *Annals of the New York Academy of Sciences, 1060*, The Neurosciences and Music II: From Perception to Performance, pp. 17–28.

—— (2002). 'Music: The Missing Humboldt System', *Musicae Scientiae*, 6: 3–21.

—— (2000). 'Synchronous Chorusing and Human Origins'. In: N. Wallin, B. Merker, and S. Brown (eds.), *The Origins of Music*, pp. 315–27. Cambridge, MA: MIT Press.

Meyer, J. (2004). 'Bioacoustics of Human Whistled Languages: An Alternative Approach to the Cognitive Processes of Language', *Annals of the Brazilian Academy of Sciences*, 76, 2: 405–12.

Minami, Y. and H. Nito (1998). 'Vocal Pitch-Matching in Infants'. In S. Woodward (ed.), *Respecting the Child in Early Childhood Learning: Proceedings of the 8th International Seminar of the Early Childhood Commission of the International Society for Music Education, Stellenbosch, South Africa*. Cape Town: International Society for Music Education.

Miranda, E. (1999a). 'The Artificial Life Route to the Origins of Music', *Scientia*, 10, 1: 5–33.

—— (1999b). 'The Role of Speech Synthesis in "Requiem per una veu perduda"', *Organised Sound*, 3, 3: 235–40.

Mithen, S. (2008). 'The Diva within', *New Scientist*, 197: 38–9.

Mithen, S. (2005). *The Singing Neanderthals*. London: Weidenfeld & Nicholson.

Morgan, O. S. and R. Tilluckdharry (1982). 'Presentation of Singing Function in Severe Aphasia', *West Indian Medical Journal*, 31: 159–61.

Morley, I (2002). 'Evolution of the Physiological and Neurological Capacities for Music', *Cambridge Archaeological Journal*, 12, 2: 195–216.

Morley, T. (1595). 'Fyer, Fyer!' In: *Il primo libro delle ballette a cinque voci*. London: Appresso Tomaso Este.

Moyle, R. (1997). *Balgo: The Musical Life of a Desert Community*. Perth, WA: Callaway International Resource Centre for Music Education.

Nattiez, J.-J. (1990). *Music and Discourse: Towards a Semiology of Music*, trans. C. Abbate. Princeton, NJ: Princeton University Press.

Ngo, T., G. Liu, A. Tilley et al. (2008). 'The Changing Face of Perceptual Rivalry', *Brain Research Bulletin*, 75: 610–18.

Nietzsche, F. (1871). *On Music and Words*, trans. W. Kaufmann. Reprinted in C. Dahlhaus (1980), *Between Romanticism and Modernism*, pp. 103–20. Berkeley and Los Angeles, CA: University of California Press.

Oubré, A. (1997). *Instinct and Revelation: Reflections on the Origins of Numinous Perception*. Newark, NJ: Gordon and Breach.

Parncutt, R. (1989). *Harmony: A Psychoacoustical Approach*. Berlin: Springer-Verlag.

Parsons, L. (2001). 'Exploring the Functional Neuroanatomy of Music Performance, Perception, and Comprehension', *Annals of the New York Academy of Sciences*, 930: 211–31.

Payne, K. (2000). 'The Progressively Changing Songs of Humpback Whales: A Window on the Creative Process in Wild Animals'. In: N. Wallin, B. Merker, and S. Brown (eds.), *The Origins of Music*, pp. 135–50. Cambridge, MA: MIT Press.

Pettigrew, J. (2001). 'Searching for the Switch: Neural Bases for Perceptual Rivalry Alternations', *Brain and Mind*, 2: 85–118.

Pinker, S. (1997). *How the Mind Works*. New York: W. W. Norton.

—— (1994). *The Language Instinct*. London: Allen Lane.

Potter, J. (2006). 'Beggar at the Door: The Rise and Fall of Portamento in Singing', *Music and Letters*, 87, 4; 523–50.

—— (1998). *Vocal Authority: Singing Style and Ideology*. Cambridge: Cambridge University Press.

Ragozzine, F. and D. Deutsch (1994). 'A Regional Difference in Perception of the Tritone Paradox within the United States', *Music Perception*, 12, 2: 213–25.

Romet, C. (1992). 'Song Acquisition in Culture: A West Javanese Study in Children's Song Development'. In: H. Lees (ed.), 'Music Education: Sharing Musics of the World', *Proceedings of the Twentieth World Conference of the International Society for Music Education, Seoul, Korea,* pp. 164–73. Christchurch, New Zealand: ISME/University of Canterbury.

Ross, D., J. Choi and D. Purves (2007). 'Musical Intervals in Speech', *Proceedings of the National Academy of Sciences, USA*, 104, 23: 9852–7.

Sachs, C. (1962). *The Wellsprings of Music*. The Hague: Martinus Nijhoff.

Sacks, O. (2007). *Musicophilia: Tales of Music and the Brain*. New York: Vintage Books.

—— (1985). *The Man who Mistook his Wife for a Hat*. London: Duckworth.

Sadie, S. (1985). *The Cambridge Music Guide*. Cambridge: Cambridge University Press.

Salgado, A. (2000). 'Voice, Emotion and Facial Gesture in Singing'. In: C. Woods, G. Luck, R. Brochard et al. (eds.), *Proceedings of the Sixth International Conference on Musical Perception and Cognition*. [CD-ROM]. Keele: Department of Psychology, University of Keele, UK.

Saltuklaroglu, T., V. Dayalu, J. Kaliowski et al. (2004). 'Say It with Me: Stuttering Inhibited', *Journal of Clinical and Experimental Neuropsychology*, 26, 2: 161–8.

Schellenberg, E. G., P. Iverson, and M. McKinnon (1999). 'Name that Tune: Identifying Poplar Recordings from Brief Excerpts', *Psychonomic Bulletin & Review*, 6, 4: 641–6.

Schenker, H. (1906). *Harmonie*, ed. O. Jonas, trans. E. M. Borgese (1954). Chicago, IL: University of Chicago Press.

Scherer, K. R. (1992). 'Vocal Affect Expression as Symptom, Symbol and Appeal'. In: H. Papousek, U. Jürgens and M. Papousek (eds.), *Nonverbal Vocal Communication: Comparative and Developmental Approaches*, pp. 43–60. Paris and Cambridge: Editions de la Maison des Sciences de l'Homme and Cambridge University Press.

Scholes, P. (1950). *The Oxford Companion to Music*, 8th edn. Oxford: Oxford University Press.

Seki, K. and M. Sugishita (1983). 'Japanese-Applied Melodic Intonation Therapy for Broca Aphasia', *No To Shinkei*, 35: 1031–7.

Sethares, W. (1993). 'Local Consonance and the Relationship between Timbre and Scale', *Journal of the Acoustical Society of America*, 94, 3: 1218–28.

Shetler, D. J. (1990). 'The Inquiry into Prenatal Musical Experience'. In: F. R. Wilson and F. L. Roehamann (eds.), *Music and Child Development*, pp. 44–62. St Louis, MI: MMB Music.

Siegel, G., M. Cooper, J. Morgan et al. (1990). 'Imitation of Intonation by Infants', *Journal of Speech and Hearing Research*, 33: 9–15.

Sloboda, J., K. Wise, and I. Peretz (2005). 'Quantifying Tone Deafness in the General Population', *Annals of the New York Academy of Sciences*, 1060: 255–61.

Soltis, J. (2004). 'The Signal Functions of Early Infant Crying', *Behavioral and Brain Sciences*, 27: 443–90.

Stefanics, G., G. Háden, I. Sziller et al. (2009). 'Newborn Infants Process Pitch Intervals', *Clinical Neurophysiology*, 120, 2: 304–8.

Stockhausen, K. (1968). *Stimmung.* Vienna: Universal Edition.

Street, A. (2003). 'Mothers' Attitudes to Singing to their Infants'. In: R. Kopiez, A. C. Lehmann, I. Wolther et al. (eds.), *Proceedings of the Fifth Triennial Conference of the European Society for the Cognitive Sciences of Music.* [CD-ROM]. Institute for Research in Music Education Monograph No. 6, Hanover, Germany.

Sundberg, J. (2001). 'Acoustics, VI: The Voice'. In: L. Macy (ed.), *Grove Music Online.* <http://www.oxfordmusiconline.com/subscriber/article/grove/music/00134pg6>.

—— (1987). *The Science of Singing.* Dekalb, IL: Northern Illinois University Press.

Tamplin, J. (2008). 'A Pilot Study into the Effect of Vocal Exercises and Singing on Dysarthric speech', *NeuroRehabilitation*, 23: 27–216.

Tarnopolsky, A., N. Fletcher, L. Hollenberg et al. (2006). 'Vocal Tract Resonances and the Sound of the Australian Didjeridu (Yidaki) I: Experiment', *Journal of the Acoustical Society of America*, 119, 2: 1194–04.

Thomas, D. A. (1995). *Music and the Origins of Language.* Cambridge: Cambridge University Press.

Tinbergen, N. (1951). *The Study of Instinct.* Oxford: Clarendon Press.

Titze, I. R. (2003). 'The Voice as a Musical Instrument: Fundamental Differences between Man-Made and Biological Designs'. In: R. Bresin (ed.), *Proceedings of the Stockholm Music Acoustics Conference*, pp. 21–4. SMAC 03, Stockholm, Sweden.

—— (2001). 'Acoustic Interpretation of Resonant Voice', *Journal of Voice*, 15, 4: 519–28.

—— (1994). *Principles of Voice Production.* Englewood Cliffs, NJ: Prentice Hall.

Trainor, L. J., C. M. Austin, and R. N. Desjardins (2000). 'Is Infant-Directed Speech Prosody a Result of the Vocal Expression of Emotion?' *Journal of the American Psychological Society*, 11: 188–95.

Traube, C. and P. Depalle (2004). 'Timbral Analogies between Vowels and Plucked String Tones', *Proceedings of the International Conference on Acoustics, Speech and Signal Processing*, 4: 293–6.

Trehub, S. E. (2001). 'Musical Predispositions in Infancy'. In: R. J. Zatorre and I. Peretz (eds.), *The Biological Foundations of Music*, pp. 11–16. *Annals of the New York Academy of Sciences, 930.*

—— (1990). 'The Perception of Musical Patterns by Human Infants: The Provision of Similar Patterns by their Parents'. In: M. A. Berkley and W. C. Stebbins (eds.), *Comparative Perception, i: Mechanisms*, pp. 429–59. New York: Wiley.

Trevarthen, C. (1999). 'Musicality and the Intrinsic Motor Pulse: Evidence from Human Psychobiology and Infant Communication', *Musica*

Scientiae, special issue: Rhythm, Musical Narrative and Origins of Human Communication, pp. 155–211.

—— (1994). 'Infant Semiosis'. In: W. Nöth (ed.), *Origins of Semiosis*, pp. 219–52. Berlin: Mouton de Gruyter.

Verdolini, K., Y. Min, I Titze et al. (2002). 'Biological Mechanisms Underlying Voice Changes due to Dehydration', *Journal of Speech, Language and Hearing Research*, 45: 268–81.

Verlaine, P. (1884/1975). *Jadis et Naguère, 'Art poétique'*. Paris: Librairie Générale Française.

Vouloumanos, A. and J. F. Werker (2004). 'Tuned to the Signal: The Privileged Status of Speech for Young Infants', *Developmental Science*, 7: 270–6.

Waarammaa, T., P. Alku, and A.-M. Laukkanen (2006). 'The Role of F3 in the Vocal Expression of Emotions', *Logopedics Phoniatrics Vocology*, 31: 153–6.

Walker, R. (1997). 'Visual Metaphors as Music Notation for Sung Vowel Spectra in Different Cultures', *Journal of New Music Research*, 26: 315–45.

Walker, A. and P. Shipman (1996). *The Wisdom of Bones: In Search of Human Origins*. New York: Alfred Knopf.

Watson, A. and D. Keating (1999). 'Architecture and Sound: An Acoustic Analysis of Megalithic Monuments in Prehistoric Britain', *Antiquity*, 73: 325–36.

Welch, G. (2001). *The Misunderstanding of Music*. London: Institute of Education, University of London.

Wishart, T. (1996). *On Sonic Art*. Amsterdam: Harwood Academic Press.

Woodward, S. (1992). 'The Transmission of Music into the Human Uterus and the Response to Music of the Human Fetus and Neonate'. Doctoral dissertation, University of Cape Town, South Africa.

Wright, A. A., J. J. Rivera, S. H. Hulse et al. (2000). 'Music Perception and Octave Generalization in Rhesus Monkeys', *Journal of Experimental Psychology: General*, 129, 3: 291–307.

Yule, H. and A. C. Burnell (1868/1979). *Hobson-Jobson: A Glossary of Colloquial Anglo-Indian Words and Phrases, and of Kindred Terms, Etymological, Historical, Geographical and Discursive*. New Delhi: Munshiram Manoharlal.

Index

Abitbol, Jean 151
Acheulean 38, 41
adaptation 90
adaptive radiations 34, 36, 39, 46
Adkins, Molly 151, 297–8
Adler, Guido 4
Africa 36, 37, 38, 41, 89, 145, 147, 156, 158, 159, 173, 174, 177, 179–86, 189, 190–1, 193, 235, 271–2, 298
anatomically modern human (AMH) 82, 100
archaeomusicology 13n, 176
Ardipithecus ramidus 35, 36
Aristotle 300
Arom, Simha 185, 264
Australia 147, 177, 299
australopithicene(s) (including *Australopithecus*) 10, 16, 37–40, 43, 51, 70–1, 74, 100, 112, 209n, 212, 234n, 271

Balter, Michael 3
Bannan, Nicholas 3, 13, 151, 269
Barthes, Roland 147, 312, 324
basicranial flexion 111–13
Baylis, J. R. 218, 222
Bekoff, Marc 273
Belyaev, Dmitri 273
Bickerton, Derek 6, 9, 322
biolinguistics 6
biomusicology 12, 176
bipedalism, bipedal 31, 39, 63, 73, 111, 113
birdsong 7
Blacking, John 4, 10, 11, 89, 91, 142, 143, 149, 178, 183, 185, 189, 265, 298–9, 319
Blackmore, Susan 325
Boer, Bart de 17, 312–13, 315, 318
Boivin, Nicole 178, 180
bonding 119, 202, 206, 208–13
bonobos 8, 264
Bowra, Maurice 7, 147
Boxgrove 96–7, 99–101
Brailoiu, Constantin 148
brain expansion (*see* encephalization) 73, 216
brain size 40, 47, 52, 59, 100, 113, 126, 228, 229, 237
Braun, Martin 16, 311–12
British Academy 3
Broca's area 72, 114–15, 120, 232–3
Brown, Donald 10, 142
Brown, Steven 13, 16, 127, 131, 305, 327
Burney, James 300–1

cantometrics 152, 154
Carlos, Wendy 312, 314, 315, 325
cave art (*see also* rock art) 88
Cheney, Dorothy 8, 264
chimpanzee(s) 8, 35–7, 43, 47, 59, 60, 62, 69–71, 111, 144, 178, 210, 211, 213, 232, 233, 243, 264, 291, 307–8, 327
choking 58, 61, 67–9
Chomsky, Noam 5, 12, 14, 289, 297, 325
cladogenesis 34
Classe, André 151, 319
cognition 12
Conard, Nicholas 14, 16, 175–7
Condillac, Étienne Bonnot de 143
conformal motive 220, 227, 241, 320
'continuity paradox' 9, 322
Cook, James 300–1
Cox, Arnie 13, 305
Crelin, Edmund 6, 66, 111
Cross, Ian 13, 14, 127, 144, 183

Damasio, Antonio 6, 87, 270
dance 10, 11, 14, 83, 86, 88–9, 94–5, 99, 130, 177, 180, 184, 191–2, 201, 208, 265, 272, 301
Dargie, David 177, 185, 188, 190, 300, 310
Darwin, Charles 4, 5, 7, 8, 15, 33, 50, 59, 72, 83, 84, 87, 100, 143, 144, 145, 175, 202, 203, 216, 263, 293, 294–5, 297, 301, 307, 309, 321–2
  Darwinist 10, 302
Daugherty, James 151

Dawkins, Richard 6, 151, 205
Deacon, Terrence 6, 8, 293–4, 319, 324, 325
Dennett, Daniel 6, 151
d'Errico, *Francesco 14, 32, 89, 177, 190*
Deutsch, Diana 12, 150, 312, 320
developmental stress hypothesis 225
Devereux, George 266
didjeridu 92, 157, 177, 310
Diehl, Randy 281
dimorphism 65, 233–4, 300
Dissanayake, Ellen 126–7
Divje Babe 16
DNA 36
Donald, Merlin 6
duetting 236, 242
Dunbar, Robin 9, 50, 51, 73, 100, 127, 130, 327
Dunsby, Jonathan 324
Durkheim, Émile 94

Edgar, Blake 6, 289
Ekman, Paul 5, 10, 125
emotion(s) 5, 12, 33, 51, 81, 83, 84, 86–8, 94, 96, 100, 101, 116, 125–6, 128–9, 132, 143, 148, 186, 268, 270, 294–5, 297, 319, 321, 324, 327
encephalization 32, 40, 228–31, 233
endocast(s) 114–16
endorphin(s) 208, 211–12
ethnomusicology 4, 5
Europe 41, 45, 89, 96, 158, 297
European Science Foundation 3
exaptation (*see also* spandrel) 49

facial expression 5, 115–16, 125, 129
Falk, Dean 9
Fernald, Alison 122, 126, 318
Fisher, Ronald 204
Fitch, W. Tecumseh 7, 10, 14, 111, 131, 215, 216, 232, 302, 305, 327
formant(s) 156, 282, 298, 300, 309, 311–16, 317, 327
Fossey, Diane 8
fossil 110, 208, 322
fossil record 109, 110, 114, 209, 245, 308
found objects 174
FOXP2 90, 209
Friberg, Anders 278, 280
Frydén, Lars 277

Garcia, Manuel 295–96, 309
Garfinkel, Yosef 14, 176
Geißenklösterle 16, 175, 272
Geissmann, Thomas 8, 130, 236
gesture(s) 8, 128–9, 133, 293–4
Gibb, Sir Duncan 145, 150
Goffman, Erving 85, 87
Goodall, Jane 8
gorilla(s) 8, 36, 71, 291
Gould, Stephen Jay 211
Grainger, Percy 11, 149, 319
grooming 9, 73, 100, 123, 124, 208–10, 212–13

Halliday, Michael 150
handicap principle 204, 224
harmonics 157, 158, 311–16, 322
harmonic series 159, 177, 185, 186, 188, 302, 307–12, 327, 328
Hauser, Marc 7, 14, 216
Helmholtz, Hermann von 4, 290, 291–3, 294, 295, 297, 302, 309, 321–2
Henry, Edward 152
Henshilwood, Charles 16, 90, 100, 272
heterophony 158, 303, 319
Hockett, Charles 5, 8, 10, 242, 297, 304, 306–7, 324
hominid 88, 272–3
hominin 31, 33, 34–46, 49, 51, 52, 59, 62, 66, 69–73, 81–2, 84, 87–8, 90, 97, 100–1, 109–11, 113, 114–15, 124, 126, 130, 181, 202, 208, 210, 211, 271, 274, 308, 322, 328
Homo 38, 40, 41, 43, 52, 83, 215, 235–40, 288, 290, 322
  H. erectus 38, 71, 83, 209, 212, 233, 234, 235–7, 243
  H. ergaster 38, 40, 41, 111, 112, 113, 115, 121, 271, 272
  H. floresiensis 16, 83
  H. habilis 38, 40, 114–15, 212, 271
  H. heidelbergensis 38, 41, 48, 62, 70, 96, 111, 113, 126, 127, 209, 271, 272
  H. helmei 39, 41, 48
  H. neanderthalensis 62, 112
  H. sapiens 35, 39, 42, 44–5, 59, 71, 100, 112, 121, 126, 132, 209, 233, 237, 244, 271
  H. sapiens sapiens 82, 271
Hornbostel, Erich von 11, 92, 143, 148, 153, 177

Humboldt, Wilhelm von 4, 231, 292, 306
Hurford, James 241
Husler, Frederick 296–7
hyoid 16, 60, 62, 65, 69–74, 111, 112

imitation 33, 121, 177, 216, 218, 220, 223, 288, 290, 293, 306, 308, 311, 320, 321, 328
infant-directed speech (IDS) 122, 126, 311, 317–18
information theory 264, 266
instruments (musical) 83, 92, 93, 99, 109, 153, 157, 174–94, 272, 289, 322, 325
International Council for Traditional Music 11
Isturitz 16, 272

Jaques-Dalcroze, Émile 291
Jenkins, Lyle 6, 289
Jespersen, Otto 4, 143, 216
Joffe, Tracey 273
Johanson, Donald 6, 289
Johnson, Anna 152
Jürgens, Uwe 13, 119–20, 217
Juslin, Patrik 12, 127–8, 286

Karmiloff-Smith, Annette 270
Keating, David 303
Kirby, Simon 17, 241–42
Kodály, Zoltan 291, 298
kulning 152
Kunej, Drago 14, 176

LaBarre, Weston 266
Laitman, Jeffrey 62, 63, 66, 111
Langer, Suzanne 266
larynx 59, 60, 62, 63–7, 73, 110, 111, 112, 131, 146, 232, 290, 295, 298
laughter 210–11
lekgodilo (flute) 187, 188
Lennerberg, Eric 5
Lévi-Strauss 84, 85
Levitin, Daniel 12
Levman, Bryan 142
Lewis-Williams, David 88, 94, 179
Lieberman, Philip 6, 60, 62, 63, 65, 66, 110, 121, 326
List, George 5
lithophone(s) 177, 178, 180
Livingstone, Frank 10
Locke, John 128, 149
Lomax, Alan 143, 152, 289
Luchsinger, Richard 151
Lucier, Alvin 303

McAllester, David 5
Mâche, *François-Bernard 7*
Malloch, Stephen 12, 15
Manén, Lucie 295, 297
Marler, Peter 7, 8, 215–16, 220, 221
Martinelli, Dario 7
material culture 84, 87, 89, 90, 91–5
mating 202, 205
  mate choice 203–4
Mayfield, G. R. 218
melodic intonation therapy 320
meme, memetics 6
Merker, Björn 8, 9, 13, 306, 317, 319, 320
Meyer, Julien 320
Meyer, Leonard 12, 268–9
Middle East 45
Miller, Geoffrey 203, 205
Miranda, Eduardo 17, 315
mirror neuron(s) 16, 17, 87, 97
Mithen, Steven 3, 9, 14, 86, 90, 299, 323–4, 327
Mode 1 technology 38
Mode 2 technology 38, 41
Mode 3 technology 38, 41, 45
monogamy 235–6, 240
monophony 302, 319
monotony 302–3, 309, 323, 328
Moyle, Richard 299
Myers, Charles Samuel 263
multilevel selection theory 206–7

Nattiez, Jean-Jacques 298
natural selection 202–3, 212, 223, 238, 243, 326
Neanderthal(s) (see also *Homo neanderthalenis*) 9, 14, 36, 39–42, 44–5, 59, 62, 66, 70–71, 73–4, 83, 89, 111, 113, 115, 271
Nietzsche, Friedrich 293–4
Nottebohm, Fernando 7, 215, 216, 231, 233

objet(s) trouvé(s) (*see also* found objects) 177–9, 187
Orrorin tugenensis 36
Ostler, Nicholas 6
overtone(s) 156, 158, 186, 310, 315

Palaeolithic 42, 43, 87, 88, 96, 99, 111, 190
Papousek, Hanus 13, 270
Papousek, Mechtild 13, 126
Parsons, Lawrence 16, 323
Payne, Katharine 7, 220, 222
Peirce, Charles Sanders 5
Pepperberg, Irene 8
performance(s) 85, 87–8, 93–101, 152, 159, 187, 205, 211, 244, 277–9, 283, 285, 288, 299, 301, 305, 307, 308, 312, 313, 315, 316, 319, 323
periaqueductal grey matter (PAG) 120
Pinker, Steven 6, 9, 49, 83, 201–2, 211, 216, 264, 265, 268, 289
Pleistocene 32, 37, 42, 44, 51
Pliocene 37
Plotkin, Henry 6
polyphony 158–9, 186, 303, 306, 310, 319
Potter, John 142, 299–300
Premack, David 8
prosody 33, 51, 52, 116, 129, 279, 314, 315, 320
proto-language 9

Reading, University of 3
recursion 14, 238, 263
respiratory tract (*see also* vocal tract) 58, 60, 72, 290, 311
Révész, Géza 11, 148, 150
Richman, Bruce 8, 9
ritual 94, 126
  ritual culture 216, 218–21, 223
rock art (*see also* cave art) 94, 179
Romet, Cheryl 150, 318–19
Rousseau, Jean Jacques 3, 4, 143

Sachs, Curt 11, 92, 143, 146, 148, 153, 175, 177, 301, 319
Sacks, Oliver 12
Sahelanthropus tchadensis 36, 39
Saussure, Ferdinand de 4, 5
Schenker, Heinrich 158, 291, 302, 312
Scherer, Klaus 12, 127, 325
Schneck, Daniel 13
selection, sexual 50, 203, 205–6, 212, 239, 263
selection, social 202, 207, 212
Sethares, William 312, 313, 315
sexy sons hypothesis 204
Seyfarth, Robert 8, 264
Shannon, Claude 264
Slater, Peter 7, 216, 220, 221, 231
Sloboda, John 4, 127–8
social selection 202, 207, 212
Société Linguistique de Paris 5, 145
song-strings 239–45
source-filter theory 65, 290, 309
spandrel 201
Spencer, Herbert 4, 143
Sperber, Dan 267, 270
Stockhausen, Karlheinz 314–15
Storr, Anthony 12
Stringer, Chris 35, 47
Sundberg, Johan 146, 151, 232, 289–90, 300, 307, 312, 326

Taylor, Hollis 7
Ternstrom, Sten 151
theory of harmony 186, 302, 327
theory of mind 186
Thorpe, William 220
timbre 118, 149, 150, 155, 156, 157, 186, 288, 290, 291, 296, 298, 300, 305, 307, 308, 311, 312, 315, 320, 321, 322, 325, 326, 327
Titze, Ingo 232, 290, 322, 323
Todd, Neil 278
tonality 309
Tracey, Hugh 185
Trehub, Sandra 150, 317
Trevarthen, Colwyn 12, 13, 15, 122, 128–9, 150, 318
tritone 281, 320
Turk, Ivan 14, 177

universal(s) 5, 10, 88, 110, 112, 122, 125, 142, 146, 147, 201, 223, 269, 285, 286, 288, 301, 302, 306, 309
  universalist 6, 144, 160
  universality 297
upright posture 31

Verillo, Vi 279
Vico, Giambattista 143
vocal learning 124, 215–46, 290, 294, 299, 304–7, 308, 320
vocal tract (see also respiratory tract) 59–74, 113, 131, 272, 290, 302, 307, 310, 322, 326
vocalization(s) 109, 110–16, 118–20, 122–32, 148–50, 160, 175, 179, 186, 288, 290–3, 295, 299, 300,

305, 307–11, 317, 320, 321, 322, 324, 326–7
vowel (vowel sounds) 110, 120–21, 125, 143, 158, 185, 282, 288, 290, 296, 300, 305, 309, 311–16, 317, 319, 321, 328

Waal, Frans de 8
Walker, Robert 327
Wallin, Nils 13
Watson, Aaron 303
Weaver, Warren 264
Welch, Graham 324
Wernicke's area 232
whales 7, 222, 305
Wilson, Deirdre 267, 270
Wilson, Edward Osborne 6
Wishart, Trevor 149, 312, 319
Wray, Alison 15, 241

Zahavi, Amotz 204, 224
Zemp, Hugo 151
zoomusicology 7